IRELAND'S SEA FISHERIES, 1400–1600

Irish Historical Monograph Series

ISSN 1740-1097

Previous titles in the series are listed at the back of the volume.

IRELAND'S SEA FISHERIES, 1400–1600

Economics, Environment and Ecology

Patrick W. Hayes

THE BOYDELL PRESS

First published 2023
The Boydell Press, Woodbridge

ISBN 978-1-78327-706-3

The Boydell Press is an imprint of Boydell & Brewer Ltd
PO Box 9, Woodbridge, Suffolk IP12 3DF, UK
and of Boydell & Brewer Inc.
668 Mt Hope Avenue, Rochester, NY 14620–2731, USA
website: www.boydellandbrewer.com

A CIP catalogue record for this title is available from the British Library

This publication is printed on acid-free paper

For Jess

Contents

Illustrations

Plates

Maps

Figures

Tables

The author and publisher are grateful to all the institutions and individuals listed for permission to reproduce the materials in which they hold copyright. Every effort has been made to trace the copyright holders; apologies are offered for any omission, and the publisher will be pleased to add any necessary acknowledgement in subsequent editions.

Acknowledgements

This book is the product of many years of work, many of which were done while I was working at the Trinity Centre for Environmental Humanities (TCEH). First, I want to thank the centre's director, Professor Poul Holm, for all his help and guidance in drafting this book and for his many valuable comments and suggestions through its many drafts. During my time at TCEH, I worked on both the NorFish and 4-Ocean projects, and I am thankful to both project teams and the European Research Council (ERD) for providing the funding to make this research possible.

I also wish to thank other members of the TCEH, including Dr Francis Ludlow, for his advice and guidance on many of the technical aspects of this work and on the oceanographic and climatic analysis. I am also indebted to Al Matthews for his advice on statistical analysis and his friendship over the whole process of writing this book. I must also thank Dr Bernard Allaire for providing me with French archival material and assisting with the transcription of manuscript documents. Lastly, I need to acknowledge the support and friendship of all my colleagues at TCEH for their company and encouragement, especially Zhen Yang and Tenaya Jorgensen.

I owe a great deal of gratitude and thanks to Peter Sowden at Boydell & Brewer for guiding me through the process of preparing this manuscript. Likewise, I extend my thanks to Professor David Hayton and my anonymous reviewer for their extremely helpful and constructive feedback on an early draft of this work.

I want to also thank the Trinity Association and Trust for assisting the publication of this work with a generous grant. They can be supported through the Bank of Ireland Trinity Affinity Credit Card (https://www.tcd.ie/alumni/services/affinity-credit-card).

I wish to also thank Robin Wiltshire at the Sheffield Archives for providing the map of Baltimore on the cover of this work. This map comes from The Wentworth Woodhouse Muniments, which have been accepted in lieu of Inheritance Tax by HM Government and allocated to Sheffield City Council. I am also extremely grateful to Dr Katharine Simms for providing a copy of the poem *Mo-chean do theacht, a sgadáin*, and I need to thank both Professor Sylvester ó Muirí and Dr Simms for translating this poem into English.

Finally, I wish to thank my parents for their endless support, and my wife Jess, whose patience, love, and support were constants throughout the writing of this book, which is why it is dedicated to her.

Abbreviations

BL	British Library
C&H	James H. Barrett and D. C. Orton (eds), *Cod and Herring: The Archaeology and History of Medieval Sea Fishing* (Oxford, 2016)
Cal. HCA 1536–1641	John C. Appleby (ed.), *A Calendar of Material Relating to Ireland from The High Court of the Admiralty Examinations 1536–1641* (Dublin, 1992)
Cal. Car. MSS	J. S. Brewer and William Bullen (eds), *Calendar of the Carew Manuscripts Preserved in the Archiepiscopal Library at Lambeth*, 6 vols (London, 1867–73)
CELT	'CELT: Corpus of Electronic Texts' (University College Cork, 1992–2020) (https://celt.ucc.ie//index.htm)
CHI vol. 1	Brendan Smith (ed.), *The Cambridge History of Ireland: Volume 1, 600–1550* (Cambridge, 2018)
CHI vol. 2	Jane Ohlmeyer (ed.), *The Cambridge History of Ireland: Volume 2, 1550–1730* (Cambridge, 2018)
CIRCLE	Peter Crooks (ed.), '*CIRCLE* – A Calendar of Irish Chancery Letters c. 1244–1509' (Dublin, 2019) (https://chancery.tcd.ie) (accessed 13 April 2020)
CSPI	*Calendar of State Papers Relating to Ireland*
CSPI (IMC)	*Calendar of State Papers, Ireland, Tudor Period, 1509–1575*, 5 vols (Dublin, 2000–17)
DIB	*Dictionary of Irish Biography* (https://www.dib.ie/)
FAO	Food and Agriculture Organization of the United Nations
FTD	Fish Trade Database
HCA	The High Court of the Admiralty
ICES	International Council for the Exploration of the Sea
IHTA	*Irish Historic Towns Atlas*
JAS	*Journal of Archaeological Science*

JRSAI	*Journal of the Royal Society of Antiquaries of Ireland*
NHI, vol. III	T. W. Moody et al. (eds), *A New History of Ireland III: Early Modern Ireland 1534–1691* (Oxford, 2009)
QDA	Qualitative Data Analysis
QFD	Qualitative Fishing Database
SP	State Papers
Statute Rolls, Ireland	Henry F. Berry and James F. Morrissey, *Statute Rolls of the Parliament of Ireland, vols II–IV* (Dublin, 1910–39)
TAC	Total allowable catch
TCD	Trinity College Dublin
TLW	Metric tonnes of live weight
TNA	The National Archives [of the UK]

Glossary of Fish Species

Common name	*Scientific name*
Blue whiting	*Micromesistius poutassou*
Cod	*Gadus morhua*
Cole	Likely, *Salvelinus colii*
Dogfish	*Scyliorhinus canicula* or *stellaris*
Eel	Likely, *Anguilla anguilla*
Flounder	*Platichthys flesus*
Gurnard	*Eutrigla gurnardus*
Haddock	*Melanogrammus aeglefinus*
Halibut	*Hippoglossus hippoglossus*
Hake	*Merluccius merluccius*
Herring	*Clupea harengus*
Horse mackerel	*Trachurus trachurus*
Ling	*Molva molva*
Mackerel	*Scomber scombrus*
Newfoundland cod	*Gadus morhua*
Pilchards	*Sardina pilchardus*
Pollock	*Pollachius virens*
Poor John	*Gadus morhua*
Salmon	*Salmo salar*
Sardines	Likely, *Sardina pilchardus*
Seal	*Halichoerus grypus* or *Phoca vitulina*
Shad	*Alosa fallax*
Sprat	Likely, *Sprattus sprattus*
Sturgeon	*Acipenser sturio*
Thornback rays	*Raja clavata*
Trout	*Salmo trutta*
Whiting	*Merlangius merlangus*

Introduction

All the joy of my heart is the herring that never was taken upon his bate.
Christopher St Lawrence, *The Book of Howth*, compiled between 1567 and 1571
(*Cal. Carew MSS*, vi, p. 153)

As an island, Ireland has a deep connection and relationship with the ocean and seas that surround it. Water has visibly and dramatically shaped the island physically and culturally for its entire history. The coastline is marked by towering cliffs, windswept beaches, abundant bays and thriving river estuaries. Some of the stormiest seas on the planet pound against the west coast, while the Irish and Celtic Seas have acted as conduits for ideas, migration, and invasion for centuries. The Atlantic Ocean also makes the country green – the gulf stream brings warm water that tempers the Irish climate: without it, the island would resemble Newfoundland, which lies on a similar latitude but experiences a far harsher climate. Water has brought wealth too. Trade has allowed port towns to flourish and grow, while fisheries have been the lifeblood of many coastal communities from prehistory to the present day. Ireland's fisheries of the fifteenth and sixteenth centuries are the focus of this book. Despite the importance of the marine space in Ireland's story, we know very little about the history of fisheries, and the subject has generated little interest among historians to date.

Ireland is not alone in this lack of knowledge; until recently, the history of the seas and oceans globally has remained largely unknown. There are many studies about things that float on water – ships carrying trade, migrants, and weapons – but few studies have been interested in what lies beneath the surface and how humans have interacted with it. Part of this may stem from older perceptions of the ocean as inexhaustible and unknowable.[1] And while there remain many unknowns, we are very aware today that the oceans are far from inexhaustible and are, in fact, frighteningly fragile in the face of unchecked human exploitation and climate change.[2] The fragility of marine life and ecosystems becomes

[1] For several examples, see Jeffrey W. Bolster, *The Mortal Sea: Fishing the Atlantic in the Age of Sail* (Cambridge, MA, 2012), pp. x, 2, 188, 191.

[2] The science is now unequivocal and there are many studies to illustrate this, but for the impact of overfishing and other human activities, see Benjamin S. Halpern et al., 'Spatial and Temporal Changes in Cumulative Human Impacts on the World's Ocean', *Nature Communications*, 6 (2015), pp. 1–7; and for the effects of climate change, see *The Ocean and*

only starker in the presence of an accurate historical perspective. For this reason, history is vital.

It was the marine scientist Daniel Pauly who first argued for the existence of a 'shifting baselines' syndrome, in 1995; he contended that a lack of memory about past ocean productivity often results in managers making decisions that keep fish populations depressed.[3] Most fisheries managers make decisions based on data from only a few decades previous; this leads to false assumptions that fisheries were in a pristine state in the recent past and does not account for centuries' worth of exploitation that may have taken place. The actual pristine state of an ecosystem and the baseline we should be aiming for may rest much further back in time. The presence of this syndrome is in itself an argument for the examination of historical fisheries and the fields of marine environmental history and historical ecology. While not every study, including the current one, can provide concrete data on the state of past fish populations, they can still provide valuable historical perspectives and observations on past ecosystems and fishing practices.[4]

The value of such observations is indubitable; in other contexts, they are crucial to increasing equity in marine fisheries, as they can help to link traditional management practices to productive and sustainable fisheries, contributing to efforts to regain Indigenous fisheries rights. For example, historical ecology showed that native populations in Hawai'i maintained fisheries' catch-levels above modern levels for more than three centuries before European contact, which serves as important evidence for both setting higher targets for fish abundance and restoring Indigenous fishing rights.[5] The situation in Ireland is overall very different, but it is also true that the sixteenth and seventeenth centuries saw large-scale English colonisation and the disconnection of older communities like the Gaelic Irish from the marine space. In that process, traditional ecological knowledge and practice were lost. Unearthing that lost knowledge and understanding how communities in late medieval and early modern Ireland interacted with the ocean could yield valuable insights for the present day.

Cryosphere in a Changing Climate: Special Report of the Intergovernmental Panel on Climate Change (Cambridge, 2022).

[3] Daniel Pauly, 'Anecdotes and the Shifting Baseline Syndrome of Fisheries', *Trends in Ecology and Evolution*, 10 (1995), p. 430.

[4] For a prominent example, see Jeremy B. C. Jackson et al., 'Historical Overfishing and the Recent Collapse of Coastal Ecosystems', *Science*, 293 (2001), pp. 629–37; for a general overview, see Heike K. Lotze and Loren McClenachan, 'Historical Ecology: Informing the Future by Learning from the Past', in M. D. Bertness et al. (eds), *Marine Community Ecology and Conservation* (Sunderland, MA, 2014), pp. 165–200.

[5] Loren McClenachan and John N. Kittinger, 'Multicentury Trends and the Sustainability of Coral Reef Fisheries in Hawai'i and Florida', *Fish and Fisheries*, 14 (2013), pp. 239–55.

Building on the earlier observations of marine scientists like Pauly, Jeffrey Bolster argued that the ocean was the next frontier for environmental historians.[6] In the almost two decades since Bolster made his call to tackle this new frontier, historians globally have turned their attention to the seas, developed new approaches to marine environmental history and given voice to changing human relationships with marine fisheries. In the North Atlantic context, a series of key research projects have pushed the field forward and provided a wealth of new insights.

At the forefront of these developments is Professor Poul Holm, who has been involved in or headed three significant projects over the last two decades. The first of these was the History of Marine Animals Population (HMAP) project, which sought to reconstruct the state of past oceans and ran from 2000 to 2010.[7] The HMAP project demonstrated that natural scientists and historians could collaborate to reach shared research goals and answer questions that neither discipline could tackle in isolation. Following on from this, Professor Holm headed the NorFish project at TCD, which ran from 2016 to 2020, and focused on the economic, cultural, and environmental consequences of the Newfoundland cod fishery for early modern Europe.[8] A major outcome of this project was the development of a long-term time series (1520–1790) of cod and herring catches for the North Atlantic, which demonstrated that historical extractions far exceeded previous estimates. Historical data on the Irish herring fishery generated for this book formed part of this broader dataset.[9]

Finally, in 2021, Professor Holm and three fellow investigators from other European institutions began the 4-Oceans project, which seeks to assess the importance of marine life to human societies during the last two millennia.[10] This is a highly ambitious research agenda, but the fact that major funding organisations such as the European Research Council are backing projects demonstrates this field's growth over the last years. Marine environmental history is now at the forefront of historical research, and these projects are helping to demonstrate

[6] Jeffrey W. Bolster, 'Opportunities in Marine Environmental History', *Environmental History*, 11 (2006), pp. 567–97.

[7] The HMAP project formed part of the more extensive Census of Marine Life research initiative; Poul Holm et al., 'Marine Animal Populations: A New Look Back in Time', in A. D. MacIntyre (ed.), *Life in the World's Oceans: Diversity, Distribution, and Abundance* (Oxford, 2010), pp. 1–24.

[8] 'The North Atlantic Fish Revolution: An Environmental History of the North Atlantic 1400–1700 (NorFish)', European Research Council Advanced Grant ERC-2014-ADG (TCD, 2016–20) (http://cehresearch.org/norfish.html).

[9] Poul Holm et al., 'Accelerated Extractions of North Atlantic Cod and Herring, 1520–1790', *Fish and Fisheries*, 23 (2022), pp. 54–72.

[10] '4-Oceans Project' (TCD, CHAM Centro de Humanidades, Norwegian University of Science and Technology, 2023) (https://www.tcd.ie/tceh/4-oceans/).

the critical role marine resources have played in human affairs and the long-term impact humans have had on the ocean.

Our knowledge of Ireland's marine history has benefited too, with the NorFish project making the current study possible. The necessity for such a study is also apparent based on the claims of some historians. In his monograph, *Merchants and Mariners in Medieval Ireland*, Tim O'Neill writes that the arrival of herring shoals and the subsequent herring fisheries was the 'greatest single economic event' of the fifteenth century.[11] In *The Cambridge History of Ireland*, Margaret Murphy reiterates this claim in her overview of the medieval Irish economy.[12] Similarly, Raymond Gillespie refers to fisheries as 'one of the most profitable areas of the Irish economy in the sixteenth century' in his outline of economic trends from 1550 to 1700.[13] While claims like these place fisheries at the forefront of Irish economic history, they have never been fully substantiated. O'Neill provides only snippets of information on the fisheries in his work and no detailed evidence. Murphy and Gillespie were providing broad overviews, so they also had no thorough, referenceable studies to back up their assertions. There has been some other work on Irish fisheries history, but none of these have been systematic or comprehensive in their approaches.[14] There is a clear need for a comprehensive examination of the fifteenth- and sixteenth-century fisheries to verify whether such high claims are, in fact, correct.

One of the most prolific writers on the history of Irish fisheries was Arthur E. J. Went, an active researcher from the 1940s to the 1970s. Went was ahead of his time. He was a marine biologist who served as president of ICES from 1966 to 1969 and also wrote extensively on the history of Irish fisheries, authoring close to thirty articles, many of which were original studies using evidence other historians had overlooked.[15] The ecology and history of salmon were his particular interest.[16] Despite Went's background in biology, he rarely melded his historical

[11] Timothy O'Neill, *Merchants and Mariners in Medieval Ireland* (Dublin, 1987), p. 131.

[12] *CHI vol. 1*, pp. 385–414, at p. 414.

[13] Raymond Gillespie, *The Transformation of the Irish Economy 1550–1700* (Dublin, 1991), p. 7.

[14] Jim Mac Laughlin, *Troubled Waters: A Social and Cultural History of Ireland's Sea Fisheries* (Dublin, 2010); John de Courcy, *Ireland's Sea Fisheries: A History* (Dublin, 1981).

[15] Christopher Moriarty, 'Arthur E. J. Went, 1910–1980', *The Irish Naturalists' Journal*, 20 (1981), pp. 309–12; A. E. J. Went, 'List of Some Historical Papers etc on Irish Fish, Fishing and Fisheries 1940–1974', *Department of Fisheries and Forestry Fishing Leaflet*, 97 (1978), pp. 1–7.

[16] Patricia M. Byrne, 'Went, Arthur Edward James' (*DIB*, 2009) (https://doi.org/10.3318/dib.008966.v1) (accessed 29 January 2023); A. E. J. Went, 'The Pursuit of Salmon in Ireland', *Proceedings of the Royal Irish Academy*, 63C (1964), pp. 191–244; 'Fisheries of the Munster Blackwater', *JRSAI*, 90 (1960), pp. 97–131; 'Fisheries of the River Liffey. Notes on the Corporation Fishery up to the Dissolution of the Monasteries', *JRSAI*, 83 (1953), pp. 163–73; 'Material for a History of the Fisheries of the River Boyne', *Journal of the County Louth Archaeological Society*, 13 (1953), pp. 18–33.

and ecological work and instead kept them very separate. This split is not surprising, as historical ecology had yet to emerge as a distinct field in his lifetime. Went's historical writings also tend toward the anecdotal, with no real attempt to synthesise the longer interconnected story of different fisheries around Ireland. Overall, Went provides some tantalising snippets into fisheries history, and was a pioneer in the subject, but new studies are needed to flesh out our understanding of the subject and connect it to the broader themes of Irish history.

This book attempts to place Irish fisheries into the broader political and military history of the sixteenth century, and the relationship between the major ethnic groups: the 'Old Irish', 'Old English' and 'New English'. The 'Old Irish' or Gaelic Irish regarded themselves as Ireland's original inhabitants before the Norman invasion. They were Catholic and Gaelic-speaking and maintained their own cultural customs and laws. By contrast, the 'Old English' identified themselves as descendants of the Norman invaders of the twelfth century. In religion, they were also Catholic, but stressed their 'Englishness' and loyalty to the English Crown. It should be noted that 'Old English' was not widely used in Ireland until the seventeenth century. Before then, 'Anglo-Irish' was the more usual description. However, the term has since been widely adopted by historians to identify this group and is employed here. Despite the differences, by the sixteenth century, there had been a considerable amount of 'intermarriage and cultural cross-assimilation' between these two communities, with the Gaelic Irish having become Anglicised and the Old English Gaelicised.[17] The connections between these groups also extended to fisheries, with the two parties cooperating and sharing organisational structures. Additionally, with English plantation in the sixteenth century, a group of Protestant 'New English' settlers immigrated directly from Britain and retained a strong connection to their home countries in the following centuries. The arrival of the New English would ultimately transform the access to and ownership of fisheries in Ireland.

There have been many histories written of fisheries in other parts of Europe, and the present work draws on these studies, both methodologically and thematically. Particularly important have been studies of Scandinavian countries. Poul Holm's work on Danish fisheries has been influential in this respect, as has J. Jónsson's research on Icelandic fisheries.[18] The scholarship of Arnved Nedkvitne and Alf Ragnar Nilelssen concerning Norway's medieval and early modern

[17] *CHI vol. 2*, p. 10.

[18] Poul Holm, 'Human Impacts on Fisheries Resources and Abundance of the Danish Wadden Sea, c. 1520 to the Present', *Helgoland Marine Research*, 59 (2005), pp. 39–44; J. Jónsson, 'Fisheries off Iceland, 1600–1900', *ICES Marine Science Symposia*, 198 (1992), pp. 3–16; Poul Holm, 'The Danish Fisheries, c. 1450–1800: Medieval and Early Modern Sources and Their Potential for Marine Environmental History', in P. Holm et al. (eds), *The Exploited Seas: New Directions for Marine Environmental History* (St. John's, Nfld, 2001), pp. 97–122.

fisheries also provided crucial context.[19] Norway was one of the major producers of dried cod, or stockfish, in the medieval and early modern periods.[20] This fish was one of the most important trading commodities of the age and found its way to Ireland as well.[21]

Elsewhere, Bo Poulsen's study of the Dutch herring fishery showed the potential for historical sources with new digital methods, and a similar approach to the use of technology was applied in this study.[22] Work on English fisheries, especially Maryanne Kowaleski's research on the south-west of the country, has helped fill in the broader history of fisheries interactions between Ireland and England.[23] While the extensive work of Wendy Childs has been critical in placing Ireland's fisheries firmly within the European perspective.[24] The writings of Michael Barkham for Spain and Laurier Turgeon for France were also crucial in this regard.[25] Further afield, Peter Pope produced the seminal text on the English

19 Arnved Nedkvitne, 'The Development of the Norwegian Long-Distance Stockfish Trade', in *C&H*, pp. 50–59; Alf Ragnar Nilelssen, 'Norwegian Fisheries, c. 1100–1850', in D. J. Starkey and James E. Candow (eds), *The North Atlantic Fisheries: Supply, Marketing and Consumption, 1560–1990* (Hull, 2006), pp. 83–112.

20 See also the work of Terje Inderhaug, 'Stockfish Production, Cultural and Culinary Values', *Food Ethics*, 5 (2020), pp. 1–40.

21 For more on the networks that facilitated the trade of stockfish, see Philippe Dollinger, *The German Hansa* (London, 1970); Justyna Wubs-Mrozewicz and Stuart Jenks (eds), *The Hanse in Medieval and Early Modern Europe* (Leiden, 2013).

22 Bo Poulsen, *Dutch Herring: An Environmental History, c. 1600–1860* (Amsterdam, 2008).

23 M. Kowaleski, 'The Expansion of the South-Western Fisheries in Late Medieval England', *Economic History Review*, 53 (2000), pp. 429–54.

24 Wendy R. Childs, 'Commercial Relations between Basque Provinces and England in the Later Middle Ages, c.1200–c.1500', *Itsas Memoria. Revista de Estudios Marítimos del País Vasco*, 4 (2003), pp. 55–64; 'The Commercial Shipping of South-Western England in the Later Fifteenth Century', *The Mariner's Mirror*, 83 (1997), pp. 272–92; 'England's Icelandic Trade in the Fifteenth Century: The Role of the Port of Hull', in Poul Holm et al. (eds), *Northern Seas Yearbook* (Esbjerg, 1995), pp. 11–31; 'Ireland's Trade with England in the Later Middle Ages', *Irish Economic and Social History*, 9 (1982), pp. 5–33; 'Irish Merchants and Seamen in Late Medieval England', *Irish Historical Studies*, 32 (2000–01), pp. 22–43; 'Medieval English Trade in the North Sea: Denmark and Jutland, c1300–1500', in Ellen Damgaard et al. (eds), *A North Sea Region: West Jutland and the World II* (Esbjerg, 1998), pp. 47–74.

25 Michael M. Barkham, 'El comercio marítimo vizcaíno y guipuzcoana con el Atlántico peninsular (Asturias, Galicia, Portugal y Andalucía) y con los archipiélagos de Canarias y Madeira al principio de la Edad Moderna', *Itsas Memoria. Revista de Estudios Marítimos del País Vasco*, 4 (2003), pp. 147–64; 'French Basque "New Found Land" Entrepreneurs and the Import of Codfish and Whale Oil to Northern Spain, c. 1580 to c. 1620', *Newfoundland Studies*, 10 (1994), pp. 1–43; 'La industria pesquera en el País Vasco peninsular al principio de la Edad Moderna: ¿una edad de oro?', *Itsas Memoria. Revista de Estudios Marítimos del País Vasco*, 3 (2000), pp. 29–75; 'The Offshore and Distant-Water Fisheries of the Spanish Basques, c.1500–1650', in D. J. Starkey et al. (eds), *A History of the North Atlantic Fisheries 1* (Bremerhaven, 2009), pp. 229–49; 'The Spanish Basque Irish Fishery and Trade in the Sixteenth-Century', *History Ireland*,

fishery in Newfoundland in the seventeenth century, which is a vital part of understanding Ireland's place in the development of fisheries in North America.[26]

In recent years, scholars of Irish maritime and fisheries history have shown a greater awareness of this broader European context and its importance to Ireland's fisheries development. In a recent article, Mark Gardiner and T. E. McNeill argued that Ulster was part of a broader trade system across Europe and the demand for fish, in particular, drove that trade.[27] These authors, and others, have also shown the advantages of combining documentary and archaeological sources to place Irish maritime history firmly within the larger cultural and economic system of the North Atlantic.[28]

To date, most marine archaeological studies of Ireland have been local in nature, and while these studies provide valuable insights, they are more difficult to base broad conclusions on or integrate with documentary evidence.[29] Colin Breen has provided a succinct overview of the development of the Irish fisheries over the medieval period and into the sixteenth century.[30] While Sheila Hamilton-Dyer provides an overview of the evidence currently available from medieval fishbone assemblages, which highlighted some limitations of zooarchaeological evidence in Ireland. For instance, many smaller fishbones, such as those from herring, have been overlooked by archaeologists and not systematically collected.[31] Far more detailed studies, and meta-analyses of marine archaeological and zooarchaeological evidence could be conducted. Although new research, such as that from the FoodCult project at TCD, promises progress on this front.[32]

9, no. 3 (2001), pp. 12–15; Laurier Turgeon, *The Era of the Far-Distant Fisheries: Permanence and Transformation (circa 1500–1850)* (St. John's, Nfld, 2005).

26 Peter Pope, *Fish into Wine: The Newfoundland Plantation in the Seventeenth Century* (Chapel Hill, NC, 2004).

27 Mark Gardiner and T. E. McNeill, 'Seaborne Trade and the Commercialisation of Fifteenth- and Sixteenth-Century Gaelic Ulster', *Proceedings of the Royal Irish Academy*, 116C (2016), pp. 229–62.

28 See, for example, Colin Breen and John Raven, 'Maritime Lordship in Late-Medieval Gaelic Ireland', *Medieval Archaeology*, 61 (2017), pp. 149–82; and Aidan O'Sullivan and Colin Breen, *Maritime Ireland: An Archaeology of Coastal Communities* (Stroud, 2007), p. 217.

29 For some examples of local studies, see Thomas McErlean et al. (eds), *Strangford Lough: An Archaeological Survey of the Maritime Culture Landscape* (Belfast, 2002); Aidan O'Sullivan, 'Medieval Fish Traps on the Shannon Estuary, Ireland: Interpreting People, Place and Identity in Estuarine Landscapes', *Journal of Wetland Archaeology*, 5 (2013), pp. 65–77; Mark Gardiner, 'The Late Medieval Port Buildings of Ardglass, County Down, in the Context of Northern European Trade', *Ulster Journal of Archaeology*, 74 (2017–18), pp. 202–12.

30 Colin Breen, 'Marine Fisheries and Society in Medieval Ireland', in *C&H*, pp. 91–98.

31 Sheila Hamilton-Dyer, 'Exploring the Contrast: Fish-Bone Assemblages from Medieval Ireland', in *C&H*, pp. 231–38, at p. 236.

32 Susan Flavin, 'FoodCult: Food, Culture and Identity in Ireland c. 1550–1650' (TCD, 2020) (https://foodcult.eu/).

Overall, the study of Irish marine history is still underdeveloped and far too many investigations to date have relied on anecdotal evidence. The present book aims to change that by utilising a range of digital humanities methods designed to collect and analyse evidence from a variety of sources systematically. All relevant information about fisheries was extracted from a wide selection of documentary primary sources and amalgamated into two central databases, namely, the Qualitative Fishing Database (QFD), which deals with qualitative evidence, and the Fish Trade Database (FTD), which contains quantitative information. All the data employed here has been made available online.[33] Details on how these databases were constructed and how data was analysed can be found in the Appendices.

Such an approach is valuable because no sources from medieval or early modern Ireland deal specifically with fisheries; the evidence is scattered throughout a range of primary sources and can be found in many different contexts. Ireland was also far from a unified or centralised state in the fifteenth and sixteenth centuries. This gave rise to several different economic systems and created varied coastal communities which had their own ways of engaging with and extracting wealth from the marine environment. This fragmentation creates difficulty for the historian, since there is far more documentary evidence from the New English and Old English parts of Ireland. We have little written material from Gaelic Ireland, and often what has survived comes from the perspective of the coloniser. This challenge can be overcome to some degree, by using archaeological evidence and the careful consideration and contextualization of the existing documentary evidence, but it remains true that we may never know as much about some fisheries, particularly those in the west of Ireland, as we do for others, such as those in Munster.

Before examining the sources employed here, it is important to clarify that this study is focused on marine fisheries, those being fisheries that occur in saltwater bodies. Marine fisheries were generally larger and more commercially developed in the sixteenth century, but this was not universally true. Some salmon river fisheries, such as the Bann, were highly economically significant and saw foreign fishers travel considerable distances to exploit them. These commercially important salmon fisheries will be discussed at length; however, most other river and lake fisheries are not covered in detail. There are abundant records relating to the granting and ownership of weirs and fishing rights on rivers, but these have not been examined.[34]

33 These have also been related online as datasets; see Patrick Hayes, 'Qualitative Fishing Dataset' (2022) (10.6084/m9.figshare.20367951) (25 July 2022) and 'Fish Trade Dataset' (2022) (10.6084/m9.figshare.20367882) (25 July 2022). A third dataset related to piracy has also been produced for this study: Patrick Hayes, 'Piracy Events, 1564 to 1603' (2022) (10.6084/m9.figshare.20367972) (25 July 2022).

34 For example, the Irish fiants and patent rolls contain many such grants.

Fisheries went from being one of the largest industries in Ireland in the fifteenth century to almost complete collapse by 1600. Tracing that change is a core aim of this work. Furthermore, only in these centuries do documentary records concerning the fisheries become numerous, making such a study as this possible. The year 1603 forms a natural terminus to the book, as it was a significant milestone in Irish history: on 24 March 1603 Queen Elizabeth I of England died, and a few days later, Hugh O'Neill surrendered to Charles Blount, 8th Baron Mountjoy.[35] These events marked the climax of a tumultuous century in Ireland, with both the conclusion of the Tudor dynasty and the end of Gaelic power. It was also a time of profound change in Ireland's fisheries.

One of the most critical source collections used is the records of the English High Court of the Admiralty. Over its lifetime from the fifteenth to the mid-twentieth century, the court generated a vast number of documentary records, although these do not become numerous until the sixteenth century.[36] Their particular significance lies in the fact that they relate to all manner of maritime affairs, from commercial disputes to crimes at sea, such as piracy.[37] And while few entries specifically relate to fisheries, their maritime focus means they hold much incidental detail.

In addition to the English court, Ireland also had its own admiralty court from the sixteenth century onwards.[38] Unfortunately, most of the records from the Irish court have not survived. We can draw some consolation from the fact that many cases related to Ireland were still heard in the English court following the establishment of the Irish court. So, despite the lack of Irish records, many documents survive relating to Irish cases. However, by their nature, cases heard in the English court are biased towards merchants from the more Anglicised parts of Ireland, as well as wealthier traders who could afford to travel to England.

The massive volume of records generated by the English Admiralty court has also made them difficult for historians to use. Given the challenge of working with these manuscripts, the current book has adopted a digital research strategy, utilising John Appleby's *A Calendar of Material Relating to Ireland from The High Court of the Admiralty Examinations 1536–1641*.[39] This extensive calendar was

[35] Colm Lennon, *Sixteenth Century Ireland: The Incomplete Conquest* (Dublin, 2005), p. 303.

[36] For a collection of the earliest records of the court, see A. Alwyn Ruddock, 'The Earliest Records of the High Court of Admiralty', *Bulletin of the Institute of Historical Research*, 22 (1949), pp. 139–51.

[37] 'High Court of Admiralty Research Guide' (http://www.nationalarchives.gov.uk/help-with-your-research/research-guides/high-court-admiralty-records/) (accessed 8 October 2019).

[38] Kevin Costello, *The Court of Admiralty of Ireland, 1575–1893* (Dublin, 2011); John C. Appleby and Mary O'Dowd, 'The Irish Admiralty: Organisation and Development 1570–1640', *Irish Historical Studies*, 24 (1985–86), pp. 299–326.

[39] *Cal. HCA 1536–1641.*

compiled from original manuscripts and contains witness depositions relating to Ireland from the Examinations and Answers series of records (HCA 13).[40]

The Irish State Papers are the most extensive documentary source collection used in this study. They mostly comprise correspondence between the English officials within Ireland and the central government in London.[41] The papers contain a wide variety of material pertaining to matters such as security, finance, trade, military campaigns, and property ownership. Few records relate directly to the operation of fisheries, but details about fisheries emerge in many other contexts. However, like any records from the early modern period, the State Papers must be used with care. In some cases, only abstracts of internal correspondence have survived, and these reflect what the chief governors in Dublin wanted officials in London to see. Additionally, as letters were often sent in batches, the dates on these relate to the day they were sent – not necessarily the day they were written.[42] The burning of the Irish Four Courts in 1922 also destroyed most of the State Papers kept in Ireland, so what remains was gathered from collections in Great Britain. However, even before 1922, little of the inward correspondence to Ireland had survived, meaning most of the extant records are outward correspondence to London.[43]

The State Papers begin in 1509 and continue for the rest of the sixteenth century and beyond. All the extant sixteenth-century State Papers from this period have been calendared in some form. However, the older calendars vary wildly in quality, with the first few volumes severely lacking in detail. So, starting in the year 2000, the Irish Manuscripts Commission and The National Archives at Kew began publishing a new series of calendars to replace the older and deficient ones.[44] Over five volumes, these new calendars cover the period 1509 to 1553, as well as 1566 to 1575.[45] While these new calendars only cover up until 1575, the original calendars for the latter part of the sixteenth century are fortunately more reliable.

In addition to these official State Papers, there are several series of letters that were kept and held in private collections. The *Calendar of Carew Manuscripts*

[40] HCA 13, The High Court of Admiralty: Instance and Prize Courts: Examinations and Answers, 1536–1826 (TNA).

[41] The National Archives of the United Kingdom, 'Introduction: State Papers Ireland 1509–1782: Government Papers on Irish Affairs' (https://www.nationalarchives.gov.uk/help-with-your-research/research-guides/state-papers-ireland-1509-1782/) (accessed 8 December 2019).

[42] R. Dudley Edwards and Mary O'Dowd, *Sources for Early Modern Irish History, 1534–1641* (Cambridge, 1985), p. 11.

[43] Edwards and O'Dowd, *Sources for Early Modern Irish History*, p. 11.

[44] The Kew archives were only involved in the first volume published in 2000, *CSPI* (IMC), 1509–47, p. xi.

[45] *CSPI* (IMC), 1509–47, 1547–53, 1566–67, 1568–71, and 1571–75.

covers the period from 1515 to 1625 and contains six volumes of correspondence and other material related to Sir George Carew's time in office. Carew was a noted antiquarian who served in a number of capacities in Ireland, including as Lord President of Munster from 1600 to 1603.[46] The *Letter Book of Sir Arthur Chichester* is another collection of primary material, and while most of the material relates to the seventeenth century, the book was searched for relevant details about fisheries.[47]

This study also makes extensive use of trade data from port books and customs accounts: records maintained in English, Welsh, Irish and Scottish ports to report on the customs duties paid on overseas trade. Unfortunately, as no sixteenth-century Irish port books have survived, we have to rely on the records of Irish imports to English, Welsh and Scottish ports. In England and Wales, the port books and customs accounts were recorded by officials known as customers or collectors, who submitted their 'particular' accounts each year to the exchequer, where they were audited and stored.[48] The primary purpose of these records was not to document the volume of trade meticulously but rather to trace the customs duties paid and owned to the Crown.[49] However, these records can be used to understand the volume of goods moving between ports, even if the quality of data differs markedly between record sets, ranging from high levels of detail and lists of every item of cargo, to simple records of generic bulk shipments. The temporal and spatial coverage is also patchy; records do not exist for every port or every year. Thus, they can only provide snapshots of trade over time. Moreover, there was also always a certain amount of customs evasion and smuggling between ports that would have gone unrecorded. According to Evan Jones, smuggling was most common for prohibited or highly taxed goods. Most other goods, including fish, were traded openly and in line with customs checks.[50]

A central aim of this book was to gather as many published and digitised port books and customs accounts as feasible and combine them in the FTD to allow for the study of broader trends in the export of fish from Ireland. In this respect, the English port of Bristol provides some of the most detailed and comprehensive customs accounts for the fifteenth and sixteenth centuries. In the 1920s Ada K. Longfield provided an early overview of the Bristol accounts, including the

[46] *NHI, vol. III*, p. iv; Edwards and O'Dowd, *Sources for Early Modern Irish History*, p. 104.

[47] Arthur Chichester and Dudley R. Edwards, 'Letter-Book of Sir Arthur Chichester', *Analecta Hibernica*, 8 (1938), pp. 3, 5–177.

[48] Evan Jones and Susan Flavin (eds), *Bristol's Trade with Ireland and the Continent, 1503–1601: The Evidence of the Exchequer Customs Accounts* (Dublin, 2009), p. xi.

[49] R. J. Hunter and Brendan Scott (eds), *The Ulster Port Books, 1612–15* (Belfast, 2012), p. xi.

[50] Evan Jones, 'Illicit Business: Accounting for Smuggling in Mid-Sixteenth-Century Bristol', *Economic History Review*, 54 (2001), pp. 17–38, at p. 36.

collection of data on the fish trade.[51] In the early 2000s, the ERC-funded *Ireland-Bristol Trade in the Sixteenth Century* project at the University of Bristol worked to fully transcribe, digitise and publish the eleven extant Bristol port books and customs accounts for the sixteenth century (up to 1601).[52] Evan Jones and Susan Flavin published the finished work in 2009, and the related data was also released online.[53] Jones and Margaret M. Condon subsequently published digital versions of Bristol's accounts for the fifteenth century.[54] The work of Jones, Flavin and Condon has been particularly influential on this study, as the structure of the FTD was based on the sixteenth-century Bristol port book dataset. Wendy Childs has also published summaries for several more fifteenth-century Bristol accounts.[55] The *Gloucester Port Book* project from the 1990s also succeeded in digitising Gloucester's trade records from 1575 to 1765.[56] However, after a detailed assessment of the data, it was decided it was poor quality and contained too few details related to Ireland to be included in the FTD.

Data from several English and Welsh ports was also extracted from printed volumes. Chester was Dublin's largest trading partner in the Irish Sea during the sixteenth century.[57] Several of the Chester customs accounts from the fifteenth and sixteenth centuries have survived, and were published by K. P. Wilson in 1969.[58] The Chester accounts do not have the same breadth of chronological coverage or level of detail as the Bristol accounts, making them far less useful for establishing trade patterns. However, they contain enough detail to understand which fish species were most frequently traded and which Irish ports engaged in trade with Chester.

Numerous Welsh port books and customs accounts have also been printed, but these only begin in 1550 and run to 1603.[59] Welsh ports had a very active trade with south-east Ireland and Bristol in the latter half of the sixteenth century, and the records illustrate well the dynamics of that trade. Again, data from the printed Welsh volume related to fish and Ireland were manually extracted and added to

51 Ada K. Longfield, *Anglo-Irish Trade in the Sixteenth Century* (London, 1929).

52 Evan Jones and Susan Flavin (eds), *Ireland-Bristol Trade in the Sixteenth Century* (www.bris.ac.uk/Depts/History/Ireland/datasets.htm) (accessed 23 April 2020).

53 Jones and Flavin, *Bristol's Trade with Ireland*.

54 M. Condon and Evan Jones (eds), *Bristol 'Particular' Customs Account Transcriptions* (http://www.bristol.ac.uk/history/research/cabot/publications/) (accessed 9 September 2020).

55 Childs, 'Ireland's Trade with England'.

56 D. Hussey et al. (eds), *Gloucester Port Books, 1575–1765*, SN: 3218 vols (http://doi.org/10.5255/UKDA-SN-3218-1).

57 *NHI, vol. III*, p. 6.

58 K. P. Wilson (ed.), *Chester Customs Accounts, 1301–1566* (Record Society of Lancashire and Cheshire, vol. CXI, Liverpool, 1969).

59 Arthur Edward Lewis (ed.), *The Welsh Port Books, 1550–1603* (Cymmrodorion Record Series, xii, London, 1927).

the database. A limited amount of data is also available from Scotland, compiled by Martin Rorke in his study of Scottish overseas trade.[60] This data was not added to the FTD due to its limited nature; however, Chapter 3 will explore what it can tell us about Ireland's fish trade with Scotland in the sixteenth century.

There are no extant Irish export records from the sixteenth century, but some records are available from the early seventeenth century. The only Irish port books that survived from this period are the Ulster port books for the years 1612 to 1615. Today, the originals are found in the West Yorkshire Archive Service, which explains why they survived compared to those records stored in the Four Courts in Dublin. They were transcribed initially by R. J. Hunter but not published in his lifetime. However, they were posthumously published by the Ulster Historical Foundation, following editing by Brendan Scott.[61] The Ulster Historical Foundation kindly provided the published data in Microsoft Excel format for this study, and this data was added to the FTD. However, this study does not offer an in-depth analysis of these records due to their limited scope and because they fall outside the study period.

In the early seventeenth century, the Irish customs system was poorly organised, so in order to generate more income, the customs were farmed out to a succession of English syndicates who took over the collection process. These operations generated statistical series that were intended to show the operators how their business was performing. These customs statistics provide our most complete data on Irish exports, including fish, from the early seventeenth century. The surviving evidence from these series was collected by David Dickson and Donald Woodward and published in 1999.[62] Unlike the data from Ulster, these accounts cover the whole of Ireland and therefore provide a useful perspective on the total scale of the fish trade that can be compared to the earlier data from the sixteenth century. For this reason, the data was added to the FTD, and Chapter 3 contains an analysis of these records.

The Bardic Poetry Database of Dr Katharine Simms and Dr Mícheál Hoyne was examined for references to fish using the website's built-in search function.[63] Additionally, Dr Simms was contacted and provided several references to poems that discuss sea fishing. Dr Simms also noted that references to sea fishing are rare in Gaelic poetry, likely because fishers were of a low social class and not a

[60] Martin Rorke, 'Scottish Overseas Trade, 1275/86–1597' (PhD thesis, 2 vols, University of Edinburgh, 2001).

[61] Hunter and Scott, *The Ulster Port Books, 1612–15*.

[62] Donald Woodward, 'Irish Trade and Customs Statistics, 1614–1641', *Irish Economic and Social History*, 26 (1999), pp. 54–80.

[63] Katharine Simms and Mícheál Hoyne (eds), 'Bardic Poetry Database' (https://bardic.celt.dias.ie/) (accessed 25 August 2020).

typical subject for poets, who focused on the ruling class.[64] The aristocracy also had contempt for agriculture and the merchant class, and likely the nature of this disdain extended to fishers and fish sellers.[65] However, a common theme in Bardic poetry of praise for a landscape's fertility under a just ruler, and fish, sometimes featured in these descriptions. For example, in a poem to Pilib Ó Raighilligh (d. 1596), the author describes 'Corn in the soil' and 'fish in the rivers'.[66] Unfortunately, references like this tell us little specific about fisheries, but they do provide some insight into what the Gaelic Irish valued in their landscape. Chapter 5 also has further discussion about what Gaelic poetry can tell us about the consumption of fish in the sixteenth century.

Overall, the reliance on printed material in this book introduces an inherent bias toward English sources, as we have relatively few written sources from the Irish perspective and those that survive had little to say about the fisheries. However, there are ways to mitigate this discrepancy and gain a more complete understanding of the variety of fishing practices around the Irish coast. The main sources used also introduce a regional bias. Certain areas, such as the south-eastern coast, are over-represented in the port books and customs accounts, for the simple reason that these regions were engaged in much international trade, but we lack records for trade in other areas like the west and north coast. Other locations, like the river Bann, could be over-represented in State Papers because the area was highly contested and therefore generated a great deal of written material.

A large selection of maps was also used in this study. These maps can impart a myriad of details concerning fisheries and maritime landscapes not attainable through other sources. The majority of sixteenth-century maps cited originate in two core online collections: first, a new collection compiled by The National Archives of the UK, which contains sixty Tudor and Stuart maps covering the period from 1558 to 1610;[67] and second, the *Hardiman Atlas* in the library of TCD, comprising seventy Tudor and Stuart maps initially collected by George Carew and rediscovered and reorganised by Galway historian James Hardiman in the early nineteenth century.[68]

64 From personal correspondence with Dr Simms. My thanks to her for providing expert insight on this topic.

65 Katharine Simms, 'References to Landscape and Economy in Irish Bardic Poetry', in H. B. Clark et al. (eds), *Surveying Ireland's Past: Multidisciplinary Essays in Honour of Anngret Simms* (Dublin, 2004), pp. 145–68, at pp. 156–57.

66 Simms, 'References to Landscape', p. 146.

67 'Irish Maps c.1558–c.1610' (TNA, MPF 1/35-102) (https://www.nationalarchives.gov.uk/help-with-your-research/research-guides/irish-maps-c1558-c1610/).

68 James Hardiman and George Carew Totnes (eds), *Hardiman Atlas* (TCD Digital Collections, Dublin) (https://www.tcd.ie/library/manuscripts/collections/maps.php).

This study also attempts to contextualise and understand historical fisheries with insights from modern marine biology. To this end, some scientific sources and language are employed. Every species of fish has both a common name and a scientific name. Scientific names are useful due to the lack of ambiguity, as several different species of fish may have the same common name. The common and scientific names of fish that are discussed throughout the text have been detailed in the Glossary of Fish Species at the beginning of this volume.[69]

This book aims to explore Irish fisheries in far greater detail than other historians have exhibited; it takes an interdisciplinary approach to the topic and applies innovative digital methods and systematic evidence collection to a wide range of sources. By employing new methods and blending the approaches of environmental history and historical ecology, we can move beyond the anecdotal and establish a firmer understanding of Ireland's fisheries history. We must also understand the key influences that acted on fisheries over time. Significant factors considered here are; war and piracy, changing consumption patterns, the rise of the Newfoundland cod fishery, and environmental change and degradation. In this process, we can abolish common misconceptions and place fisheries firmly in the broader history of Ireland and Europe in the fifteenth and sixteenth centuries.

[69] In some cases, with generic names like 'salted fish' or 'ray' we cannot be certain what species is being described.

1

The Development of Fisheries, 1400–1500

All was fish that came to net.
Proverb, in H. C., *A Book on the State of Ireland*, 1599

The origins of intensive or commercial marine fishing in Ireland are poorly understood, but we can say that it did not start in the fifteenth century. Zooarchaeological evidence has allowed researchers to pinpoint the birth of concentrated marine fishing in England, and the wider European region, to the beginning of the tenth century CE. Around 1000, there was a shift away from freshwater fishing and a rapid increase in sea fishing, an event that has been coined as the 'fish event horizon' by researchers.[1] Herring was the first species to be widely exploited in the first millennium; cod and hake fisheries followed in the thirteenth and fourteenth centuries.[2]

As yet no detailed zooarchaeological study of fishbones or fish genetics has been conducted for Ireland, but Sheila Hamilton-Dyer has undertaken an overview of the available fishbones evidence from existing archaeological reports.[3] Her analysis shows that the intensity of cod fishing increased throughout the medieval period in Ireland, matching what has been observed in other parts of Europe.[4] However, the presence of larger fish might be overstated in the data, as, for the most part, these assemblages were hand-collected, which means that smaller bones from species like herring could have been overlooked owing to a lack of sieving.[5] Overall, much more research is required before definitive conclusions can be drawn from Irish zooarchaeological evidence. Fortunately, new research projects promise to increase our range of knowledge in this area through the application of cutting-edge methods to the study of Irish archaeological material.[6]

1 James H. Barrett et al., 'The Origins of Intensive Marine Fishing in Medieval Europe: The English Evidence', *Proceedings of the Royal Society B: Biological Sciences*, 271 (2004), pp. 2417–21.

2 Barrett, 'Origins of Intensive Marine Fishing', p. 2420.

3 Sheila Hamilton-Dyer, 'Exploring the Contrast: Fish-Bone Assemblages from Medieval Ireland', in *C&H*, pp. 231–38.

4 Hamilton-Dyer, 'Exploring the Contrast', pp. 236–37.

5 Hamilton-Dyer, 'Exploring the Contrast', p. 236.

6 Susan Flavin, 'FoodCult: Food, Culture and Identity in Ireland c. 1550–1650' (TCD, 2020) (https://foodcult.eu/).

The records of the Irish Chancery are one of the most comprehensive printed documentary sources extant for medieval Ireland. The records run intermittently from 1244 to 1509.[7] These records provide the first quantitative evidence of intensive, commercial marine fisheries in Ireland. However, they also have an inherent regional bias, relating mostly to the areas in which the Norman and then English state had the most significant control following the arrival of the Normans in Ireland in the twelfth century. To balance this, a selection of the most extensive Irish Annals have also been surveyed. Entries from the Annals after 1500 will be covered at greater length in later chapters, but the present discussion includes some consideration of what they can tell us about fishing before 1500. All dates from the Annals have been kept in their original formats.[8] Although these various sources only provide snippets of information, they can, taken together, tell us a great deal about the nature of Ireland's marine fisheries in the late medieval period.

Within the Chancery records, the first spike of records relating to fishing occurred in 1258. That year, the King's son, Lord Edward (the future Edward I), was fighting in Wales, and several shipments of fish were sent from Ireland to his troops.[9] The most common species were herring, hake, eel, and cod. There was a further spike in fishing-related entries in the Chancery records in 1298, driven by Edward I's war in Scotland.[10] Almost all the fish exports from Ireland at this time went to Carlisle in the far north of England. Later, in 1323, more army provisions were sent from Ireland, this time to a small town near Carlisle called Skinburness, as the wars continued there.[11] While none of these shipments constituted regular commercial activity, they demonstrate the capacity of fisheries in the English part of Ireland to meet the demand for large orders of sea fish.

We can also map out these trade connections, the better to understand the wider network at play. Map 1.1 indicates the routes with lines coming from the locations in Ireland from whence the fish originated to the respective destinations. These lines are not meant to indicate the routes taken by ships but merely link origin and destination. The thickness of the lines also indicates the number of mentions for each connection. For some entries, no specific ports were mentioned, and only countries of origin were discussed; in these cases, the centre point of each country was used to map the network. The map includes all mentions of fish exports in Chancery records, and the influence of wars in Wales and Scotland is

[7] Peter Crooks (ed.), '*CIRCLE* – A Calendar of Irish Chancery Letters c. 1244–1509' (Dublin, 2019) (https://chancery.tcd.ie).

[8] In some cases, the dates in the Annals may be a few years different than modern dates.

[9] *CIRCLE*, Henry III, no. 16, 17, 18, 19, 20.

[10] *CIRCLE*, Edward I, Close Roll 26, no. 22, 23, 28, 34, 36, 40; Close Roll 27, no. 2, 3, 4, 5, 6, 23, 38, 40, 43, 47, 48, 49.

[11] *CIRCLE*, Edward II, Close Roll 17, no. 7.

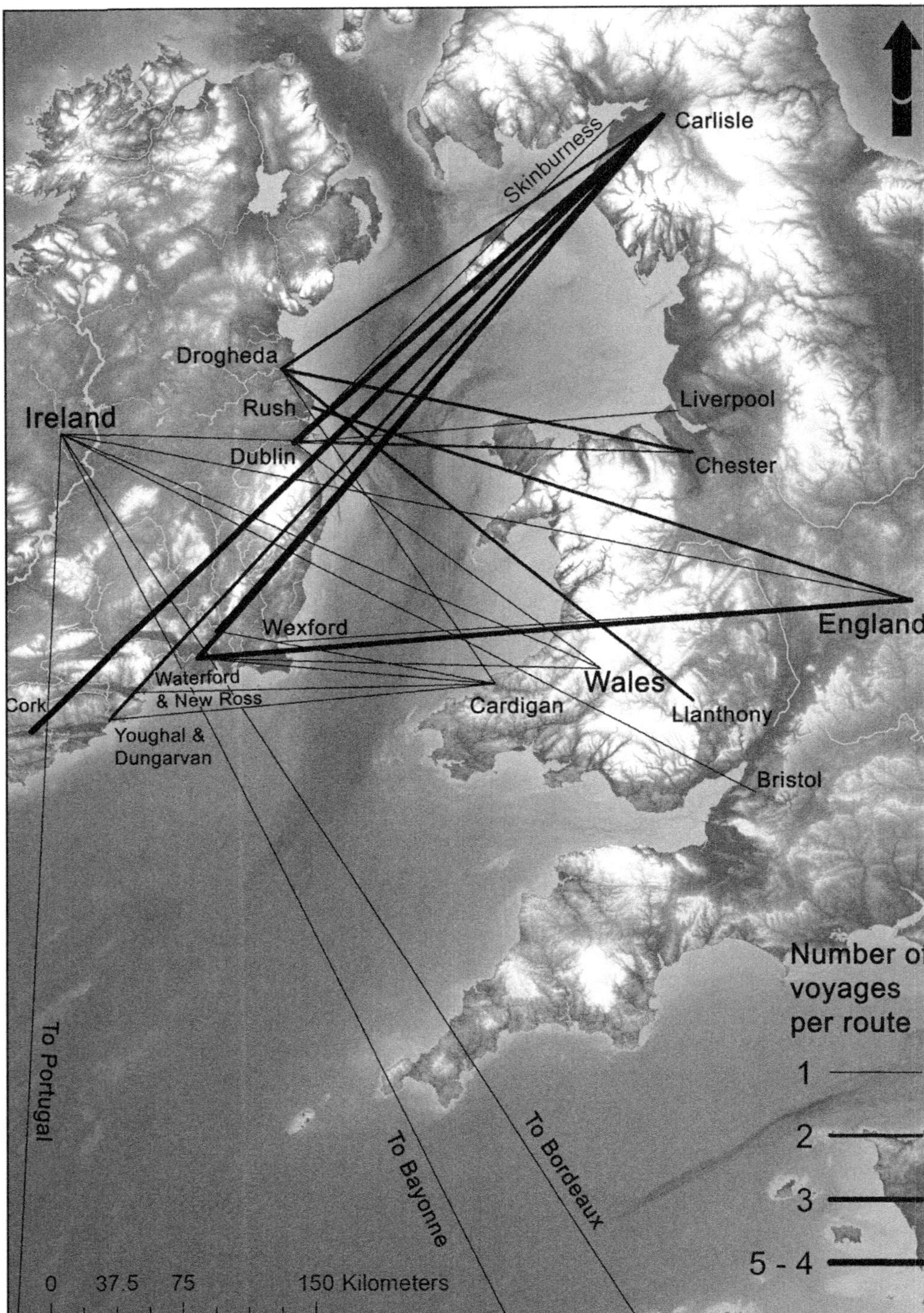

Map 1.1. Export of fish from Ireland, 1258–1403.

clearly visible. And while most fish shipments went to England and Wales, a few went further afield to France and Portugal in the medieval period.

Compared to the Chancery records, there are very few mentions of sea fish in the Irish Annals in the early medieval period. In general, river fish, especially salmon, are mentioned more often. For instance, the Annals of Tigernach report that in the year 564 a host of Aodh Ó Conchobhair raided Munster and there they 'burned Killaloe and demolished the weir of Kincora, and they ate up the salmon that lived in the well'.[12] Another entry from 1113 reports that a salmon 'twelve feet in length, twelve hands in breadth without being split' had been caught in Clonmacnoise that year.[13] It is likely that, having been written in ecclesiastical settings, the Annals also over-represent salmon, because before the dissolution of the monasteries many freshwater weirs were controlled by religious houses.[14]

Some entries in the Annals are more fanciful in nature. In 564, the Annals of Tigernach reported a fisherman caught a 'muirgelt' (mermaid) on the strand of 'Ollarba'.[15] In 1118, both the Annals of Ulster and Loch Cé reported fishers caught two mermaids, one in Osraige (roughly the area of modern Kilkenny) and another in Port Láirge (Waterford).[16] It is difficult to know if these reports of mermaids correspond to the sighting of a particular fish or marine mammal, or are purely fictional: reports of a mermaid in other contexts have been related to the discovery of manatees by Europeans.[17]

The Annals also explicitly mention whales being washed ashore. The number and content of these entries suggest they were rare events. An entry in the Annals of the Four Masters from 739 states, 'The sea cast ashore a whale in Boirche, in the province of Ulster. Everyone in the neighbourhood went to see it for its wondrousness.'[18] These strandings were special events that could save whole communities during winters or times of famine. A later entry from the Annals of Connacht in 1246 records that, 'A whale was stranded in Carbury, at Cuil Irra,

[12] Donnchadh Ó Corráin (ed.), *The Annals of Tigernach (Gearóid Mac Niocaill)* (CELT, 2010), 564.1, p. 144.

[13] Ó Corráin (ed.), *The Annals of Tigernach*, 1113.6, p. 32.

[14] A. E. J. Went, 'Fishing Weirs of the River Erne', *JRSAI*, 75 (1945), pp. 213–23; Newport Benjamin White, *Extents of Irish Monastic Possessions, 1540–1541, from Manuscripts in the Public Record Office* (London and Dublin, 1943).

[15] Ó Corráin (ed.), *The Annals of Tigernach*, 564.1, p. 147.

[16] Donnchadh Ó Corráin and Mavis Cournane (eds), *The Annals of Ulster AD 431–1201* (CELT, 2020), 1118.7, p. 563; William M. Hennessy (ed.), *Annals of Loch Cé 1* (CELT, 2008), 1118.9, p. 111.

[17] Cristina Brito, 'On Mermaids and Manatees: A First Approach to the Evolution of Natural History Images in Early Modern Times', paper presented at the Report from the European Cetacean Society Conference Workshop (Setúbal, 2013).

[18] John O'Donovan and Emma Ryan (eds), *Annals of the Four Masters, Volume 1* (CELT, 2002), 739.6, p. 342.

which brought great relief and joy to the countryside.'[19] Similarly, a pod of pilot whales washed ashore near Dublin in 1331. There was a famine at the time, and 200 whales were harvested from the beach, giving great relief to the city.[20] There is also evidence to suggest that in the eleventh century the Irish hunted young whales using harpoons.[21] Another entry from the Annals of Ulster speaks of 'a great slaughter of porpoises on the coast of Ard Cianachta by the foreigners'.[22] Ard Cianachta is in modern-day Louth, and 'foreigners' refers to the Norse.

An entry in the Annals of Loch Cé in 1217 has the first concrete details about sea fishing in these sources. It reports that fishers from Port-Lairge (Waterford) and along the coast from Loch-Carman (Wexford) to Doire Choluim Chille (Derry) all travelled to Manainn (the Isle of Man) to fish. It says they 'committed violence' there and 'were all slain in punishment for their violence in Manainn'.[23] If this entry is not an exaggeration, it suggests that fishers from both the Norman and Gaelic communities were capable of travelling to the Isle of Man for a seasonal fishery. This entry also provides an earlier date for the start of intensive marine fishing in Ireland than the Chancery records (1217 compared to 1258). The most likely species targeted was herring. Modern research has shown that spawning herring populations fluctuate between the Irish Mourne coast and the Isle of Man on a regular basis.[24] A similar change of spawning location in the thirteenth century could have driven medieval fishers from their fishing grounds in Ireland to seek out the herring in the Isle of Man.

Unfortunately, this is the last solid entry about sea fishing in the Annals before 1500, but there is more information to be found on the topic in the Chancery records. From the mid-fourteenth century onwards, the Chancery records mostly relate to the granting of fishing rights and the unregulated export of fish. Several entries also relate to complaints about merchants exporting fish without a licence.[25] The fourteenth century was a period of turbulence and tragedy across the whole of Europe, which Ireland did not escape. The island was hit severely by the Great European Famine of 1315–21 and the cattle murrain that followed.[26]

[19] The Cúil Irra peninsula is in modern-day Sligo; Pádraig Bambury (ed.), *Annals of Connacht* (*CELT*, 2008), 1246.2, p. 88.

[20] Margaret Murphy and Michael Potterton, *The Dublin Region in the Middle Ages* (Dublin, 2010), p. 394.

[21] Colin Breen, 'Marine Fisheries and Society in Medieval Ireland', in *C&H*, pp. 91–98, at p. 91.

[22] Ó Corráin and Cournane (eds), *The Annals of Ulster AD 431–1201*, 828.3, p. 285.

[23] Hennessy (ed.), *Annals of Loch Cé 1*, 1217.2, p. 257.

[24] Mark Dickey-Collas et al., 'The Location of Spawning of Irish Sea Herring (Clupea harengus)', *Journal of the Marine Biological Association of the UK*, 81 (2001), pp. 713–14.

[25] See, for example, *CIRCLE*, Edward III, Patent Roll 10, no. 4; Close Roll 29, no. 39, 44; Patent Roll 32, no. 141.

[26] Bruce M. S. Campbell, 'Nature as Historical Protagonist: Environment and Society in

Then the Black Death first struck the country, causing demographic desolation.[27] The impact of these catastrophes on British agriculture has been studied in detail, but no comparative work has been done on fisheries.[28] The evidence suggests that most cities suffered commercially and demographically as the population was decimated, so fisheries must have been impacted as well.[29] Increasing objections to fish exports may be related to the broader effort to secure domestic food supplies in Ireland during this time of crisis.

A letter patent from January 1356 gives some indication of the impact of the plague on Irish towns. The letter details the customs charged on goods entering and leaving the port of Drogheda, which 'had been impoverished both by the great and horrible pestilence and because of merchandise of the burgesses accidentally lost at sea'. The customs were being levied to pay for a new wall to protect the town from attack. Included on this list were herring, salmon and generic 'dried fish'.[30] The same fish are also listed in general customs duties related to the whole of Ireland that the Irish Parliament enacted in 1382.[31] The presence of these species again indicates their commercial importance in this period.

In 1355, Thomas de Rokeby, Justiciar of Ireland, sent a letter to the mayor and bailiffs of Dublin in which he referred to the 'frequent' complaints of Dublin's inhabitants regarding people forestalling the fish market and exporting fish from the Pale without a licence.[32] Forestalling was the practice of buying goods before they reached the markets to avoid city taxes.[33] De Rokeby writes that 'loders or braggers take fish in the ports of Dublin, Howth, Malahide, Rogerston, Portraghlyn, Russhe, Holmpatrick, Dalkey and elsewhere, and buy them and by night and dawn or twilight bring them to the city and salt them secretly in their houses, or put them in casks and convey them by night to places where the king is at war'.[34] Exporting fish to conflict zones could enable forestallers to obtain prices far above the domestic market value of the fish. The purpose of the laws against forestalling and export was to control market prices and ensure a steady supply of food to urban centres like Dublin. Preserving prices, food security, and

Pre-Industrial England', *Economic History Review*, 63 (2010), pp. 281–314, at p. 287.

27 *CHI vol. 1*, p. 410.

28 Such as, Campbell, 'Nature as Historical Protagonist'.

29 Maria Kelly, '"Unheard-of Mortality": The Black Death in Ireland', *History Ireland*, 9, no. 4 (2001), pp. 12–17.

30 *CIRCLE*, Richard II, Patent Roll 4, no. 2.

31 *CIRCLE*, Richard II, Patent Roll 5, no. 247

32 *CIRCLE*, Edward III, Close Roll 29, no. 39, RCH 59/39.

33 Timothy O'Neill, *Merchants and Mariners in Medieval Ireland* (Dublin, 1987), p. 31; Wendell Herbruck, 'Forestalling, Regrating and Engrossing', *Michigan Law Review*, 27 (1929), pp. 365–88.

34 *CIRCLE*, Edward III, Close Roll 29, no. 39, RCH 59/39.

famine prevention were constant concerns for the authorities in all medieval and early modern states.[35]

However, despite the concern with preserving home supplies, the Chancery records show that a select few merchants were granted licences to export their fish. An export licence from 1386 permitted cod, herring, and salmon to be sent to the priory of Llanthony, near Gloucester in England.[36] A grant of a licence from 1387 gave Robert Crull the right to load grain and fish on his ship and transport it to 'England, Wales, Bayonne, Bordeaux and Portugal'.[37] But overall, this trade was somewhat limited, and there is strong evidence to show that Ireland was far less economically productive in the fourteenth century than it had been a century earlier. It is estimated that the Irish exchequer collected an average of £6,300 a year in 1278–99, and only £2,512 from 1368–84.[38]

In the fifteenth century, mentions of fish in the Chancery records become very scarce, but that does not mean that fisheries were in decline; other evidence indicates the opposite. But before exploring that evidence, we can look at a macro view of Chancery records to see what they tell us about the broader pattern of fisheries in medieval Ireland. Map 1.2 shows both freshwater and sea fishing sites mentioned in Chancery documents.[39] These records are of course inherently biased towards the English parts of Ireland, so do not show the full pattern of fisheries. For example, the Aran Islands is the only sea fishing site on the west coast. Understanding the bigger pattern of fishing sites for Gaelic Ireland in the medieval period is far more difficult: sources are limited and what is available, such as the Annals, rarely contains specific details. Fortunately, it is far easier to map out the extent of Gaelic fisheries in the sixteenth century, the results of which will be outlined in the following chapter.

Although limited, the Chancery data shows that there were many smaller fishing ports on the south and east coast, as we might expect. The cities of Limerick and Galway also feature prominently; both were established Norman trading towns by the thirteenth century. The most frequently mentioned location was the Shannon estuary, which generated many records relating to granting fishing rights. The highest concentration of individual locations was around Dublin, with Holmpatrick (Skerries), Rush, Rogerstown estuary, Portrane, Malahide, Howth, Dublin city, Chapelizod and the River Liffey all being discussed as fishing sites.

[35] For more on the Irish case, see Raymond Gillespie, 'Harvest Crises in Early Seventeenth-Century Ireland', *Irish Economic and Social History*, 11 (1984), pp. 5–18; Margaret E. Crawford (ed.), *Famine: The Irish Experience 900–1900: Subsistence Crises and Famines in Ireland* (Edinburgh, 1989).

[36] *CIRCLE*, Richard II, Close Roll 9, no. 64.

[37] *CIRCLE*, Richard II, Patent Roll 10, no. 163.

[38] *CHI vol. 1*, p. 411.

[39] The first site is mentioned in 1258 and the last in 1497.

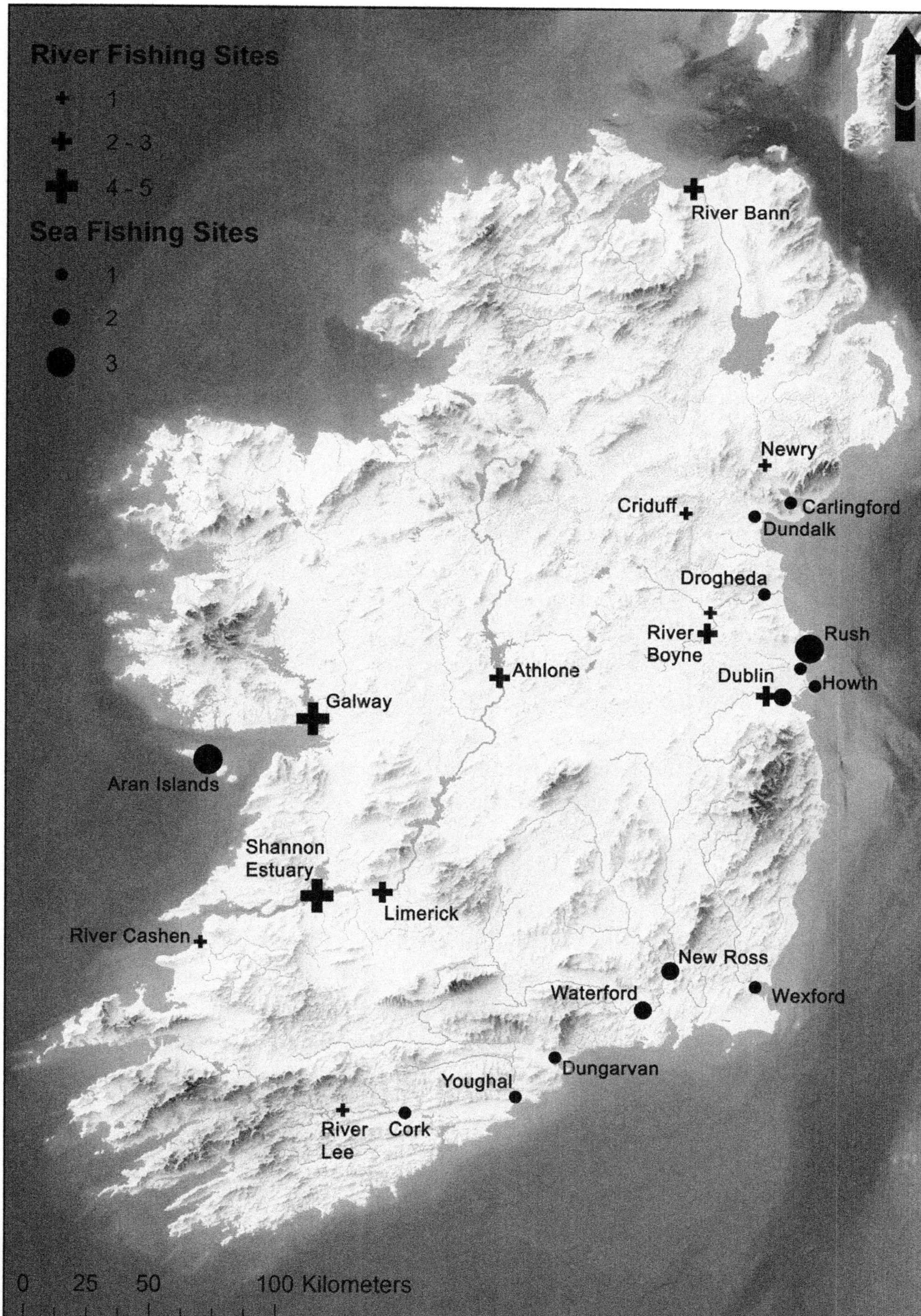

Map 1.2. Fishing locations in Ireland, 1258–1497.

The river Bann is the only fishing site on the north coast appearing in these records, and as we will see later, the river continued to host a major salmon fishery in the following centuries.

The decline in the number of Chancery records in the fifteenth century does not indicate that fisheries were in decline; rather, it reflects the fact that the English state had lost much of its power in Ireland by this time, and in its place, there had been a revival of Gaelic power. With this came general economic growth, in which fisheries may have played a part.

Tim O'Neill has claimed that the arrival of the herring shoals on the Irish coast was the most significant economic event of the fifteenth century.[40] The ecological basis for this claim is questionable, but it is true that fisheries then experienced a period of rapid growth, which may have contributed to a general increase of prosperity across Ireland. An estimated 3,000 tower houses were built by Gaelic and Old English lords over the late fourteenth and fifteenth centuries.[41] These buildings could signal a greater need for defence, but they were also centres of commerce and trade, and towers built on the coast were heavily involved in controlling and regulating fisheries.[42]

There was also a marked increase in the building and refurbishment of monastic houses and friaries in the fifteenth and sixteenth centuries. Many of these construction projects took place in the west of Ireland and were sponsored by Gaelic lords; religious orders such as the Franciscans benefited particularly from the building boom.[43] Trade data from the fifteenth century also attests to the rapid growth of the fishing industry around 1400. In 1404/5 £2,091 worth of Irish goods were imported into Bristol, of which fish accounted for £1,754 (84%).[44] The export of fish would continue to grow and reach even higher levels in the late fifteenth century.

The growth of Irish fisheries also relied on broader trends across Europe. Many fishers who worked in Ireland during the fifteenth century were not Irish, but came from England, France, Spain and Scotland amongst other places. The growth of foreign fisheries relied on the parallel development of the fishing industry in these visitors' home countries. There was a general growth of fisheries in south-west England in the fifteenth century; these English fishers came to

40 O'Neill, *Merchants and Mariners*, p. 131.

41 *CHI vol. 1*, p. 411.

42 Colin Breen, 'Maritime Connections: Landscape and Lordship along the Gaelic Atlantic Seaboard of Scotland and the North of Ireland during the Middle Ages', *Maritime Communities of the North Atlantic Arc in the Early Modern Period, Journal of the North Atlantic*, Special Volume 12 (2019), pp. 3–15, at p. 9.

43 O'Neill, *Merchants and Mariners*, p. 36; *CHI vol. 2*, p. 538.

44 Wendy R. Childs, 'Ireland's Trade with England in the Later Middle Ages', *Irish Economic and Social History*, 9 (1982), pp. 5–33, at p. 15.

Ireland, but they also fished within home waters and went as far as Iceland.[45] In this way, Ireland was part of a wider Europe-wide trend, which saw fishers willing to travel much further from home waters to make catches.

These visitors combined with Ireland's domestic fishers to make the industry reach its height in the late fifteenth and early sixteenth centuries. The cooperation between different fishing groups came to define fisheries in this period. The following chapter will explore these visiting fisheries in detail, but it is important to note that it was probably the decline of English power and the Gaelic revival of the fifteenth century that allowed these fisheries to prosper. The political situation in fifteenth-century Ireland was ripe to welcome visitors; the absence of a central controlling or domineering state meant that the Gaelic Irish, Old English and many port towns were free to establish their own fisheries and keep much of the profit. They could also develop fishing and trading relationships with each other and with visitors from across Europe. The Tudor state's supremacy in Ireland would end much of this.

The income from these fisheries could help explain the fifteenth-century building boom, but we must also acknowledge that this was a general period of economic growth.[46] Agricultural production rose too, with cattle-herding increasing markedly in Ireland and across Europe.[47] Establishing the relative importance of different industries like fisheries or livestock is not easy, but overall, agriculture has received more attention from historians. A recent survey of economic trends in medieval Ireland outlined cattle-herding and grain production in some detail, but consigned fisheries to a single paragraph.[48] This is not surprising, since the extent of the industry has never been fully delineated. While this book cannot conclusively establish the importance of fisheries relative to cattle-herding or other branches of agriculture, by establishing the scale and nature of fisheries in fifteenth- and sixteenth-century Ireland, we can gain a greater understanding of the true economic significance of the industry.

To that end, the next section will offer a snapshot of fisheries around their most prosperous period in the early years of the sixteenth century. Subsequent chapters will further trace how and why fisheries changed over the sixteenth century, thus providing a portrait of fisheries around their high-water mark which will provide a valuable baseline to demonstrate later changes.

[45] M. Kowaleski, 'The Expansion of the South-Western Fisheries in Late Medieval England', *Economic History Review*, 53 (2000), pp. 429–54, at p. 430.

[46] *CHI vol. 2*, p. 538.

[47] *CHI vol. 1*, p. 412.

[48] *CHI vol. 1*, p. 414.

Overview of Fisheries around 1500

The natural place to start this tour of the Irish fisheries around the year 1500 is in Dublin, the capital of the Pale and the administrative centre of the English regime in Ireland.[49] Dublin city lies in the fold of a sizeable bay where the River Liffey meets the sea; prior to the modern sea walls' construction, the town could only be reached through a narrow channel running between shallow beaches and shifting sandbanks. In part due to the inherent difficulty of navigating into Dublin port, several smaller satellite harbour towns sprang up around the city, acting as additional landing-places for fishing boats that fed fish to the city. In royal correspondence from the medieval period, the phrase 'all harbours from Holmpatrick in the North to Dalkey in the south' was used to describe the catchment area of fish brought into Dublin.[50]

It seems that Dublin was more consumer-focused than towns like Waterford, which operated a much stronger export trade in the medieval and early modern periods. Unfortunately, there are no population estimates for Dublin in 1500, but the city likely had around 5,000 inhabitants in the year 1600, which made it by far the largest settlement in Ireland.[51] The whole island of Ireland probably had a population of only between 0.5 and 1 million people in 1500.[52] Dublin did export some fish and had a strong trading relationship with Chester, but its size meant that Dublin needed a stable supply of locally caught seafood, leaving a relatively small surplus for export.

The next major fishing town north of Dublin was Drogheda, which shared many of the same characteristics. It was a Norman town in the Pale with a strong trade connection with England.[53] Drogheda lies at the mouth of the river Boyne, which has a long history as a salmon-fishing river, going right back into Irish mythology.[54] Comparatively little is known about the town's sea fisheries. The customs list from the Chancery records discussed earlier indicated that the town had a thriving international trade from at least the fourteenth century.

[49] H. B. Clarke, 'Dublin, Part I, to 1610', *IHTA*, 11 (2002), pp. 1–36, at p. 1; *NHI, vol. III*, p. 5.

[50] Murphy and Potterton, *The Dublin Region in the Middle Ages*, p. 399.

[51] Raymond Gillespie, *The Transformation of the Irish Economy 1550–1700* (Dublin, 1991), p. 28.

[52] Gillespie, *Transformation*, p. 13; L. M. Cullen, 'Population Trends in Seventeenth-Century Ireland', *Economic and Social Review*, 6 (1975), pp. 149–65, p. 163.

[53] Donald Woodward, 'The Trade of Elizabethan Chester' (PhD thesis, University of Hull, 1970).

[54] A. E. J. Went, 'Material for a History of the Fisheries of the River Boyne', *Journal of the County Louth Archaeological Society*, 13 (1953), pp. 18–33; Chiara Buldorini, 'Drogheda as a Case Study of Anglo-Norman Town Foundation in Ireland, 1194–1412' (PhD thesis, TCD, 2009).

The Irish Parliament of 1460 was held in Drogheda, and a statute then enacted provides further evidence for the town's trade connections.[55] The Act describes various Drogheda merchants freighting a ship for a voyage to 'Westmony' in Iceland.[56] Upon arrival in Iceland, they found 'the stock of fish sold to other merchants of England and this land, so that they could not vend their merchandise'.[57] It was decided that one of the merchants, Nicholas Abbot, would remain in Iceland with his servant to sell all the goods over several months. However, Abbot's partners never returned, and the Icelanders eventually confiscated his goods. The statute reviewed the case and decided if Abbot was owed compensation. The English fishery in Iceland in the fifteenth century is well understood, but we have little evidence of Irish involvement, which makes this statute a rare glimpse into this network.[58]

Just north of Drogheda is the small port of Clogherhead. Another statute from the Irish Parliament of 1475–76 refers to rebuilding the town's fishing pier. To pay for the reconstruction, every boat using the port had to pay 4*d.* and every ship 8*d.* Additionally, every boat fishing for herring had to pay a mease of herring (600 fish) per year.[59] This reference suggests that there was an active herring fishery operating here during the late fifteenth century.

Further north is the market town of Dundalk. There was fishing in the town from at least 1408 when John Rowe was granted access to the waters.[60] The town had been a free borough since the fourteenth century, and a renewal of that right in 1465 granted residents 'fishing of the water of Dundalk as amply they had ever enjoyed'.[61] Dundalk was well positioned to exploit spawning herring populations in this part of the Irish Sea, but we know that Carlingford, just to the north, was more heavily involved in that activity.

Carlingford, and the bay with the same name, was the last outpost of the Pale and served as a significant centre for herring fishing during the fifteenth and sixteenth centuries. The town initially grew around King John's Castle, built before 1200. Today the ruins of the tholsel and the mint (an elaborate merchant's

55 *Statute Rolls, Ireland*, vol. II, 38 Henry IV, A.D. 1460, pp. 697–99.

56 Westmony is an Anglicised name for the port of Vestmannaeyjar on the Icelandic south coast.

57 *Statute Rolls, Ireland*, vol. II, 38 Henry IV, A.D. 1460, p. 697.

58 Wendy R. Childs, 'England's Icelandic Trade in the Fifteenth Century: The Role of the Port of Hull', in Poul Holm et al. (eds), *Northern Seas Yearbook* (Esbjerg, 1995), pp. 11–31; O'Neill, *Merchants and Mariners*, p. 37.

59 Mease was a contemporary term that usually referred to 5 long hundreds, or 600 fish. *Statute Rolls, Ireland*, vol. IV, 15 & 16 Edward IV, A.D. 1475–76, p. 389.

60 *CIRCLE*, Henry IV, Patent Roll 10, no. 18.

61 Anngret Simms et al., 'Dundalk', *IHTA*, 16 (2006), pp. 1–24, at p. 2; *CIRCLE*, Edward IV, Patent Roll 5, no. 2.

residence), which were built in the fifteenth century, still survive.[62] These remains suggest the fifteenth and early sixteenth centuries were extremely prosperous periods for the town, and it is likely much of that prosperity came from the herring fishery. Today the town is known for its oyster beds, and there is evidence that oysters were harvested in sixteenth-century Carlingford as well.[63] Carlingford was granted to the earl of Kildare in 1506 but was confiscated following the failed rebellion of 1534–35. In 1535, the English army visited the town under William Brabazon, then Treasurer at War. He reported that 600 English fishing boats were taking part in the seasonal herring fishery,[64] providing solid evidence that Carlingford hosted a sizeable herring fishery in the early sixteenth century.

Just north of Carlingford is the small port town of Kilkeel, which was referred to as 'a key for fisher boats' in 1497.[65] Up the coast from Kilkeel is the harbour of Dundrum, which was said to host 'small boats and barks' in 1497.[66] Dundrum's castle was under the control of the Gaelic Irish in the early sixteenth century but was recaptured by the English in 1537.[67] Archaeological remains from the townland of Rossglass on the shore of Dundrum Bay also provide insight into how small-scale local fisheries operated in the medieval and early modern periods. A stone fish trap used to capture fish using the tides has been uncovered in the bay, one of many found in County Down.[68] Evidence suggests that these traps were still in use during the sixteenth century and would have provided locals with fresh fish for consumption.[69]

Like Carlingford, the town of Ardglass, just east of Dundrum, experienced a herring fishing boom in the fifteenth century. The town used the revenue from the fishery to build a raft of new structures in the early fifteenth century, including a complex with shops for merchants and a communal dining room for sailors, as well as a new tower house.[70] From 1515, the earl of Kildare controlled

62 Anngret Simms et al., 'Carlingford', *IHTA*, 23 (2011), pp. 1–15, at p. 3.

63 Aidan O'Sullivan and Colin Breen, *Maritime Ireland: An Archaeology of Coastal Communities* (Stroud, 2007), p. 223.

64 *Cal. Car. MSS*, 1515–74, no. 70, p. 85; D. B. Quinn, 'Anglo-Irish in Ulster in the Early Sixteenth Century', in Arthur Deane (ed.), *Proceedings and Reports of the Belfast Natural History and Philosophical Society* (Belfast, 1935), pp. 56–78, at pp. 61–64.

65 *CSPI*, 1601–03, no. 696, pp. 676–77.

66 List of, and notes on, the Havens of Ireland, 1497 (TNA, SP 63/214, f.36).

67 Quinn, 'Anglo-Irish in Ulster', p. 65.

68 Colin Breen, 'A Medieval Fishery at Rossglass, Co. Down', *Ulster Journal of Archaeology*, 58 (1999), pp. 153–55.

69 Breen, 'A Medieval Fishery at Rossglass', p. 155.

70 Mark Gardiner and T. E. McNeill, 'Seaborne Trade and the Commercialisation of Fifteenth- and Sixteenth-Century Gaelic Ulster', *Proceedings of the Royal Irish Academy*, 116C (2016), pp. 229–62, at p. 243; O'Sullivan and Breen, *Maritime Ireland*, p. 188. For more on these buildings, see Mark Gardiner, 'The Late Medieval Port Buildings of Ardglass, County Down, in the Context of Northern European Trade', *Ulster Journal of Archaeology*, 74 (2017–18), pp. 202–12.

the fisheries in Ardglass; Kildare's customs records show that herring and white fish were commonly landed by fishing boats at that time.[71] In 1524, the customs revenues of Ardglass and Strangford were leased for £150; D. B. Quinn writes that this was an 'extensive' amount for the period.[72] During the fifteenth century, several tower houses were constructed in the town, signalling the prosperity it gained from the fishery.

The next major fishing location along the Ulster coast was Strangford Lough. As with Ardglass, the earl of Kildare was granted the fishing rights here in the early sixteenth century.[73] An extensive archaeological survey of the lough found twenty examples of stone and wooden fish traps, mostly dating from the medieval period. There is little visual or documentary evidence relating to fish traps, but a map of south-east Ulster from 1568 has an area named 'pascafdry ennomies' on the east coast of the lough, which suggests that traps were in use at that time.[74] In a description of the Ards peninsula written in 1684, William Montgomery explained how the stone fish traps used the tides to catch fish, but also wrote that 'since fish days were neglected, those yards [fish traps] decayed'.[75] Thomas McErlean and Aidan O'Sullivan have observed that Montgomery's account is solid evidence that the traps 'were abandoned by the end of the sixteenth century', and as we will see later, the fisheries in Strangford were not alone in their decline at this time.[76]

Gaelic Irish and Scottish clans controlled most of the coastline of Ulster in the early sixteenth century, although many of the Scottish families had become indistinguishable from the Irish.[77] Only Carrickfergus stood as an English outpost in Belfast Lough.[78] There are few records detailing fishing operations in sixteenth-century Carrickfergus, but we know that the town had an active fishery in the eighteenth century;[79] and while Belfast is the largest city in Ulster today, in 1500, it was only the site of a small outpost owned by the O'Neills.[80]

71 Quinn, 'Anglo-Irish in Ulster', p. 64.

72 Quinn, 'Anglo-Irish in Ulster', p. 63.

73 Quinn, 'Anglo-Irish in Ulster', p. 64.

74 Thomas McErlean et al. (eds), *Strangford Lough: An Archaeological Survey of the Maritime Culture Landscape* (Belfast, 2002), pp. 146–47. The authors write that this map is from 1568 and was based on a survey from 1567–68. However, the National Archives catalogue dates the map from 1580: see 'Map of the coast of Ireland from Carrickfergus (County Antrim) to Dundrum Bay (County Down)', 1580 (TNA, MPF 1/87).

75 McErlean et al., *Strangford Lough*, p. 149.

76 McErlean et al., *Strangford Lough*, p. 148.

77 O'Neill, *Merchants and Mariners*, p. 31.

78 Quinn, 'Anglo-Irish in Ulster', p. 69.

79 Philip Robinson, 'Carrickfergus', *IHTA*, 2 (1986), pp. 1–16, at p. 6.

80 The O'Neill base was destroyed by the earl of Kildare in 1503; Quinn, 'Anglo-Irish in Ulster', p. 70.

The fishery on the river Bann was one of the most important in Ulster, and the whole of Ireland, during the sixteenth century. The English Crown acquired the rights to the fishery in 1386–87 and leased it to various owners over the following centuries. A fifteenth-century proverb shows English mercantile involvement in the Bann fishery: 'Herring of Sligo and Salmon of Bann, had made in Bristol many a rich man'.[81] A list of Irish havens from 1497 in the State Papers describes the Bann as having 'salmons leaping and with nets taken afore they fall'. It also reported that ships of 40 to 60 tons were engaged in the fishery and usually took two months to get a full cargo of salmon.[82]

The State Papers also tell us that the Bann fishery failed to generate any income in the years 1485–86 and 1495–96. The earl of Kildare was then granted the fishery after 1496, and it subsequently generated a substantial income for him. The earl leased the rights of the river to Otwell Corbett in 1519 for £60. However, Corbett did not pay the full rent and even received compensation, suggesting that he suffered attacks from the native Irish, who also wanted to control and exploit the river's resources.[83] While the English Crown nominally owned the fishing rights to the river, in reality, the O'Neill, O'Cahan, MacQuillan and O'Donnell families actually controlled the Bann fisheries in the early sixteenth century.[84] As we will see, the river remained a site of conflict throughout the rest of our period.

Within Ulster, the two most powerful Gaelic families in 1500 were the O'Neills and O'Donnells. The O'Neills of Tyrone controlled most of central Ulster, while the O'Neills of Clandeboye held the coastlines of modern Down and south Antrim. Salmon was part of the O'Neill identity; even the family crest included a salmon under the red hand of Ulster.[85] Most of north Antrim was controlled by the MacQuillans, whose ancestors had come from Scotland. Another Scottish family, the MacDonnells, controlled the Glens of Antrim on the east coast of Ulster. The majority of modern County Londonderry was in the hands of the O'Cahan clan, while the O'Dohertys controlled the Inishowen peninsula. The remainder of the coast, roughly the area of modern County Donegal, was controlled by the O'Donnells and their MacSweeney vassals.[86] The MacSweeney

[81] O'Neill, *Merchants and Mariners*, p. 35. This quote has been corrected to modern English.

[82] *CSPI*, 1601–03, p. 676; list of, and notes on, the Havens of Ireland, 1497 (TNA, SP 63/214, f.36).

[83] Quinn, 'Anglo-Irish in Ulster', p. 67.

[84] Audrey J. Horning, 'On the Banks of the Bann – the Riverine Economy of an Ulster Plantation Village', *Historical Archaeology*, 41 (2007), pp. 94–114, at p. 95.

[85] A sixteenth-century example of this can be seen in a letter from Hugh, earl of Tirone to the Privy Council, 31 October 1591 (TNA, SP 63/160, f.126).

[86] *NHI, vol. III*, pp. 15–17; Simon Egan, 'By Land and by Sea: The Role of the Maritime Sphere in the Expansion of O'Donnell Power ca. 1380–1500', *Maritime Communities of the North Atlantic Arc in the Early Modern Period, Journal of the North Atlantic*, Special Volume 12 (2019),

clan also traced their heritage to Scotland but had essentially become Gaelic Irish lords by the sixteenth century.[87]

Gaelic Ireland did not have urban centres comparable to the Anglo-Norman settlements of Ireland, but it did have tower houses that acted as centres of power, administration, and trade. Colin Breen and Aidan O'Sullivan have argued that tower houses need to be viewed as 'working buildings, oriented towards marine economies'.[88] In that respect, the placement of these tower houses can help us unravel the marine resources valued by coastal communities. Marine resources do seem to have been important to the economy of the O'Donnells in north Donegal. An assessment of tower house placement in the O'Donnells' land suggested the majority were positioned near the sea or navigable rivers.[89] The plantation town of Derry would later become a major exporter of salmon, but in the sixteenth century, an O'Donnell tower house stood to keep control over the river.[90] The same is true in other parts of the coast; for example, the tower house at Carlingford was positioned towards the sea to facilitate the storage of herring catches.[91]

A later English survey from 1608 catalogues the fisheries along the coast from Derry to Ballyshannon, showing a diverse range of fishing locations in this area with salmon, herring and ling as the main targets.[92] What is today the town of Donegal was the site of the largest O'Donnell castle, constructed in 1474–75 and reported in 1608 to hold a weekly market. Closer to the shore of Donegal Bay, the O'Donnells had another castle, described in 1601 as the place 'where the ships used to ride'.[93] In many respects, the O'Donnell infrastructure in Donegal acted as a town, but unfortunately, much of the written material that describes it is retrospective in nature.

It seems likely too that Ballyshannon was a centre for fishing. Records from later in the sixteenth and seventeenth centuries tell us that the river Erne held an extensive salmon and eel fishery.[94] There is evidence that merchants, familiar with the area and language, came to O'Donnell territory in the sixteenth century. In 1539 a Galway merchant who had travelled to O'Donnell's country with a ship

pp. 16–31; Colin Breen and John Raven, 'Maritime Lordship in Late-Medieval Gaelic Ireland', *Medieval Archaeology*, 61 (2017), pp. 149–82.

87 Breen, 'Maritime Connections', p. 9.

88 O'Sullivan and Breen, *Maritime Ireland*, p. 164.

89 Máire Ní Loingsigh, 'An Assessment of Castles and Landownership in Late Medieval North Donegal', *Ulster Journal of Archaeology*, 57 (1994), pp. 145–58, at p. 148.

90 R. J. Hunter and Brendan Scott (eds), *The Ulster Port Books, 1612–15* (Belfast, 2012); Avril Thomas, 'Derry-Londonderry', *IHTA*, 15 (2005), pp. 1–47, at p. 2.

91 O'Sullivan and Breen, *Maritime Ireland*, p. 164.

92 *Cal. Car. MSS*, 1603–25, pp. 13–22.

93 Gardiner and McNeill, 'Seaborne Trade', p. 254.

94 Went, 'Fishing Weirs'.

full of wine and was relaying intelligence about an impending attack by O'Neill in the Pale, reported that a ship from Bristol was currently at Assaroe and would 'stay two months getting salmon'.[95] Assaroe was the name of the local abbey and waterfall near Ballyshannon.[96] The merchant reported that the master of the Bristol ship also spoke Irish, which suggests this was a regular practice for him.

West of Ballyshannon, across Donegal Bay, the O'Donnells held another castle at Killybegs, known as McSwyne Banagh. Today Killybegs is by far the largest fishing port in Ireland.[97] Some of Ireland's most important commercial species, such as mackerel and horse mackerel, are caught in large numbers on the continental shelf edge off the west coast and landed in Killybegs.[98] Species on the continental shelf were out of the reach of early modern fishers, but large herring spawning areas close to shore in the west and north of Donegal were accessible to fishers in the sixteenth century.[99]

The earliest report of herring fishing in the region comes from 1543. In that year, Manus O'Donnell granted away several rights to his lands and marine resources to Henry VIII, including a promise that 'if a multitude of herrings and other fish called "garbhushe"[100] should resort to the country of Lord O'Donnell or to his sea, such as now comes into Arayn or Ynisnycadyrn, one half thereof shall belong to the king and the other to O'Donnell'.[101] This suggests that herring were not always present on the Donegal coast, but when they were available, they were targeted by fishers. Based on this account, we can say that the herring fishery was more active in 'Arayn', or the Aran Islands, at this time.

South of Ballyshannon is the town of Sligo, located at the junction between Ulster and Connacht. This position meant it was a flashpoint for conflict over the centuries, changing hands between the O'Donnells and the O'Connors of Sligo, who controlled the town in 1500.[102] Sligo built its first quay in 1423 and covered the expense by collecting customs taxes.[103] Goods exported from the town

95 *CSPI* (IMC), 1509–47, no. 470.2, p. 236.

96 Went, 'Fishing Weirs', p. 213.

97 H. D. Gerritsen and E. Kelly (eds), *Atlas of Commercial Fisheries around Ireland* (3rd edn, Oranmore, 2019), p. 6.

98 Gerritsen and Kelly (eds), *Atlas of Commercial Fisheries around Ireland*, p. 42; L. Dransfeld et al., *North Western Waters Atlas* (3rd edn, Oranmore, 2014).

99 David O'Sullivan et al., 'An Inventory of Irish Herring Spawning Grounds', *Marine Institute, Irish Fisheries Bulletin*, 42 (2013), pp. 1–31, at p. 21.

100 'Garbhushe' is an Irish name for shad; J. Murray writes, 'Shad was, by the older naturalists, styled the mother of herrings': J. Murray (ed.), *State Papers, King Henry the Eight, Part III – Continued, 1538–1546* (London, 1834), p. 590.

101 *Cal. Car.* MSS, 1515–74, no. 182, pp. 208–09.

102 Fióna Gallagher and Marie-Louise Legg, 'Sligo', *IHTA*, 24 (2012), pp. 1–27, at p. 1.

103 Gallagher and Legg, 'Sligo', p. 1.

in the fifteenth century included herring and other generic 'fish'.[104] The English saying 'Herring of Sligo and Salmon of Bann' further suggests an active herring fishery and a trading connection to Bristol. In 1400 a licence was issued for a Bristol merchant to travel to Sligo with 20 tons of old wine and exchange it for salmon.[105] And in the 1450s we have a record of a ship going from England to Lisbon to collect salt and then to Sligo to pick up hides.[106] This entry makes no explicit mention of fish, but we could speculate that the salt was for processing fish, although it could also have been for curing animal hides.

Unfortunately, there is little evidence for fisheries in Sligo during the sixteenth century, and it seems clear that the town's prosperity was damaged as it became increasingly embroiled in conflict.[107] A later English survey of the coast of Connacht from 1614 reported that a ship of 200 tons could anchor in Sligo Bay, and that the area had 'excellent fishing for cod and linge without, and salmon within'.[108] Export figures from 1621/2 also show small amounts of salmon and pilchards exported from the town.[109]

The coastline directly south of Sligo was controlled by the Gaelic Irish O'Dowds. Colin Breen and John Raven have suggested that because this coastline had limited landing-places but rich inland agriculture, the O'Dowds may not have engaged with the sea as much as other Gaelic families.[110] The northern coast of Mayo was dominated by the Old English Lower MacWilliam Burke family, while the Gaelic Irish O'Malleys occupied the south coast.[111] We know very little about the maritime activities of the Burkes, but the O'Malleys had a strong maritime tradition and a reputation for piracy.[112] The O'Malley tower houses on Clare Island and in Clew Bay were both well positioned to control access to the sea, which suggests this is where the clan's focus lay.[113]

As the soil in this part of the Mayo coast is poor, it is not surprising that the O'Malleys looked to the ocean for resources. The area is also dotted with many islands, including Achill, Inisturk and Inishbofin. The 1614 English survey of Connacht described Inishbofin as 'good fishing for cod and ling, and it is much frequented by his Majesty's subjects and others'. Similarly, Inisturk is described as

[104] Gallagher and Legg, 'Sligo', p. 1.

[105] O'Neill, *Merchants and Mariners*, p. 39.

[106] E. M. Carus-Wilson (ed.), *The Overseas Trade of Bristol in the Later Middle Ages* (Bristol Record Society, vii, Bristol, 1937), pp. 106–07.

[107] Gallagher and Legg, 'Sligo', p. 2.

[108] *Cal. Car. MSS*, 1603–25, no. 152, p. 297.

[109] Donald Woodward, 'Irish Trade and Customs Statistics, 1614–1641', *Irish Economic and Social History*, 26 (1999), pp. 54–80.

[110] Breen and Raven, 'Maritime Lordship', p. 169.

[111] *NHI, vol. III*, p. 14.

[112] O'Sullivan and Breen, *Maritime Ireland*, p. 165.

[113] Breen and Raven, 'Maritime Lordship', p. 169.

having fishing for cod and ling, and 'much haunted by the English and Dutch'. While Clew Bay was described as having very good fishing, Inishkea South was said to have the 'best and most plentiful fishing of all the west of Ireland'.[114]

This 1614 account must be problematic in some ways because of the date of composition, and it is not clear how well the situation described in the seventeenth century reflected the situation in 1500. Furthermore, the point of the survey was to promote and encourage English plantations in Connacht, with the author, Sir Oliver St John, providing suggestions on where forts could be strategically placed to achieve control of the province. It was to St John's benefit to exaggerate the fishing potential of the area, to give his scheme a better chance of becoming policy. However, his description shows there was potential for fisheries on this coast, and it is certainly feasible that the O'Malleys exploited these waters in the sixteenth century and taxed fishers who travelled from elsewhere to exploit them too. As we shall see later, the facilitation and taxation of visiting fishers were common practices among other Gaelic Irish families in the sixteenth century.

South of Mayo, the independent O'Flaherty family controlled most of the modern coast of County Galway. The O'Flahertys were known for their naval strength in the medieval period, but their prowess at sea had waned by the sixteenth century.[115] Breen and Raven note that the O'Flahertys had two main concentrations of tower houses, one inland and one along the coast. The coastal towers were positioned in mostly hard-to-reach inlets, which suggests the clan were expert navigators and would have increased the defensive capabilities of the sites.[116] In terms of fishing, we know little about the O'Flahertys, but given their skills at sea, we can assume they did engage in some fishing or perhaps welcomed foreign fishers to their shore.

Of all the west coast islands, the strongest evidence for fisheries is on the Aran Islands, located just outside Galway Bay. In 1379 the king granted the rights to fish on the Aran Islands to an English noble, Theobald Verdun.[117] As Norman power declined in the following centuries, the islands came back under Irish control and were in the hands of the O'Brian family in 1500. They were strategically important for controlling trade entering and leaving the port of Galway, and in the early sixteenth century, the O'Brians kept the bay clear of pirates in exchange for a yearly tribute of wine from Galway.[118]

Like other locations on the west coast, the islands were visited by continental fishers in the early modern period. In 1511, a ship left the Basque country and, having picked up salt at La Rochelle, travelled to the Aran Islands, where it used

[114] *Cal. Car. MSS*, 1603–25, no. 152, p. 297.

[115] Breen and Raven, 'Maritime Lordship', p. 157.

[116] Breen and Raven, 'Maritime Lordship', pp. 167–68.

[117] *CIRCLE*, Richard II, Patent Roll 2, no. 9.

[118] O'Neill, *Merchants and Mariners*, p. 127.

six smaller boats for 'fishing hake and one for killing sardines'.[119] The mention of sardines is interesting, as the fish are not always present on the Irish coast.[120] The record from 1543 mentioned earlier also tells of herring present off the islands in that year.[121] Sir Oliver St John also wrote in 1614 that 'English, Bretons and Portugalls (in times past) had a great trade of fishing here for cod, ling, hake, and conger' on the islands.[122] Thus, in this case, St John's report is supported by other evidence, and it does seem likely that foreign fishers visited the islands.

East of the Aran Islands, at the head of Galway Bay, lies Galway town. Galway was an important trading post and an independent Old English town in 1500, perhaps the most detached from the central government in Dublin. It was the second-largest city in Ireland, with our earliest population estimates starting in 1600 at 4,000 residents.[123] Galway had strong trade connections to continental Europe in the sixteenth century, and owing to generous charters, the town kept most of the revenue from its trade.[124] Among the records of the English Admiralty court, we find merchants from Galway engaging in voyages to every part of Europe.[125] These do not always originate from Galway, but they nonetheless show that Galway had strong links to continental European ports. A customs list from Galway in 1395 contains a wide array of exotic goods and several fish species, including cod, herring, eel, whiting and salmon by the individual fish and pipe.[126] It is also likely that Galway was a supply hub for English (and possibly Irish) ships that were heading to the Icelandic cod fishery.[127]

The available archaeological evidence indicates there was a wide array of marine species present in medieval and early modern Galway. These include shark, ray, herring, cod, and gurnard.[128] The town also had an important freshwater fishery in both Corrib lough and river, providing an abundance of salmon, trout and eel

[119] Michael M. Barkham, 'The Spanish Basque Irish Fishery and Trade in the Sixteenth-Century', *History Ireland*, 9, no. 3 (2001), pp. 12–15, at p. 14.

[120] Liam Downey and Ingelise Stuijts, 'Historical Fisheries: Irish Pilchard Fishery and Historical Fish Harvesting and Processing Methods, Fish Products and Consumption', *Kerry Archaeological & Historical Journal*, 2 (2014), pp. 1–35; A. E. J. Went, 'The Irish Pilchard Fishery', *Proceedings of the Royal Irish Academy*, 51B (1946), pp. 81–120.

[121] *Cal. Car.* MSS, 1515–74, no. 182, pp. 208–09.

[122] *Cal. Car.* MSS, 1603–25, no. 152, p. 296.

[123] Anthony Sheehan, 'Irish Towns in a Period of Change, 1558–1625', in Ciaran Brady and Raymond Gillespie (eds), *Natives and Newcomers: Making of Irish Colonial Society, 1534–1641* (Dublin, 1986), pp. 93–119.

[124] *NHI, vol. III*, p. 13; Sheehan, 'Irish Towns in a Period of Change', p. 95.

[125] *Cal. HCA 1536–1641*, see for example, no. 16, p. 3; no. 55, p. 11; no. 122, p. 23; no. 247, p. 49; no. 254, p. 52; no. 260, p. 53; no. 275, p. 57; no. 278, p. 58; no. 407, p. 88; no. 429, p. 94; no. 721, p. 166; no. 722, p. 166.

[126] *CIRCLE*, Richard II, Patent Roll 19, no. 4.

[127] O'Neill, *Merchants and Mariners*, pp. 36–37.

[128] Hamilton-Dyer, 'Exploring the Contrast', p. 235.

that were sold in the market three times a week. The salmon fishery was important from the fourteenth century when Chancery records refer to its granting to Richard Panys, a burgess of Bristol.[129] Shortly afterwards, the king ordered two burgesses native to Galway to stop interfering in the fishery, indicating its importance to the Crown.[130] The rights to the salmon fishery continued to be granted and rented out in the sixteenth century, with three fishing spots being rented to Jenet Lynch in 1521 for 10*s*. a year.[131] Barnaby Goog's map of Galway in 1583 shows salmon fishing taking place from the town's bridge, with the fishers employing both nets and spears.[132]

Despite Galway's thriving salmon fishery, there is little evidence of a large-scale export of fish; perhaps, as in Dublin, the local population consumed much of the fish caught. No customs accounts or port books have survived for sixteenth-century Galway. Later customs statistics from 1621–22 show a small export of salmon and pilchards, but far less when compared to other ports like Limerick or Wexford.[133] Galway clearly had an active salmon fishery and consumed a wide variety of fish, but too few records remain to estimate the scale of these activities.

South of Galway, the imposing coast of Clare provides few access points to the ocean. However, the Gaelic O'Loughlin Burren built a series of tower houses along the shoreline. The tower house near Pouldoody was positioned by a prominent oyster fishery, which suggests once again that the physical infrastructure of the O'Loughlins was in part placed to control marine resources.[134] Further south along the coast is the port town of Limerick, located close to where the river Shannon runs into the 113 km long Shannon estuary.[135] Originally a Viking town, Limerick was further developed by the Normans, and by the sixteenth century, it was, like Galway, an independent port town surrounded by Irish and Old English lords. Limerick was the third largest town in Ireland in the sixteenth century, with a population of 3,000 by 1600.[136] A map from 1587 clearly shows the maritime character of the town, with many ships illustrated, sailing and anchored in the river.[137] Archaeological evidence

[129] *CIRCLE*, Richard II, Patent Roll 10, no. 32, RCH 131/29; Richard II, Patent Roll 10, no. 104, RCH 133/85.

[130] *CIRCLE*, Richard II, Patent Roll 10, no. 105, RCH 133/85.

[131] M. D. O'Sullivan, *Old Galway* (Cambridge, 1942), p. 409.

[132] Arthur Went has also thoroughly explored the methods used in the Galway salmon fishery: A. E. J. Went, 'The Galway Fishery. An Account of the Ownership of the Fishery', *Proceedings of the Royal Irish Academy*, 48C (1942), pp. 233–53; 'The Galway Fishery. An Account of the Modes of Fishing Together with Notes on Other Matters Connected with the Fishery', *Proceedings of the Royal Irish Academy*, 49C (1943), pp. 187–219.

[133] Woodward, 'Irish Trade and Customs Statistics', p. 70.

[134] Breen and Raven, 'Maritime Lordship', p. 167.

[135] Eamon O'Flaherty, 'Limerick', *IHTA*, 21 (2010), pp. 1–59, at p. 1.

[136] Sheehan, 'Irish Towns in a Period of Change', p. 97.

[137] Francis Jobson, 'The Plotte of the Citie of Limbrick', 1587 (TNA, MPF 1/96).

further suggests the town experienced a period of prosperity and a building boom in the fifteenth century.[138] And while we cannot be sure, this growth may have been related to the expansion of fisheries in the same period.[139]

Like Galway and the Corrib, Limerick had an important salmon fishery on the Shannon. Most of the surviving records relating to fisheries concern the granting and protection of the city's extensive network of weirs. In 1496, the earl of Desmond was granted all profits of the city's 'fishing weirs' and 'fisheries'.[140] A similar grant was re-issued in 1497.[141] The town does not often appear in the English or Welsh customs accounts or port books, as it likely had better trading connections to continental Europe. Records from the late sixteenth century indicate a small trade of salmon to Spain and elsewhere. In 1598 the *Phoenix* was freighted in Limerick with various goods, including one barrel of salmon, for a voyage to Spain.[142] Customs records from 1616–17 show that the only fish the town exported in that period was 77 tons of salmon.[143]

While Limerick had few marine fisheries, the Shannon estuary was a major fishing centre in the medieval and early modern periods. The river and its estuary were the most frequently mentioned site in the Chancery records. Archaeologists have found the remains of wooden posts and wattle fish traps in the estuary dating from the fifth century at the earliest, although the use of these structures seems to have peaked from the eleventh to the fourteenth centuries. These worked by funnelling fish into baskets that could be collected by hand. According to Aidan O'Sullivan, the traps could be used to make large catches of 'salmon, sea trout, lampreys, shad, flounder and eels'.[144] The catches made in the estuary could have supplied the local population with fish, but they could also have been preserved and sold at markets in Limerick or exported through the town's port.[145]

There is also evidence of active sea fisheries in the area. A statistical survey of County Clare from 1808 refers to a grant by Queen Elizabeth to the city of Limerick of the rights to a herring fishery on Scattery Island, which lies near the mouth of the Shannon estuary. It stipulated that the city would get 'from every herring boat,

138 O'Flaherty, 'Limerick', p. 4.

139 O'Neill, *Merchants and Mariners*, p. 36.

140 *CIRCLE*, Henry VII, Patent Roll 12, no. 1.

141 *CIRCLE*, Henry VII, Patent Roll 12, no. 4.

142 *Cal. HCA 1536–1641*, p. 94, no. 431.

143 Woodward, 'Irish Trade and Customs Statistics', p. 68.

144 Aidan O'Sullivan, 'Medieval Fish Traps on the Shannon Estuary, Ireland: Interpreting People, Place and Identity in Estuarine Landscapes', *Journal of Wetland Archaeology*, 5 (2013), pp. 65–77, p. 70. Arthur Went has also outlined the use of head weirs in the Shannon Estuary: A. E. J. Went, 'Historical Notes on the Fisheries of the Estuary of the River Shannon', *JRSAI*, 111 (1981), pp. 107–18, at pp. 107–09.

145 O'Sullivan, 'Medieval Fish Traps', p. 71.

500 herrings once a year'.[146] Unfortunately, the original document has not survived, and the 1808 text provides no reference.[147] However, this record suggests that there was an active herring fishery in the estuary during the sixteenth century.

Limerick and the Shannon lie in the province of Munster, which was the most densely populated province in sixteenth-century Ireland and possessed some of the best marine resources.[148] In addition to major ports and fishing locations, the Munster coastline is dotted with a multitude of smaller bays, islands, and landing-places that ships could use. A rough map of Munster drawn by an English hand in 1587 illustrates just how many sea connections the province possessed. It shows twenty major landing-places along the coast, together with a brief description of the conditions and water depth at each place.[149] For example, Dunmanus Bay in County Cork was '2 faddom ½ at low water, good ground, good water', that is to say the landing-place had 2.5 fathoms, or 4.5 metres, of water at low tide.

In terms of political control, most of north-west Munster was dominated in the early sixteenth century by the powerful Old English earls of Desmond. English surveys of land taken in the later sixteenth century show a wide range of fishing sites on the Desmond lands, ranging from salmon fisheries at Askeaton, the Desmond capital, to sea fisheries on Valentia Island.[150]

The town of Dingle lay at the westward extent of the Desmond territory. It had an active merchant community and strong trading connections to continental Europe in the sixteenth century.[151] The local Gaelic families also welcomed visiting fishers: in 1482 the earl of Desmond settled a dispute between two competing families in the area. He gave one family the right to charge for anchorage in Ventry Bay and split between them the rights to charge visiting fishers to use the shore, salt fish and dry nets.[152]

A record from Bordeaux in 1534 gives details about a fishing voyage to Dingle. The *Isabella de Losquirec* was chartered in Bordeaux for a voyage carrying wine. On reaching Dingle, the wine was unloaded, and the ship set about fishing on the coast before sailing back to La Rochelle with its catch. Many ships sailing to Ireland were Breton and combined trading and fishing. Like the *Isabella de Losquirec*, they would carry cargo to Ireland and then spend time fishing on the Irish coast before returning home. It is difficult to say how extensive the continental trade was in

[146] Went, 'Fisheries of the Estuary of the River Shannon', p. 113; Hely Dutton, *Statistical Survey of the County of Clare: With Observations on the Means of Improvement* (Dublin, 1808), p. 338.

[147] Dutton, *Statistical Survey of the County of Clare*, p. 338.

[148] Sheehan, 'Irish Towns in a Period of Change', p. 95.

[149] 'Map of the coast of Munster from Kinsale (County Cork) and Dingle (County Kerry)', 1587 (TNA, MPF 1/100).

[150] John A. Murphy (ed.), *The Desmond Survey* (CELT, 2013).

[151] *NHI, vol. III*, p. 12.

[152] O'Neill, *Merchants and Mariners*, p. 34.

Dingle and the rest of the south coast. Notarial records from the early sixteenth century in Bordeaux show that 150 Irish merchants (from several different towns) and 75 Gascon merchants participated in trade to Ireland. There is no indication of the number of individual voyages made. The most common cargo carried to Ireland by far was wine, while hides were the most frequent export. Fish featured prominently, with herring and hake being the most common.[153] Most Irish ports outside Ulster featured in the notarial records, but the most frequented ports were Waterford and Youghal, followed by Cork and Kinsale.[154] Dingle also had connections to Spain, with some of the older architecture of the town said to be in the Spanish style.[155] In 1567, John Denton wrote that the town had 'plentiful fish and the great commodity the Spaniards there have, by which, the queen loses her customs'.[156] Again this could have been an exaggeration, but it does suggest the Spanish were using that part of the coast for fisheries in the sixteenth century.

Gaelic Irish lords controlled much of the remaining coast of southern Munster. In the south-west of Munster, the dominant family was the MacCarthy Mór. As with the west coast, the soil quality was poor, so the lords looked to the sea for wealth. Today Castletown-Bearhaven is Ireland's second largest fishing port by volume,[157] located on the southern coast of the Beara Peninsula just south of Bere Island. In the sixteenth century, the area was known as Berehaven and was controlled by the O'Sullivan Beare clan, vassals of the MacCarthy Mór.[158]

The prominence of Berehaven as a fishing location stretches back into the sixteenth century, as it was one of the most frequently mentioned fishing locations across the sources analysed for the QFD. Colin Breen has written on the Gaelic lordship of O'Sullivan Beare and used documentary evidence to highlight the vital role played by visiting fishers in the area's maritime economy.[159] He has noted that the tower houses at Dunboy, Whiddy and Ardea, around Berehaven, all overlook fishing grounds out at sea, indicating the importance of the sea to their economy.[160] A map of Berehaven and Bantry Bay from 1558 further illustrates the maritime activity in the area (see Plate 1.1). Galleys are pictured around Berehaven, with larger sailing ships further out in the bay. The O'Sullivan Beare used the galleys to control and protect the visiting fishers who frequented their waters.

[153] Jacques Bernard, 'The Maritime Intercourse between Bordeaux and Ireland, c. 1450–1520', *Irish Economic and Social History*, 7 (1980), pp. 7–21, at pp. 13–15.

[154] Dublin, Drogheda, and Limerick were further behind; Bernard, 'Maritime Intercourse', p. 16.

[155] Richard Hitchcock, 'Dingle in the Sixteenth Century. With an Introduction and Notes', *Transactions of the Kilkenny Archaeological Society*, 2 (1852), pp. 133–43, at pp. 133–34.

[156] *CSPI* (IMC), 1566–67, no. 345.1, p. 155.

[157] Gerritsen and Kelly (eds), *Atlas of Commercial Fisheries around Ireland*, p. 6.

[158] *NHI, vol. III*, p. 11.

[159] Colin Breen, *The Gaelic Lordship of the O'Sullivan Beare* (Dublin, 2005), pp. 113–21.

[160] Breen, *O'Sullivan Beare*, p. 116.

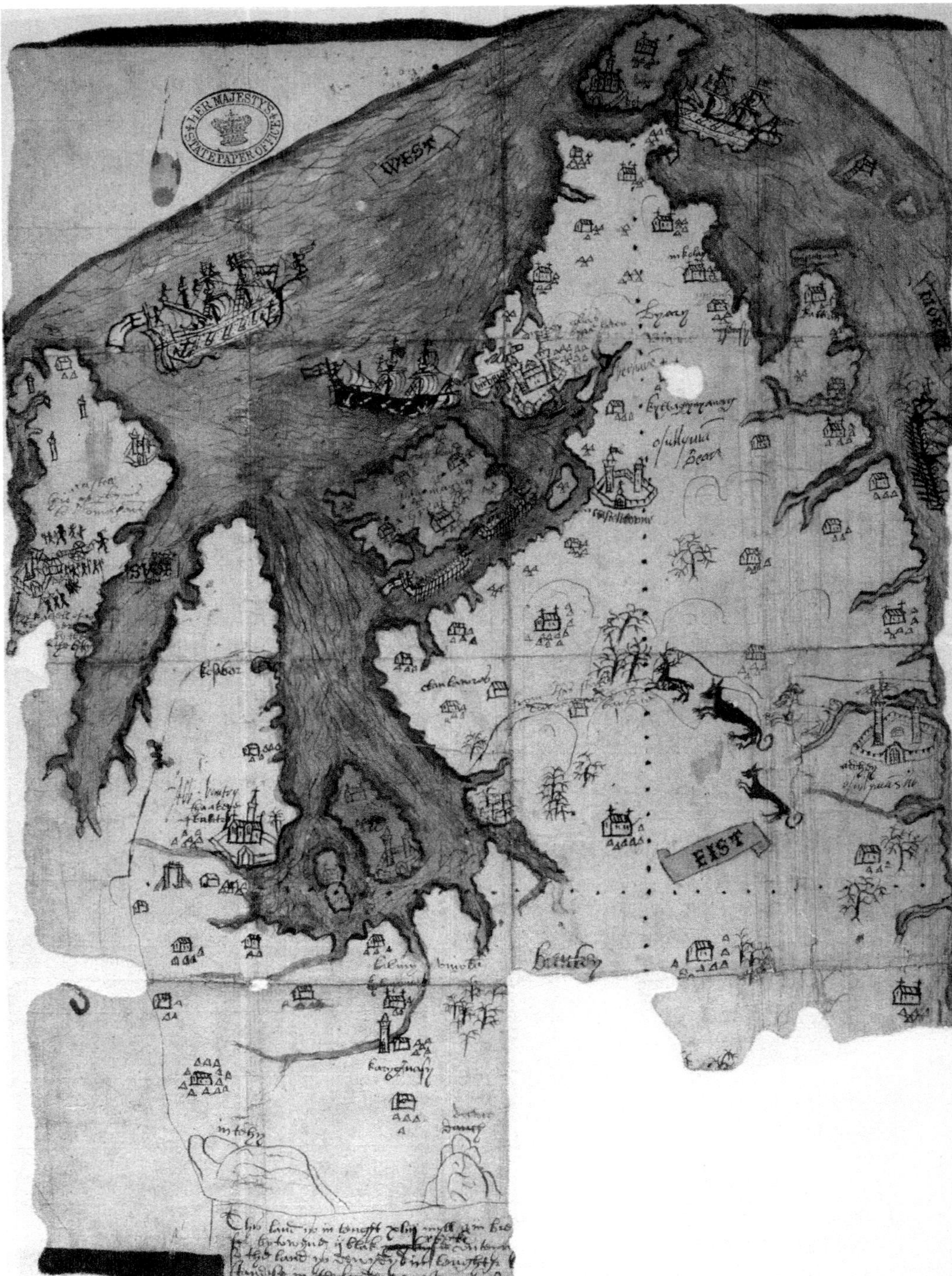

Plate 1.1. Bantry Bay, County Cork and County Kerry, 1558.

The next major fishing location in south Munster was Baltimore, which was controlled by the Gaelic Irish O'Driscolls, vassals of the MacCarthy Reagh.[161] The O'Driscoll castle on Sherkin Island gave the area its name, 'baile an tí mhóir' (Baltimore) or 'town of the great house'.[162] The castle itself was firmly positioned to control the bay and its marine resources.[163] Like Berehaven, Baltimore was a major centre for both local and distant water fisheries, and early evidence for the presence of foreign fishers in Irish waters can be found in a statute from 1450. In that year, citizens from Wexford, Waterford, Youghal, Cork, Kinsale, 'or any other liege people', were banned from fishing at Baltimore because it was claimed that Flynn O'Driscoll, the lord of Baltimore, had killed some English subjects.[164] The Act does not state how or why O'Driscoll committed these crimes, but it does show that a wide array of fishers from elsewhere in Munster travelled to the O'Driscolls' territory to make catches. And not only did Irish fishers visit the area, there is also ample evidence of the presence of continental fishers, mainly from Spain and France.[165] The connection to Spain is evidenced by the fact that the island directly north of Baltimore is called Spanish Island.[166] The complex system by which the O'Driscolls generated income from these visiting fishers is outlined in an inquisition taken in 1609.[167] Overall, the sophistication of the system indicates that the clan were economically well adapted to their surroundings. This ability to 'adapt and conform' to their environment allowed the O'Driscolls to remain independent for longer than many other Gaelic clans.[168]

The stretch of coastline from Baltimore to Wexford was probably Ireland's most active fishery region. The reasons are a mixture of political, geographical, and ecological. This region has the closest sea connections to important ports in both England and the continent, facilitating trade to those regions. Most of the port towns had Norman roots, which meant they possessed the technology and connections to trade further afield. That Norman heritage has also meant that

[161] *NHI, vol. III*, p. 10.

[162] Cindy Valler, 'Baltimore, Ireland 20 June 1631' (http://www.cindyvallar.com/Baltimore.html) (accessed 1 June 2020).

[163] For a local history of the O'Driscolls, see Éamon Lankford, *O'Driscolls Past and Present* (Cork, 2005).

[164] *Statute Rolls, Ireland*, vol. II, 28 Henry VI, A.D. 1450, X, pp. 191–93.

[165] Barkham, 'The Spanish Basque Irish Fishery'.

[166] Fiontar & Scoil na Gaeilge and The Placenames Branch [Department of Culture, Heritage and the Gaeltacht], 'Placenames Database of Ireland' (https://www.logainm.ie/en/) (accessed 8 March 2020).

[167] John O'Donovan (ed.), *The Genealogy of Corca Laidhe* (*CELT*, 2011), pp. 100–07.

[168] Connie Kelleher, *The Alliance of Pirates: Ireland and Atlantic Piracy in the Early Seventeenth Century* (Cork, 2020), p. 71; 'The Gaelic O'Driscoll Lords of Baltimore', in Linda Doran and James Lyttleton (eds), *Lordship in Medieval Ireland, Image and Reality* (Dublin, 2007), pp. 130–59, at pp. 139–42.

we have comparatively more quantitative information about trade in this region because of the survival of English port books and customs accounts. Today large catches of cod, haddock, hake, herring, and ling are made in or near this section of the coast.[169] The area also has the highest concentration of herring spawning grounds in Ireland.[170] The confluence of geography and ecology made this coastline ideal for early modern fisheries: fish were abundant close to shore and in reach of smaller craft that could not fish in the open sea.

Along the coast from Baltimore, Kinsale was the next major fishing location in the sixteenth century. The fishing enterprises of the two towns were also connected, as they jointly elected a fishing admiral every year in the fifteenth and early sixteenth centuries.[171] However, in contrast to Baltimore, Kinsale was not Gaelic Irish but a chartered English town. It was also closely connected to Cork and acted as a satellite port for the city, while also possessing a strong mercantile community of its own.[172] The council book of the town has survived, and although it covers mostly a later period, there are some details about port operations from the fifteenth and sixteenth centuries. From these we can see that Kinsale was a centre for the collection of customs and enjoyed trading connections to the continent.[173] Several entries refer to the presence of foreign fishers in the town in the later sixteenth century.[174]

Further up the coast, Cork was one of the largest conurbations in Ireland in the sixteenth century, with an estimated population of 2,400 in 1600.[175] An English mercantile port, it had received its charter directly from the English Crown.[176] The city is located at the end of one of the best natural harbours along the Irish coast. For its size, we know comparatively little about Cork or its fisheries from the early sixteenth century. Some of the early Chancery records from the thirteenth century reference the purchasing of fish in Cork to supply the English army in Scotland,[177] showing that there was an active market and a close connection to England in that period. A document in the State Papers from 1549 also details the trade in fish. It requests that the customs searchers from the town do not interfere with a Spanish ship freighted in Cork with 'hides, fish, frieze mantles, flocks and a little flesh, butter and tallow, to bring back Spanish wines'.[178] While this is just

[169] Gerritsen and Kelly (eds), *Atlas of Commercial Fisheries around Ireland*.

[170] O'Sullivan et al., 'Irish Herring Spawning Grounds'.

[171] O'Donovan (ed.), *Genealogy of Corca Laidhe*.

[172] *NHI, vol. III*, p. 10.

[173] Richard Caulfield (ed.), *The Council Book of the Corporation of Kinsale* (Guildford, 1870), see for example Henry V, 24 July 1415, p. xiii.

[174] Caulfield (ed.), *Corporation of Kinsale*, p. xx.

[175] Sheehan, 'Irish Towns in a Period of Change', p. 96.

[176] *NHI, vol. III*, p. 10.

[177] *CIRCLE*, Edward I, Patent Roll 27, no. 3.

[178] *CSPI* (IMC) 1547–53, no. 191, p. 54.

one example, it is indicative of a larger trade that may well have existed, and about which few records remain. Cork also had strong trade connections to Bristol in the early sixteenth century, as did many of the ports on the south-east coast.

Further along the coast from Cork was the comparatively smaller but important fishing and trading port of Youghal, which lay under the influence of the earls of Desmond in the early sixteenth century. Youghal lies at the mouth of the river Blackwater, giving it deep inland connections.[179] The Blackwater is also one of Ireland's most abundant salmon rivers.[180] Concerning the town's international trade, there is a record of a ship carrying cod and ling valued at 52*s.* from Youghal to Poole in 1504.[181] While this is only one voyage, it suggests a wider, unrecorded trade. Another entry from the State Papers in 1548 mentions the town's fishing fleet being attacked by pirates on its journey home.[182] This confirms the existence of Youghal's fleet (though not its size), and indicates the precarious position of many fishers on the water. The danger faced by fishers would only increase over the course of the sixteenth century as new conflicts drew more pirates to the Irish coast.

Further along the coast from Youghal is the port of Dungarvan, nominally a royal outpost but in fact contested in the early sixteenth century between the earl of Desmond and lords of the Decies.[183] During the Kildare rebellion, Dungarvan castle was seized, and during its recapture, the English forces were 'supported at sea by fishermen of Devonshire and Cornwall'.[184] That intervention suggests that fishers from those parts of England were already familiar with the town.[185]

At the north-eastern edge of Munster, Waterford was probably the most important exporter of fish in sixteenth-century Ireland.[186] The town is located where three rivers reach the sea – the Barrow, Nore and Suir – affording it river access to a multitude of inland towns. Through the Barrow, Waterford was connected to New Ross; by the river Nore, it was connected to Kilkenny and Thomastown; and by the Suir, it had access to Carrick-on-Suir and Clonmel.[187] The town also had an excellent harbour that could hold ships of up to 400 tons.[188] Waterford was surrounded by the semi-Gaelicised Power family, with whom it was often in conflict. Further inland lay the lands of the Butlers, earls of Ormond,

179 *NHI, vol. III*, p. 10.

180 Went, 'Fisheries of the Munster Blackwater'; A. E. J. Went, 'Fisheries of the Munster Blackwater (Continued)', *JRSAI*, 91 (1961), pp. 19–41.

181 O'Neill, *Merchants and Mariners*, p. 37.

182 *CSPI* (IMC), 1547–53, no. 27, pp. 9–10.

183 *NHI, vol. III*, p. 9.

184 *CSPI* (IMC), 1509–47, no. 143, p. 70.

185 See more on this in the following chapter.

186 Evan Jones and Susan Flavin (eds), *Bristol's Trade with Ireland and the Continent, 1503–1601: The Evidence of the Exchequer Customs Accounts* (Dublin, 2009).

187 *NHI, vol. III*, p. 8.

188 Ada K. Longfield, *Anglo-Irish Trade in the Sixteenth Century* (London, 1929), p. 33.

one of the most powerful families in early modern Ireland, whose principal seat of power lay in Kilkenny.[189]

Waterford was founded by the Vikings and was an important trading port after the Norman conquest. The town features prominently in Chancery records from the thirteenth century onwards. Licences granted to merchants of Waterford in the late fourteenth century gave them the right to export herring to England.[190] Another licence granted in the year 1400 gave the right to all the citizens of Waterford 'to transport meat, fish, salmon and other victuals to England and Wales'.[191]

The fifteenth and early sixteenth centuries seem to have been a time of marked prosperity and stability in Waterford and much of the south-east. This was in part owing to the many privileges Irish towns received from the English Crown, meaning that there were few barriers to trade for the towns' merchants. Additionally, by 1463, almost all the independent towns were free to trade with the Gaelic Irish and to welcome foreign merchants and fishers.[192] This freedom opened opportunities that benefited all coastal communities in this part of Ireland.

The small town of Ballyhack, just outside Waterford, also seems to have been a centre for fishing. The name Ballyhack itself may be derived from the hake fishery that took place there.[193] It appears often as a home port in the Welsh port books from 1550–1603, suggesting it was home to many mariners.[194] A survey of the area in 1541 showed there were nine fishers living in the area, who owned two boats between them. In addition, any foreign fishers landing at 'le key' had to pay one cod, hake or 1*d*., depending on what kind of fish they were targeting. This shows that local fishers not only engaged in their own activities, but also exploited visitors for taxation.[195]

Wexford was also an extremely important fishing port in the early sixteenth century. The town was independent but surrounded by lands owned by the English earl of Shrewsbury.[196] Like Waterford, Wexford had strong trading connections to England, and fish was the most valuable commodity exported from the town in the late fifteenth and early sixteenth centuries. The most

[189] *NHI, vol. III*, p. 8.

[190] *CIRCLE*, Richard II, Patent Roll 10, no. 234, RCH 136/203.

[191] *CIRCLE*, Henry IV, Patent Roll 1, no. 85, RCH 157/80.

[192] Susan Flavin, *Consumption and Culture in Sixteenth-Century Ireland: Saffron, Stockings and Silk* (Woodbridge, 2014), p. 23.

[193] A. E. J. Went, 'The Irish Hake Fishery, 1504–1824', *Journal of the Cork Historical and Archaeological Society*, 51 (1946), pp. 41–51, at pp. 42–43.

[194] Arthur Edward Lewis (ed.), *The Welsh Port Books, 1550–1603* (Cymmrodorion Record Series, xii, London, 1927).

[195] Went, 'The Irish Hake Fishery', p. 42, quoting from White, *Extents of Irish Monastic Possessions*.

[196] *NHI, vol. III*, p. 7.

prominent coastal towns that lay between Wexford and Dublin in the sixteenth century were Arklow, Wicklow and Bray. Unfortunately, we have no evidence of significant fishing or maritime trade in these towns in the early modern period. This may have been in part due to geography, as the coastline north of Wexford is dotted with tricky sandbars which make navigation very difficult.[197]

This survey has illustrated the great diversity of fisheries that existed around the coast of Ireland in the early sixteenth century. During the fifteenth and sixteenth centuries Ireland's seas and oceans played host to a wide variety of fishing fleets, and the coast was a tapestry of small fishing ports, fish-full bays, and industrious, semi-independent towns. There were local fishing operations in every province, some commercialised and others purely subsistence. Fishers from across Ireland, England and Europe travelled long distances to exploit rich waters around Ireland. As the next chapter will show, certain Irish lords and families became specialised in facilitating and taxing these foreign fishers. Fish was an important domestic product too, with strict laws limiting the unlicensed sale of fish in Dublin and the other English parts of Ireland. A few species dominated the sixteenth-century fisheries, with herring the most prominent, followed by several white fish species like hake and cod. Salmon was by far the most valued river fish, but eel was also commonly caught.

It is important to note that the fifteenth and early sixteenth centuries were periods of relative peace in Ireland, which contributed significantly to the success of the fisheries and the economy in general.[198] While conflict within the fisheries did occur, there was an underlying element of cooperation and trust between the different parties. Over the course of the sixteenth century this would begin to change, with many regions experiencing conflict, which severely damaged fishing operations.

Ultimately, fisheries developed differently on the Irish coast owing to the unique political, cultural, and geographic conditions existing in different regions. The environment and its human inhabitants shaped fisheries, as did internal and international relationships. Establishing a baseline of who had access to fisheries in 1500 is also important because, in no small degree, the type of change that each region would experience was dependent on its underlying political makeup. Old English fisheries experienced a different trajectory of change compared to those of the Gaelic Irish. As the rest of this book will illustrate, a number of forces moved to shift fisheries in Ireland radically over the sixteenth century, and very little of what existed in 1500 would still be the same in 1603.

[197] R. H. Buchanan, 'The Irish Sea: The Geographical Framework', in Michael McCaughan and John C. Appleby (eds), *The Irish Sea: Aspects of Maritime History* (Belfast, 1989), pp. 1–12, at pp. 4–5.

[198] O'Neill, *Merchants and Mariners*, p. 130.

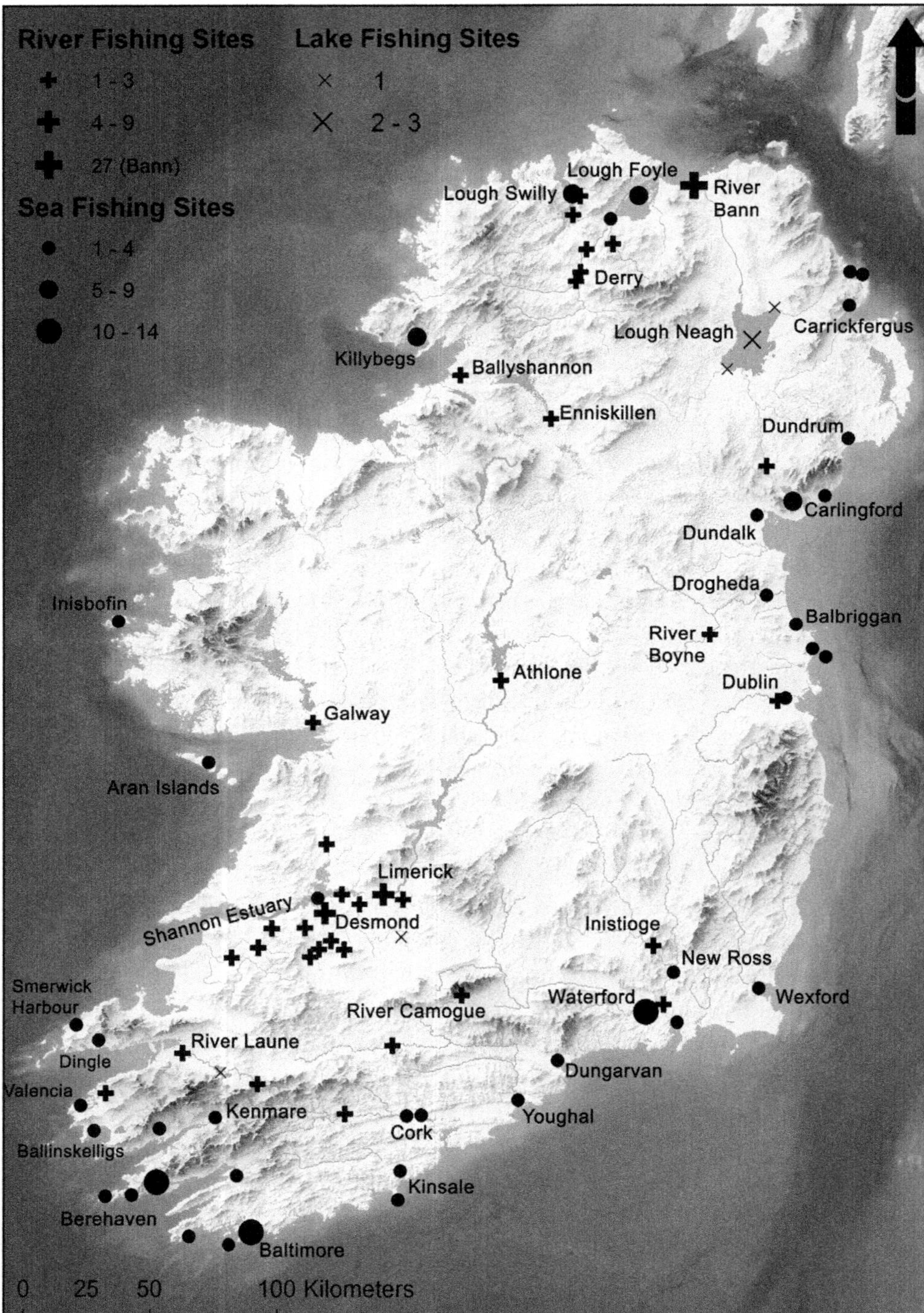

Map 2.1. Fishing locations in Ireland, 1496–1603.

2

Diversity and Cooperation in Sixteenth-Century Fisheries

> Gathered about his soft locks are the heroes of the Tuatha, the warriors of Fanad, the kingly youth of Inishowen, delightful hosts from whom the sea is easy to fish.
>
> Tadhg Dall Ó Huiginn, *The Battle of Drumleene* (composed between 1550 and 1591)

Macro View of Fisheries, 1500–1603

For the sixteenth century at least, it is possible to explore and answer many of the fundamental questions about fisheries. For example, where they operated, where fishers came from, how large the fisheries were, what species were caught and what methods were used to do the catching. Answering these basic questions is vital to understanding the diverse and distinctive nature of Irish fisheries in the sixteenth century. Furthermore, in the answers to these questions lie many of the factors that contributed to the economic significance of the industry and its ultimate decline by the end of the sixteenth century.[1]

One of the most fundamental questions we can answer about fisheries around Ireland during the sixteenth century is where they took place. To this end, Map 2.1 shows the site of every fishing location mentioned in the sources from 1496 to 1603.[2] Written accounts that explicitly discussed the location of fishing sites or fishing towns and centres were analysed to generate this map. These sites do not, therefore, in every case represent the exact location where fish was caught but do indicate the locations around which fisheries were based and where the majority of the catch was landed. For example, we have records that discuss Killybegs hosting a fishery, but it is likely that fishers made catches along the coast in the general vicinity of the town. In this instance, not all catches were made at Killybegs, but it was the central landing point.

[1] The maps and figures shown in this chapter were generated primarily from the Irish State Papers and Admiralty court records; a full description of the sources and methods used is outlined in Appendix 1.

[2] The data runs from 1496 onwards in order to include a few more relevant entries from the late fifteenth century.

The map includes river fisheries and lake fisheries, but the majority of sites recorded are sea fisheries. Table 2.1 also lists the ten most frequently mentioned sea and river fishing sites. Here locations that were very close but had different names have been combined in some cases into one central location. The spatial pattern of these locations fits into the general regions established in the previous chapter. There is a concentration of smaller fishing sites on the east coast between Dublin and Carlingford. On the Ulster coast, the English outpost of Carrickfergus stands out, as does the diverse fishery on the Bann. The remainder of the Ulster coast from Lough Foyle to Ballyshannon was in the hands of the Gaelic Irish, with the O'Donnell clan being the most dominant family. The islands Inishbofin and Aran were the most prominent fishing locations on the west coast; however, there was a lack of documentary evidence covering this region overall. The sources also frequently mention Galway in the context of its salmon fishery; the same was true of Limerick city. The south coast was host to several Gaelic Irish-controlled fisheries, with Berehaven and Baltimore being by far the most prominent. Then the south-east coast was peppered with Old English-controlled fishing towns, Waterford being the most prominent and the most mentioned sea fishing site overall.

Table 2.1. Ten most mentioned sea and river fishing sites.

Sea fishing	Number of mentions	River fishing	Number of mentions
Waterford	14	River Bann	27
Berehaven	14	Askeaton	9
Baltimore	13	Limerick	5
Lough Foyle	9	River Laune	3
Carlingford	8	River Feale	3
Killybegs	6	Corkan Isle	2
Youghal	4	Athlone	2
Lough Swilly	6	Camogue River	2
Cork	3	Galway	2
Dungarvan	3	River Galey	2

Overall, there were far fewer freshwater fishing sites collected in this data, but despite this, some noteworthy patterns emerge. The river Bann was the most frequently mentioned fishing site of any type, with twenty-four references in total. As we have seen, the Bann held a major salmon fishery in the medieval and early modern periods, making it a valuable economic asset. However, the number of records generated relating to the Bann was also driven by the constant conflict over it. At various times the English Crown, Gaelic Irish clans, Old English lords and travelling English fishers all sought to control the river. The legal and physical disputes generated many records in the Irish State Papers in particular. The case of the Bann highlights that the data collected do not necessarily show how proportionally important each fishing location was, but rather how often it was discussed in the sources.

Lake fishing, too, although not featuring much in the sources, still shows up intermittently. Lough Neagh in Ulster was the most frequently mentioned lake fishery, with five references. The cluster of freshwater fishing sites around Limerick and northern County Kerry come from the Peyton and Desmond land surveys, conducted by the English in the aftermath of the Desmond rebellions in the late sixteenth century.[3]

Where fishers originated also had a significant bearing on the type of fishing that took place. Irish waters played host to a variety of different fishing fleets. The most apparent distinction between fishers was between those who operated close to where they lived and landed their fish in home markets, and those who travelled long distances to fish and brought a portion of their catch to foreign markets. Within the data collected for this chapter, a 'fishing party' was assigned to each location to describe the cultural identity of those fishing there. If it was not clear who was fishing at a particular location, the identity of those who controlled the fishing site was recorded instead. Table 2.2 shows the overall count of origin and ownership.

Table 2.2. Fisher origins and fisheries control, 1496–1603.[4]

Fishing party	Number of mentions
Gaelic Irish	36
Old English	32
English	32
Spanish	18
Foreign general	3
French	3
Scottish	1

[3] John A. Murphy (ed.), *The Desmond Survey* (CELT, 2013).

[4] Three entries only discussed foreign fishers in general, without reference to their exact origin.

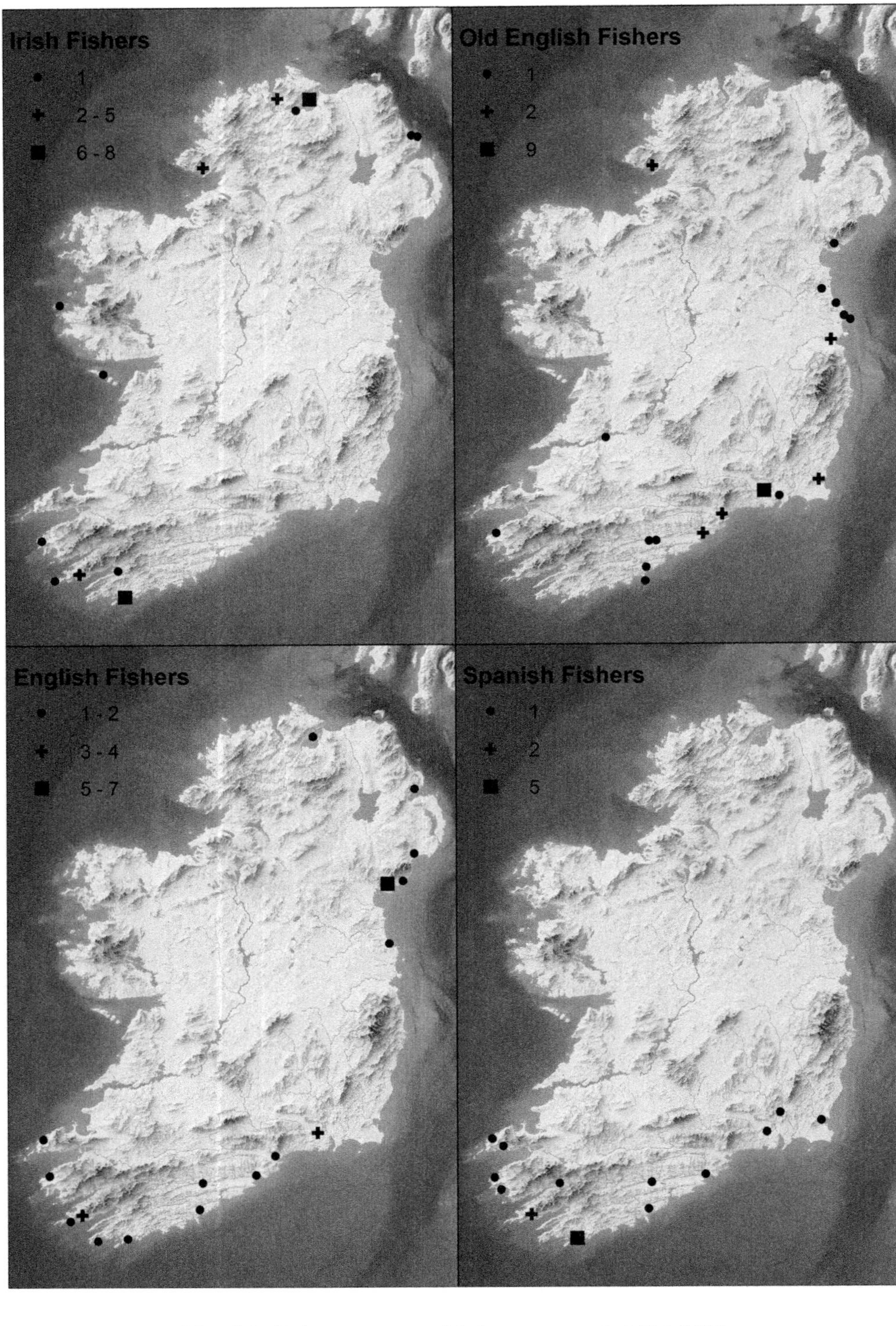

Map 2.2. Fisher origins and fisheries control, 1496–1603.

With this data, we can also see the spatial distribution of different fishers across Ireland; Map 2.2 shows the top four groups. The one Scottish mention, which is not on the map, was related to Islandmagee, County Antrim. There were also two mentions of French fishers at Baltimore and one at Berehaven, both in County Cork. The map was limited to sea fisheries because these were far more likely to have external fishers present, as Irish or Old English landowners controlled most inland fisheries. The map is split into four panels: the first shows Gaelic Irish fishing locations; the second Old English; the third English; and the fourth Spanish, French and Scottish.

The Gaelic Irish were the most frequently mentioned group overall, and their fishing sites were highly concentrated in a few key areas, such as Baltimore, Berehaven and Lough Foyle. The Irish mode of fishing involved both local domestic fishers and the hosting of foreign visitors. Overall, the largest group of visiting fishers to Ireland were the English, who operated in Irish-controlled fishing locations like Berehaven and completely controlled seasonal fishing locations at places like Carlingford, County Louth. The Old English fishers generally operated from the Norman towns found on the east and south-east coast, but they engaged in distant water fishing within Ireland, travelling to Gaelic Irish and English-controlled areas like Donegal and Carlingford. The Spanish were the largest group of continental fishers on the Irish coast, frequenting similar locations to English fishers but with little presence in Ulster and focusing their efforts on the south Munster coast. There were several reasons, not least the geographical proximity of that coastline to Spain.[5] While fewer in number, the French also concentrated their fishing efforts on the south coast. Lastly, a single reference to Scottish fishers at Islandmagee was found. There may well have been a more substantial connection between Ulster and Scotland, but the surviving sources have inadequate coverage for this area.

The Bann was left out of this map because it was a river fishery, but if we examine the Bann in isolation, we see that there were fifteen mentions of English distant water operations on the river and one mention of Old English fishers travelling to fish there. There were a further nine mentions of the Gaelic Irish fishing the river; these would have been locals from the Irish families who owned land along the riverbanks.

Temporally, the data shows that there were fewer and fewer recorded instances of continental fishers visiting Ireland as the sixteenth century went on. The last Spanish fisher was reported in 1580. Only Irish, Old English and English fishers were mentioned in the later decades. There is also a significant spike at the end of the century in the number of Irish fishers being referenced, not driven

5 A report from 1589 suggested that Spain was only 'four days sail' from the Munster coast; *CSPI*, 1588–92, no. 34, p. 252.

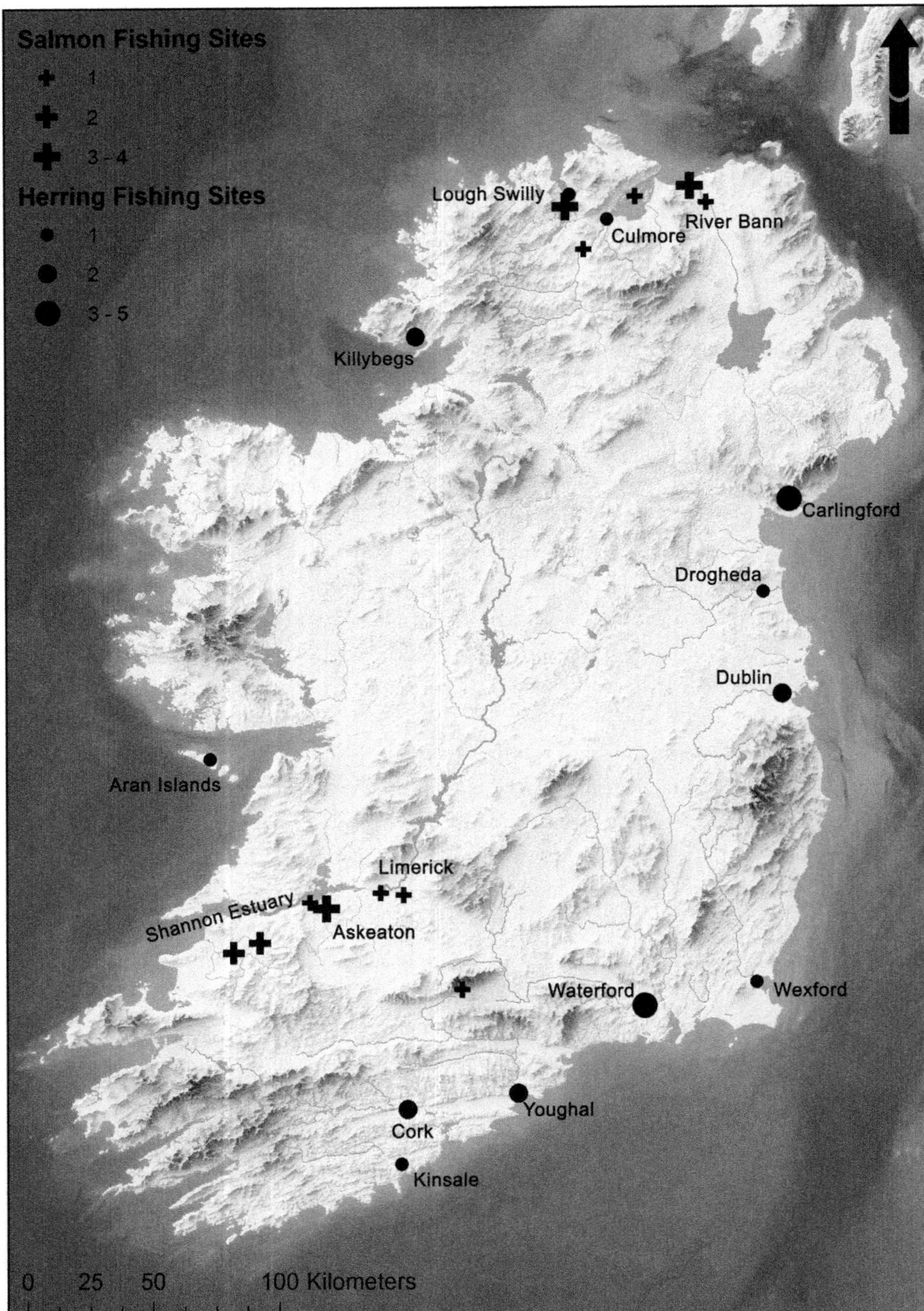

Map 2.3. Salmon and herring fishing locations, 1496–1603.

Table 2.3. Mentions of species around Ireland, 1496–1603.

Species	Number of mentions
Herring	25
Salmon	22
Eel	3
Hake	2
Trout	2
Shad	1
Pilchards	1
Flounders	1
Oyster	1
Cod	1
Mussels	1

by an increase in fishing activity but by the encroachment of the English into Irish territory. The English desire to control and incorporate these fisheries into plantation schemes led to them being recorded with higher frequency.

This illustrates a particular trait of the State Papers and other English sources. When fisheries were operating normally, they usually went unnoticed and unrecorded. If things went wrong, for example when fisheries were involved in a conflict or attracted the attention of the authorities or lawmakers, they would be mentioned in the sources. This does not invalidate any conclusions being suggested, but should make us aware of the limitations of the sources and mindful that much more went unrecorded and unsaid about the day-to-day operation of fisheries, particularly those in Gaelic Ireland.

The species of fish available and targeted at each location also determined the nature of the fisheries. Table 2.3 shows the frequency with which different species were mentioned at fishing locations around Ireland. The list includes fresh and saltwater fish. The species targeted were not reported for every site, but overall, herring was by far the most common sea fish, with hake a long way behind in second place, with only two mentions. Salmon was the most common freshwater fish, followed by eels.[6]

The spatial distribution of the two most dominant species can be seen in Map 2.3. Waterford, Carlingford and Killybegs (County Donegal) emerged as the places most frequently associated with herring fishing. The river Bann and the

6 Technically both are amphihaline species, meaning they live in both marine and freshwater. But as they were primarily caught in freshwater that is how they are recorded here.

Shannon estuary were the most common sites for salmon fishing. By itself, this suggests that herring was the most exploited sea fish but that there was a diversity of other fish and molluscs, albeit much smaller numbers. Other fish, such as hake, were commercially important species. So, it is unclear why hake are mentioned to a much lesser extent than herring and salmon. It is possible that hake was targeted more by distant water fishers in Ireland and was not consumed locally as much as other fish.

One of the most pronounced features of fishing in sixteenth-century Ireland was the prominence of distant water fishers who travelled from other places within Ireland or abroad to make their catch. Within the qualitative sources, four records were found that explicitly listed the origin and destination of internal Irish fishing voyages. All originated from Old English towns, with two visiting Donegal, another voyaging to the Bann and another to the herring fishery at Carlingford. These four voyages took place later in the sixteenth century, but we know from a statute that fishers from Wexford, Waterford, Youghal, Cork and Kinsale were fishing at Baltimore from at least 1450.[7] This Act specifically banned fishers from travelling to the Gaelic Irish area to fish, probably to prevent the income this brought the Irish. Old English fishing in Gaelic-controlled waters was not a phenomenon that began in the sixteenth century. Instead, it became more contentious in this period, and increasing levels of conflict in Ireland meant this activity was more often recorded. In 1571, boats from Dublin and Drogheda went fishing to Killybegs and reported to the English authorities that there were Scottish ships in the area.[8] There would be no record of this voyage if the fishers had not sent in their report. Likewise, the 1596 trip from Drogheda to Killybegs is entered into the State Papers because the Drogheda fisher reported the presence of an enemy Spanish vessel in the O'Donnells' waters.[9]

A 1588 voyage to the Bann was recorded in the context of the English attack on Macosquin Abbey just outside modern Coleraine. In this case, Francis Stafford wrote that his forces killed the local Gaelic chief, Fordorough O'Cahan. He also reported that 'the merchants of Dublin, which have used to fish the Bann', attested to 'outrageous exactions' imposed by Fordorough 'when they came to fish the Bann'.[10] This makes it clear that the Dublin merchants had a long tradition of fishing at the Bann, and the recent conflict there was the reason that the practice was being reported. As will become evident later, the practice of Gaelic Irish chiefs charging taxes for the use of fishing grounds was common in sixteenth-century Ireland.

7 *Statute Rolls, Ireland*, vol. II, 28 Henry VI, A.D. 1450, X, pp. 191–93.

8 *CSPI* (IMC), 1571–75, no. 84.4, p. 53.

9 The English and Spanish were at war from 1585 to 1604.

10 *CSPI*, 1586–88, no. 36, p. 505.

The data mined from qualitative sources indicates that the English were the most significant group involved in distant water operations in Ireland. Maryanne Kowaleski's research has shown that most of these fishers hailed from the English West Country. As well as coming to Ireland, these English fishers also travelled within their own country to towns like Rye, and from the middle of the fifteenth century onwards, went further afield to Iceland to catch cod. Nevertheless, Ireland seems to have been their most popular destination.[11] Later, in the sixteenth and seventeenth centuries, these fishers would also sail in large numbers to Newfoundland and New England.[12]

The surviving customs accounts from the English West Country provide compelling evidence for the extent of fishing operations in Ireland. Kowaleski has tabulated data from a variety of ports from the years 1492/3 and 1497/8, providing a snapshot of the fish trade in the 1490s.[13] The fish recorded in these accounts had to come from overseas, as fish caught locally was not registered, so we can be confident most of it originated in Ireland.[14] However, the catches recorded are only a fraction of the total catch brought into the ports. Kowaleski estimates that 80 per cent of all catches were transported via coastal trade or caught locally and therefore not included in the customs accounts.[15] Fortunately, the percentage caught locally is irrelevant to the study of Irish imports. By combining all Kowaleski's figures, we get a total value of £816 worth of fish imported in 1492/3 and 1497/8. Map 2.4 shows the spatial distribution of these combined imports.[16]

The map indicates that Minehead was the largest importer of fish. Given the geographic proximity to Ireland and the types of fish imported, it is likely that English vessels caught most of this imported fish off the Irish coast. Hake was the most significant import and export from the West Country ports, followed by herring, with pilchards a distant third. Hake is widely distributed along the Irish coast, with significant nursery grounds in the Celtic Sea and in the Bay of Biscay. It is around these grounds that most modern catches take place.[17] Some of the most popular English fishing sites in the sixteenth century (see Map 2.2) were also near the hake nursery grounds, and it seems likely that early modern fishers followed a similar pattern to their modern counterparts.

[11] M. Kowaleski, 'The Expansion of the South-Western Fisheries in Late Medieval England', *Economic History Review*, 53 (2000), pp. 429–54, at p. 442.

[12] Kowaleski, 'South-Western Fisheries', p. 442.

[13] The data is presented in a somewhat confusing way, but due to the limitations of the sources, records for each port were only available for either 1492/3 or 1497/8. The data was then combined to provide a single 'snapshot' of the trade during the 1490s.

[14] Kowaleski, 'South-Western Fisheries', p. 442.

[15] Kowaleski, 'South-Western Fisheries', p. 432.

[16] Kowaleski, 'South-Western Fisheries', p. 437.

[17] 'Species Dashboard, Hake (Merluccius merluccius)' (https://shiny.marine.ie/speciesdash/) (accessed 30 September 2020).

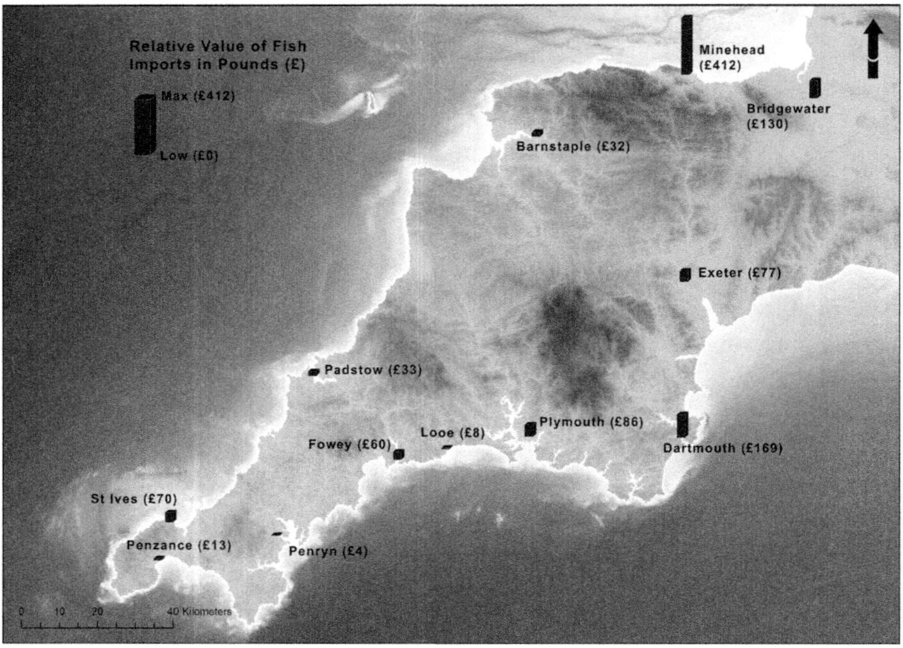

Map 2.4. Value of fish imports to south-western English ports, 1490s.

Peak spawning for hake generally occurs from January to March in the Bay of Biscay, and from May to July in the Celtic Sea.[18] Many hake also spawn on the continental shelf and migrate to nursery areas on the Irish coast over the season.[19] However, there is also much inter-annual variability in the intensity of hake spawning in the Bay of Biscay and Celtic Sea.[20] In some years, spawning populations are far more substantial in one region than in the other. This variability may have influenced fisheries in the fifteenth and sixteenth centuries. English evidence shows that hake fishing boats left south-west England in late May and early June and generally returned to their home ports in late August and

[18] Maria Korta et al., 'European Hake (Merluccius merluccius) in the Northeast Atlantic Ocean', in Hugo Arancibia (ed.), *Hakes: Biology and Exploitation* (Chichester, 2015), pp. 1–37, at p. 8; Paula Alvarez et al., 'Distribution and Abundance of European Hake Merluccius merluccius (L.), Eggs and Larvae in the North East Atlantic Waters in 1995 and 1998 in Relation to Hydrographic Conditions', *Journal of Plankton Research*, 26 (2004), pp. 811–26, at p. 819; Hilario Murau, 'Reproductive Fundamentals for the Estimation of Egg Production of the European Hake, Merluccius merluccius in the Bay of Biscay' (PhD thesis, Universidad del País Vasco, 2006), p. 24.

[19] Alvarez et al., 'Distribution of European Hake', p. 812.

[20] ICES, *Report of the Working Group on the Assessment of Southern Shelf Stocks of Hake, Monk and Megrim* (Copenhagen, 2013), p. 53.

September.[21] This timing largely coincides with the northward movement of the spawning hake. Thus it appears likely that fishers knew the hake would be more abundant on the Irish coast during the summer and early autumn.

The fishery drew ships from all along the shores of Devon, Cornwall and Somerset. It is difficult to estimate the total number of vessels going to Ireland, but we know that in 1497/8, thirty ships sailed from Plymouth to Ireland with salt as their only cargo.[22] The fishers used the salt they carried to process their catch and bring it back to England. A portion of the fish was then re-exported. An exchequer inquiry from the Devon ports in 1474–75 reveals that vessels from eleven different ports in Devon shipped Irish fish onwards to ports in Kent, Hampshire, Devon, Somerset and Gloucestershire.[23]

There is also evidence for Welsh fishers in Irish waters. A survey of ships in Pembroke in 1566 recorded sixteen different fishing vessels, reporting that most of them 'useth commonly to Ireland afishinge and upp Severne'.[24] An Admiralty case from 1577 provides details on how English and Welsh fishing voyages were organised. In June the previous year, the *John* of Plymouth left on a fishing voyage to the coast of Ireland. The owner, John Langford, funded the voyage and was to receive a sixth share of all the fish caught. The master, Matthew Hore, was responsible for supplying the ship and paying wages: he would be allowed to keep the rest of the fish, except for 'one young man of Langford's' who was also allowed to keep the fish he caught.[25] This is an excellent example of how distant water operations were funded and how the catch and profits were shared. Unfortunately, on its return voyage the ship 'put into St Ives where she was castaway and perished'.[26]

In total, 283 ships from south-west England engaged in overseas fishing during the 1490s.[27] There is no way to know how many came to Ireland each year, but other records make clear that Ireland was a significant destination for south-west English fishers. Kowaleski credits the expansion of the English south-west fisheries to several factors, including the success of the Dutch herring fishery in the North Sea, which drove east coast English fishers out of the market and created an opportunity for their counterparts on the west coast.[28] The south-west of England was also geographically favourable to exploit the vibrant waters off the Irish coast, and there were many ports in England where they could sell their

[21] Kowaleski, 'South-Western Fisheries', p. 443.

[22] Kowaleski, 'South-Western Fisheries', p. 443.

[23] Kowaleski, 'South-Western Fisheries', p. 443.

[24] Arthur Edward Lewis (ed.), *The Welsh Port Books, 1550–1603* (Cymmrodorion Record Series, xii, London, 1927), pp. 311–13.

[25] *Cal. HCA 1536–1641*, no. 214, p. 42.

[26] *Cal. HCA 1536–1641*, no. 215, p. 42.

[27] Kowaleski, 'South-Western Fisheries', p. 432.

[28] Kowaleski, 'South-Western Fisheries', p. 450.

catch. Furthermore, English fishers were well positioned to exploit new markets and sources of salt in southern Europe.

It must also be stated that all aspects of maritime activity expanded in the late fourteenth century, not just English fisheries in Ireland. As Kowaleski writes, 'naval ventures, privateering and piracy, coastal and overseas trade, the sea-carrying trade, and fishing' all increased in this period, so what we observe in Ireland was also part of a broader expansion of maritime activity.[29]

In terms of destinations, Carlingford was one of the most popular places for English fishers in Ireland. As we have seen, William Brabazon reported a fleet of 600 English fishing boats engaged in the herring fishery there in 1535. Carlingford Castle and Green Castle on the other side of the bay were in ruins, but the fishing continued would generate 'as much rent as ever was paid, and the residue of the profit will, within a short time, make both the castle of Carlingford and the Green Castle to be wardable'.[30] Brabazon made his observation because the English fishers at Carlingford offered themselves as a mercenary force of 3,000 soldiers should they be needed. Because these figures had strategic military importance, we should expect them to be accurate and thus provide clear evidence of a substantial fishery operating at Carlingford at this time.

If divided evenly, these figures would mean each fishing boat at Carlingford had a crew of five, which implies that they were small boats, probably pinnaces. Kowaleski has suggested that most of the fishers at Carlingford came from Somerset and north Cornwall.[31] A later extent of land taken in 1540 refers to Carlingford as 'a very safe harbour for ships that arrive in the months of September and October' for 'an abundance of herring', with a fishing fleet which 'rivals in size anything in England'.[32] The report of fishing in autumn matches with the modern spawning period of herring in the north Irish Sea, although marine scientists have observed that these spawning periods fluctuate yearly.[33] Irish fishers visited the area too; in 1554, John Dympsy, a fisher from Dublin, reported to the Admiralty court that he had been 'fishing in Carlingford haven for herrings' at Michaelmas the year before.[34]

The evidence also suggests that the south-east coast was a frequent location for English fishers targeting herring and hake, and Map 2.2 shows a concentration of English fishing activity in this area. In 1535 lord deputy Skeffington had reported

29 Kowaleski, 'South-Western Fisheries', p. 451.

30 *Cal. Car. MSS*, 1515–74, no. 70, p. 85.

31 Kowaleski, 'South-Western Fisheries', p. 442.

32 Translated from the original Latin; Gearóid Mac Niocaill (ed.), *Crown Survey of Lands 1540–41* (Dublin, 1992), p. 75.

33 D. Brophy et al., 'Temporal Trends in Stock Origin and Abundance of Juvenile Herring (Clupea harengus) in the Irish Sea', *ICES Journal of Marine Science*, 66 (2008), pp. 1749–53.

34 *Cal. HCA 1536–1641*, no. 46, p. 9.

that fishers from Devon and Cornwall assisted his forces in the siege of Dungarvan Castle in County Waterford.[35] The assault took place in September, which suggests that the English fishers may have been at the tail-end of the hake fishing season. Alternatively, they could have been arriving for the herring fishing that began in that month,[36] but the town was renowned for hake, probably the fish the English were after.[37] This account, combined with the evidence relating to Carlingford, may indicate that English fleets were operating in both areas in 1535.

English fishing around Waterford town was reported most frequently in the later sixteenth century, in part owing to an escalation in conflict in the period, which brought the fishery, and the danger it faced, into greater focus. In 1587, Sir John Perrot observed that if the Spanish had attacked Ireland in that winter, they would have captured '200 sails of bark' and 1,200 mariners from England, come for the fishing. Perrot went on to say that the capture and destruction of the fishing fleet would cause significant economic damage and that, in previous years, the fishers were 'accustomed to come thither a fishing two or three thousand'.[38] If that figure refers to the number of persons, it matches the reported number from Carlingford earlier in the sixteenth century. A letter from later in 1596 again warned that the English fishery at Waterford was in danger of attack and stated that the fishery usually took place in November and December.[39] This chimes with the modern observed spawning period of herring there, again suggesting that the timing of the fisheries was based on spawning patterns.[40]

The English also fished with single larger ships operating independently. A case heard before the Admiralty court in 1579 gives an insight into how these vessels worked. In the summer of 1577, two merchants, Peter Smyth and Gilbert Bodleighe, outfitted a ship called the *Emerye* in London for fishing off the coast of Ireland. The voyage was financed by the shipowner, who remained in London. The crew loaded 'salt in barrels and other empty casks, hoops, nets and other necessaries for fishing'.[41] The *Emerye* then travelled to Cork, Kinsale and Waterford and made a catch of 192 barrels of herring, which the crew cured with 27 casks of salt.[42] They then transported it to Le Havre in Normandy to sell. Thus we can see that single vessels operating alone came to the Irish coast equipped to

35 *CSPI* (IMC), 1509–47, no. 143, p. 70.

36 Kowaleski, 'South-Western Fisheries', p. 443; David O'Sullivan et al., 'An Inventory of Irish Herring Spawning Grounds', *Marine Institute, Irish Fisheries Bulletin*, 42 (2013), pp. 1–31, at p. 27.

37 A. E. J. Went, 'The Irish Hake Fishery, 1504–1824', *Journal of the Cork Historical and Archaeological Society*, 51 (1946), pp. 41–51, at p. 42; *Cal. Car. MSS*, 1603–25, no. 101, p. 175.

38 *CSPI*, 1586–88, no. 35, p. 443.

39 *CSPI*, 1596–97, no. 53, p. 153.

40 O'Sullivan et al., 'Irish Herring Spawning Grounds'.

41 *Cal. HCA 1536–1641*, no. 220, p. 43.

42 The original reports a catch of 16 lasts, which were equal to 12 barrels each.

catch and process their catch self-sufficiently. It is also notable that the *Emerye*'s catch was brought to a market outside both Ireland and England to be sold.

Distant water operations for salmon were also popular with English fishers. The most prominent salmon fishing location frequented by the English was the Bann. In 1519 the river was leased to Otwell Corbett, an English merchant. In 1524, it was again let to Corbett and two other merchants from Chester. The river continued to be granted to different English subjects across the century.[43] As the 1588 case illustrates, the local Irish continued to exert influence over the river in the late sixteenth century. In 1542 the English captured the O'Cahan castle on the Bann, which they called 'an obstacle to the fishing there'.[44] Later, in 1543, the O'Cahans signed an agreement with the lord deputy to settle the conflict, the first article of which was a promise to stop extracting 'ransoms, or tributes from farmers and other subjects of the king repairing to the Banne for the purpose of fishing'.[45] These events illustrate that there was an active English fishery in the period and that establishing a peaceful river in which to fish was an important issue for the English authorities.

There is no way to know how many English vessels visited the Bann each year, but individual examples provide us with details about how the fishery worked. An Admiralty case from 1555 describes how a merchant in London, Thomas Kent, purchased a pinnace to send to the Bann with other fishing ships owned by various London merchants, the profits of this venture to be shared among the merchants when the ships returned.[46] Bristol merchants were also heavily invested in the Bann fishery. In 1566, an English official, Valentine Browne, and a consortium of Bristol merchants were competing for the rights to the river. The lord deputy favoured Browne, but in 1569, the Lord Treasurer of Ireland was advised to collect tax from the Bristol merchants who traded 'in the north, for salmon and also hides, yarn and tallow'.[47]

English fishers and merchants visited other rivers in Ulster to fish for salmon. In 1539, a Galway merchant reported that a ship from Bristol was at Assaroe, near Ballyshannon, County Donegal, and would 'stay two months getting salmon'.[48] Salmon voyages were thus similar to those for hake and herring. Fishers would travel long distances and remain in the fishing grounds for an extended period. In 1539, Assaroe was controlled by the O'Donnells, showing that at this time

43 David B. Quinn, 'Anglo-Irish in Ulster in the Early Sixteenth Century', in Arthur Deane (ed.), *Proceedings and Reports of the Belfast Natural History and Philosophical Society* (Belfast, 1935), pp. 56–78, at p. 67.

44 *CSPI* (IMC), 1509–47, no. 611, p. 339.

45 *CSPI*, 1586–88, no. 177, p. 202.

46 *Cal. HCA 1536–1641*, no. 61, p. 12.

47 *CSPI* (IMC), 1566–67, no. 292, p. 132; *CSPI* (IMC), 1568–71, no. 379, p. 154.

48 *CSPI* (IMC), 1509–47, no. 470.2, p. 236.

English fishers and merchants could still operate in Gaelic Irish territory without consequence. However, the same merchant wrote of an impending Irish attack on the English Pale, a harbinger of the conflict that would engulf Ulster later in the century.[49]

Such examples illustrate the complex systems of ownership and finance that lay behind many of these voyages to Ireland. The majority of the larger fishing fleets originated from south-west England. However, we also have evidence for English ships travelling from Chester, London and elsewhere to fish in Ireland. It is difficult to trace the exact scale of these operations, but it seems clear that they were at their most frequent during the late fifteenth and early sixteenth centuries. In 1587 Sir John Perrot estimated that in previous years the number of English fishers at Waterford had numbered 2,000 to 3,000.[50] If we assume that the report in 1535 of 3,000 fishers at Carlingford referred to a separate fishing fleet, there could have been more than 6,000 English fishers operating in Ireland at various times during the first half of the century. If all 6,000 fishers operated in crews of five, that would mean 1,200 small boats. This figure is difficult to verify, but is the best estimate we can establish with the available evidence.

While the English seem to have been the most numerous, French and Spanish fishers were also frequent visitors to Irish waters, particularly Gaelic Irish regions. For the most part, the Gaelic Irish did not engage in their own extensive sea fisheries but instead had systems in place to cater for, protect, and tax visiting fishers from other areas. Spanish and French voyages to Ireland were happening from at least the middle of the fifteenth century. It was also during this period that the English authorities began attempting to limit and tax these voyages. In 1465 a statute ordered that 'every vessel of other lands coming to the said land of Ireland fishing' over 12 tons had to pay 13*s*. 4*d*., while every vessel under 12 tons had to pay 2*s*. per year to the king to be allowed to fish in Ireland. It also stated the money was to be used to fund the king's wars in Ireland.[51] Fishers who left Ireland without paying would forfeit their vessels and goods. However, the statute declared that every ship north of Wicklow was exempt from the charge. This focus on the area to the south suggests that the English authorities were most concerned with restricting the income visiting fishers provided to the Gaelic Irish.

At Baltimore, the Gaelic Irish O'Driscolls maintained a complex system of taxes to charge visiting fishers. An inquisition from 1609 described how this system had worked.[52] Any ship wishing to anchor there had to pay 4*d*. sterling to the lord of Baltimore, who also had the right to purchase any goods before they

49 *CSPI* (IMC), 1509–47, no. 470.2, p. 236.

50 *CSPI*, 1586-88, no. 35, p. 443.

51 *Statute Rolls, Ireland*, vol. III, 5 Edward IV, A.D. 1465, p. 355.

52 John O'Donovan (ed.), *The Genealogy of Corca Laidhe* (*CELT*, 2011), pp. 100–07; see also John O'Donovan (ed.), *Miscellany of the Celtic Society* (Dublin, 1849), p. 9.

went to market and would receive a discount of one in 20*s.* from the total price. If merchants or fishers sold their goods at the market, the lord would receive 8*d.* out of every 2*s.* One of the few goods explicitly mentioned in the rules was wine, with the lord receiving a tribute of one butt per four gallons of wine landed at Baltimore. The lord would also get a discount of 2*d.* on every additional gallon bought for personal use. The O'Driscolls were not alone in their appetite for wine; many Gaelic Irish lords imported wine for their personal use in the sixteenth century.[53] Indeed, almost all wine imported to Ireland in this period came from France or Spain.[54] The demand for wine in Ireland and the desire for fish among the continental fishers likely created an exchange that benefited both parties.

Every boat coming to Baltimore to fish had to pay a base tax of 9*s.* 2*d.* and a collection of foodstuffs three times a week.[55] Every fisher who wanted to use a seine net had to get a special licence.[56] Additionally, the fishers paid 13*s.* for the use of a rock to dry their fish. Different rates were also applied to fishers using other parts of the coastline, with the areas 'between Fastness and the Stagges' being half price. In the early sixteenth century, the earl of Kildare levied similar dues for drying fish on rocks at Ardglass, County Down.[57] The process of drying fish on rocks sounds very similar to the production of 'klipfish' – Norwegian cod that was split, salted and dried on rocks.[58] Given that the Norwegian method was used to process cod, it seems likely it was employed on white fish in Ireland too, with hake and cod the most likely candidates. Plate 2.1 shows a typical scene of klipfish production in Norway from the twentieth century. The method of preservation would have gone virtually unchanged since the early modern period.

Plate 2.2 is an image of rocks in a small bay at Trafraska, just outside Baltimore Bay. These rocks have a similar shape to those depicted in the photograph from

[53] Susan Flavin, 'Consumption and Material Culture in Sixteenth-Century Ireland' (PhD thesis, University of Bristol, 2011), pp. 148–53; for a specific study on the importance of wine to Gaelic Irish culture, see Katharine Simms, 'Guesting and Feasting in Gaelic Ireland', *JRSAI*, 108 (1978), pp. 67–100.

[54] H. F. Kearney, 'The Irish Wine Trade, 1614–15', *Irish Historical Studies*, 9 (1954–55), pp. 400–42, at p. 405.

[55] This was paid on Wednesday, Friday, and Saturday. The food required was 'a barrel of flour, a barrel of salt, a hogshead of beer, and a dish of fish'; O'Donovan (ed.), *Genealogy of Corca Laidhe*, p. 104.

[56] A seine was a long net used to surround fish and envelop them; see Plate 2.9 for an illustration of how this worked.

[57] Mark Gardiner and T. E. McNeill, 'Seaborne Trade and the Commercialisation of Fifteenth- and Sixteenth-Century Gaelic Ulster', *Proceedings of the Royal Irish Academy*, 116C (2016), pp. 229–62, at p. 237; the dues can be found in BL, Harleian MS 37556, f.94v: rental and survey of the property of Gerald fitz Gerald, ninth earl of Kildare.

[58] Charles L. Cutting, *Fish Saving* (London, 1955), p. 145; Alison Mary Locker, 'The Role of Stored Fish in England 900–1750AD; the Evidence from Historical and Archaeological Data' (PhD thesis, University of Southampton, 2000), p. 52.

Plate 2.1. Klipfish production, Kristiansund, Norway, 1922.

Plate 2.2. Rocks at Trafraska, near Baltimore.

Norway. They are also flat and black, which means that they would have absorbed more heat to help dry the fish, although drying fish on rocks was very weather dependent, and the process in Ireland may have required more salting than the Norwegian equivalent.[59] It is plausible that similar scenes to those in Norway could have been seen on rocks near Baltimore, Ardglass and elsewhere in Ireland during the sixteenth century. Perhaps in Ireland, too, the work would have been carried out by a mixed workforce of women and men, as it is in the Norwegian image. Unfortunately, no records remain to inform us about the structure of the workforce in Baltimore. The practice of charging to dry fish was not unique to Ireland either; at Lydd, in Kent, visiting fishers in 1462 were charged by the local authorities for the privilege of drying their fish and nets.[60]

There were separate charges for separate species at Baltimore too. If any fisher found a halibut they were required to give it to the lord in exchange for a barrel of butter, concealment carrying a fine of 40*s*. Clearly the halibut, a large flatfish, must have been prized by Gaelic Irish, presumably because of its rarity and impressive size. Furthermore, every boat that came into the harbour had to pay 6*s*. 8*d*. per 100 white fish brought in. For every barrel of herring and pilchards sold on the open market the lord would receive 8*d*. from the seller. These charges indicate what fish were caught in Baltimore and what species held special significance.

Several estimates from sixteenth-century sources approximated the size of the Spanish fishery in Ireland. In April 1543, lord deputy Anthony St Leger reported to the king that '200 or 300' fishing boats sailed to Baltimore each year.[61] In 1569, one John Corbine warned secretary of state William Cecil: 'The Spaniards will have good harbours on the southwest coast, at Berehaven, Crookhaven, Baltimore, O'Sullivan's river, Ballinskellig and Valentia.' He went on to say that 'The area is already fruitful to the Spaniards. Every year 200 boats fish here and take away beef, hides and tallow, paying no dues to the queen.'[62] This figure seems similar to that provided by St Leger, but Corbine was trying to persuade Cecil to attack the Gaelic Irish at this time, so it was also in his interest to exaggerate.

A document from 1572 provides the perspective of an English colonist, Humphrey Gilbert, on the Spanish fisheries. He wrote that '600 sail of ships and barks for fishing only' visited the south coast of Ireland each year. The Spanish 'lay aground' at Baltimore during the fishing season, and both 'Spaniards and Biscayans' used Great Blasket Island as a base for fishing. Gilbert's figures must be taken with caution; he was arguing for the plantation of Munster and for the 'trade with Spain to be destroyed', so may have inflated the figures to his

59 Locker, 'Stored Fish', pp. 52–3.

60 Kowaleski, 'South-Western Fisheries', p. 441.

61 *CSPI* (IMC), 1509–47, no. 627, p. 350.

62 *CSPI* (IMC), 1568–71, no. 362., p. 145.

advantage, but the locations he provides for the Spanish fishing are plausible and consistent with other evidence.[63]

Later, in 1583, Flynn O'Driscoll, lord of Baltimore, reported that 100 fishing boats were heading to Baltimore that year.[64] Much later, in 1605 Donal Cam O'Sullivan Beare informed Philip II of Spain that at least 500 Spanish fishing boats had formerly visited Berehaven each year.[65] O'Sullivan wrote while in exile in Spain, so was probably overestimating the size of the fishery to gain more support from the Spanish king. There are, however, more reliable figures for the size of the fleet in the Spanish sources. An event in 1537 also indicates the size of the O'Driscolls' own fleet. Ships from Waterford had attacked Baltimore, burned the O'Driscoll castle on Sherkin Island, captured the galley of the O'Driscoll chief, seized 'above three or four score pinnaces' and burned another fifty.[66] Even with some exaggeration, this is a sizeable fleet of small boats that the O'Driscolls could have used for transport and inshore fishing. It is also possible these boats were rented out to visiting fishers who arrived in larger ships unsuitable for fishing in the small bays of the Cork coast.

Baltimore is the most detailed example of taxes being imposed on foreign fishers, but a plethora of other evidence shows that this was a common practice in many Gaelic and Old English areas. A report from 1580 describes the O'Sullivans imposing similar taxes at Berehaven. Here larger boats, like skiffs, were charged 8*s.* 17*d.* for the right to fish, while a 'small boat' paid 6*s.* 8*d.* There were also charges for anchorage and for 'places to dry their fishe'.[67] Thus the system of taxation on the Munster coast seems to have been relatively consistent among the Gaelic families who controlled the best fishing spots. Nor was the practice confined to Munster. In the 1540s the O'Cahans signed an agreement with the English government promising to refrain from extracting tributes from visiting fishers on the Bann,[68] and in 1557, lord deputy Sussex reported that James Mac Donnel 'desires peace' and had offered to allow the English to fish the Bann as long as fishers paid him the 'accustomed droits'.[69]

Protection was a crucial part of the service offered by the Gaelic lords in exchange for the dues paid by fishers.[70] In 1531, when an English vessel captured a Spanish fishing boat off Dursey Island, near Berehaven, Dermot O'Sullivan, chief of Berehaven, took his ships out and captured both the English and Spanish

[63] *Cal. Car. MSS*, 1515–74, no. 290, pp. 422–23.

[64] Flynn O'Driscoll to Lord Burghley, 1583 (TNA, SP 63/104, f.238).

[65] Flavin, 'Consumption and Material Culture', p. 92.

[66] O'Donovan (ed.), *Genealogy of Corca Laidhe*, pp. 94–95.

[67] Fulk Greville to his cousin Secretary Walsyngham, 2 July 1580 (TNA, SP 63/74, f.4).

[68] *CSPI*, 1586–88, no. 177, p. 202.

[69] A droit is a legal right or due. *CSPI*, 1509–73, no. 28, p. 136.

[70] Colin Breen, *The Gaelic Lordship of the O'Sullivan Beare* (Dublin, 2005), p. 116.

vessels. 'The English Captain he hanged and set the other at liberty.'[71] This incident is a clear example of the Gaelic Irish protecting fishers in their waters. There were other benefits for the foreign fishers; they received a safe and suitable place to dry their fish and anchor their boats, and access to markets where they could trade commodities like wine, for fish. However, we should also keep in mind that the visitors were at the mercy of the local lords. The Gaelic Irish positioned their settlements and fortifications to control the best fishing locations, so visiting fishers who did not pay would not have access to the best fishing grounds. However, despite the taxes, these distant water voyages must have been profitable.

This was not a practice limited to the Gaelic Irish, and Old English lords generated income from visiting fishers, employing similar methods to the Gaelic Irish. In 1537, the residents of Dungarvan collected a custom of 6*s*. 8*d*. and one hake from every fishing boat that landed in the town.[72] A survey taken in the 1540s records that the town of Ballyhack, just outside Waterford, welcomed 'all fishing boats of strangers', upon the donation of one 'large fish', or one hake if they had been fishing for them. Boats also paid 'one penny-fish' if they had been catching smaller fish. The survey suggests the town relied more on the visiting fishers for income than those who were residents. At the time the survey was taken, the taxes and rent of the town were being paid to the countess of Ormond.[73] Additionally, in the early sixteenth century, the earl of Kildare charged customs on the different types of ships and boats that landed at Ardglass. Fishers wishing to dry their fish also had to pay the earl. The visitors were required to give 'a chief fish to the castle' every time they came into port. Every herring boat also had to donate a mease of herring (600 fish) to the castle.[74] Ardglass and Strangford were leased out by the Crown in 1524 for £150, a high rent for the period.[75] Therefore, this was not a practice limited to the Gaelic Irish, and Old English lords generated income from visiting fishers too, even in a relatively small port like Ardglass.

Before the dissolution of the monasteries in the 1530s and 1540s, monastic houses also charged visiting fishers to use their infrastructure. A statute from 1493 imposed a new custom on ships and fishing boats using the new pier at Skerries constructed by Holmpatrick priory. The priory charged a poundage tax on any boats 'coming to those parts for fishing, for flask fishing, cod and herrings'.[76] In

[71] A. E. J. Went, 'Foreign Fishing Fleets along the Irish Coasts', *Journal of the Cork Historical and Archaeological Society*, 54 (1949), pp. 17–24, at p. 17.

[72] Went, 'The Irish Hake Fishery', p. 42.

[73] Newport Benjamin White, *Extents of Irish Monastic Possessions, 1540–1541, from Manuscripts in the Public Record Office* (London and Dublin, 1943), p. 355.

[74] Quinn, 'Anglo-Irish in Ulster', p. 64.

[75] Quinn, 'Anglo-Irish in Ulster', p. 63.

[76] Philomena Connolly (ed.), *Statute Rolls of the Irish Parliament, Richard III – Henry VIII* (Dublin, 2002), Statute roll, 8 Henry VII, p. 141.

this instance, 'flask fishing' might refer to pot fishing for Nephrops, which we have few other historical references to from this period.

The elaborate nature of the duties in many Gaelic Irish and Old English areas indicates the importance of the visiting fishers to the local economy. There are several possible reasons why the Gaelic Irish may have used this system rather than catching and selling the fish themselves. They may have lacked the trained workforce to conduct a large enough fishery, or not have possessed the requisite knowledge (although this seems unlikely). Moreover, the visiting fishers were clearly a good source of money and brought exotic goods, such as wine. The economic model used in Gaelic areas allowed the fish caught at places like Baltimore to reach a broader market that may have been otherwise inaccessible to the Irish.

We know that the Irish did take part in subsistence fishing, with archaeological remains of stone traps at Rossglass and other places illustrating the occurrence of small-scale fishing in the early modern period.[77] However, if Gaelic Irish clans had intended to make large catches of fish above their subsistence needs, they lacked local supplies of salt to preserve it.[78] Almost all early modern techniques for preserving fish required salt, and any fish that was not consumed locally needed to be preserved before being transported.[79] Salt could have been imported, but this would have added to the cost and perhaps made the industry less commercially viable.[80]

In contrast, foreign fishers brought salt with them to Ireland, which cut out the need for local supplies. Furthermore, the Irish could exchange fishing rights for salt that could be used on other products such as hides or for preserving meat and fish. Ultimately, the exchange of goods and money for the right to fish benefited both parties. The Irish had ample fishing resources and could trade access for both luxury goods and mundane necessities like salt. As Colin Breen has written, the presence of foreign fishers shows that the Gaelic Irish were not economically backward but 'instead had comprehensive economic management systems in place' to profit from their physical and cultural environment.[81] However, as the sixteenth century progressed, the English authorities and their allies in Ireland became increasingly concerned with the income visiting fishers brought to the Gaelic Irish and to rebellious Old English lords. The English were also concerned

[77] Colin Breen, 'A Medieval Fishery at Rossglass, Co. Down', *Ulster Journal of Archaeology*, 58 (1999), pp. 153–55.

[78] Ada K. Longfield, *Anglo-Irish Trade in the Sixteenth Century* (London, 1929), p. 163.

[79] See, Locker, 'Stored Fish'.

[80] Aidan O'Sullivan and Colin Breen, *Maritime Ireland: An Archaeology of Coastal Communities* (Stroud, 2007), p. 227.

[81] Breen, *O'Sullivan Beare*, p. 115.

by the number of Spanish and French fishers visiting Ireland and the potential military and religious consequences.

The work of Michael Barkham in the Spanish archives provides an insight into this activity from the Spanish perspective: most of the Spanish fishers who travelled to Ireland for fishing came from the Basque country in northern Spain.[82] The reasons why Basque fishers came to Ireland are manifold. The region itself was mountainous with poor agriculture, but it did have an abundant supply of oak and iron to make ships.[83] Given this, the Basques looked outward to the ocean to expand their economic activity. They did not just come to Ireland but travelled to fish in Galicia, England and elsewhere.

Additionally, Basque whalers hunted in the Bay of Biscay from at least the twelfth until the fifteenth century, when they began to go further out into the Atlantic.[84] These distant voyages were necessary due to centuries of overhunting that reduced whale populations closer to Spain.[85] From the early decades of the sixteenth century, Basques also began to travel across the Atlantic in increasing numbers to fish and hunt whales in Newfoundland and Labrador, respectively.[86] The Basques were some of the first European fishers to exploit these waters. Their activity in Ireland played a role in preparing them for later voyages across the Atlantic. However, we know substantially more about the Basque Newfoundland fishery than we do about the Irish fishing that predated it.

Evidence from Spanish ports suggests that the Basques were fishing in Ireland from at least the fourteenth century. This activity continued into the sixteenth century, and according to Barkham, the main targets of the Spanish fishery in Ireland were hake, conger eel and herring or pilchards (depending on which was present).[87] These target species are very similar to those sought by the English, apart from conger eel. The Spanish ships that travelled to Ireland left for the fishery in June or July and would often not return to Spain until at least October or November. The timing of these voyages suggests they came to Ireland initially for the hake fishery, with some vessels perhaps staying longer to catch the early

[82] Michael M. Barkham, 'The Offshore and Distant-Water Fisheries of the Spanish Basques, c.1500–1650', in D. J. Starkey et al. (eds), *A History of the North Atlantic Fisheries 1* (Bremerhaven, 2009), pp. 229–49; 'The Spanish Basque Irish Fishery and Trade in the Sixteenth-Century', *History Ireland*, 9, no. 3 (2001), pp. 12–15; 'La industria pesquera en el País Vasco peninsular al principio de la Edad Moderna: ¿una edad de oro?', *Itsas Memoria. Revista de Estudios Marítimos del País Vasco*, 3 (2000), pp. 29–75.

[83] Barkham, 'Fisheries of the Spanish Basques', p. 229.

[84] Jean-Pierre Proulx, *Basque Whaling in Labrador in the 16th Century* (Ottawa, 1993), p. 12.

[85] Alex Aguilar, 'A Review of Old Basque Whaling and Its Effect on the Right Whales (Eubalaena glacialis) of the North Atlantic', *Report – International Whaling Commission*, 10 (1986), pp. 191–99.

[86] Barkham, 'Fisheries of the Spanish Basques', p. 241.

[87] Barkham, 'Fisheries of the Spanish Basques', p. 231.

autumn spawning herring. These periods are similar to those of the English fishers reported earlier, though the English left Ireland slightly earlier. Barkham suggests that Spanish fishers mostly frequented the south coast between Baltimore and Valencia, and Killybegs in Donegal, an assertion supported by the data collected for this study.[88]

The journey of the *Santa Maria* in 1511 provides a detailed example of a typical Spanish voyage to Ireland. The charter party of the ship records that by 29 June 1511, the master was meant to have loaded 'six pinnaces for hake and one for killing sardines' onto the ship. Two of these pinnaces had a crew of five men, and four had a crew of four men. These figures would give the ship a minimum crew of twenty-six. These figures are very similar to the five-men crews described for the English herring fishery at Carlingford.[89] The ship was to cross the Bay of Biscay from Lekeitio in the Basque country to La Rochelle, where it would buy salt to cure the fish. From La Rochelle, the ship was meant to navigate to the Aran Islands to make its catch. The ship was to stay there until Christmas Day and then travel to Cadiz to sell the fish. Though, if the ship could not sell its catch in Cadiz, it was allowed to travel further to a town near Rome to sell the fish.[90] This example shows the complex nature of these voyages, involving ships from the north of Spain, French salt, and markets in the central Mediterranean.

Concerning the size of the Spanish fleet, Barkham writes that in peaceful years the Spanish Basque fishery in Ireland 'consisted of 20–30 small and medium-sized ships crewed by a total of 300–500' fishers. A census of ships conducted in the province of Gipuzkoa in the Basque country from September 1534 recorded that seventy-six ships operated from the port that year, and of these, fifteen travelled to Ireland with a total crew of 250 to 300 fishers. The fleet consisted of eleven caravels and four small craft, and the ships were expected back in Spain by mid-November. It is likely they were targeting herring if they were to remain on the coast that long. Barkham estimates that another similar number of ships would have travelled from other Basque ports, which were not included in the census.[91] The total estimate is therefore thirty ships, with an upper estimate of 600 crew. If the fishers operated in crews of five, that would mean 120 pinnaces with each large ship carrying four pinnaces, which is within the number carried by the *Santa Maria*. These figures are again an upper estimate, but given the evidence, they seem plausible.

Additionally, the figure of 120 small boats seems more reasonable than most estimates for the size of the Spanish fishery from the English and Irish sources

88 Barkham, 'Fisheries of the Spanish Basques', p. 234.
89 *Cal. Car.* MSS, 1515–74, no. 70, p. 85.
90 Barkham, 'The Spanish Basque Irish Fishery', pp. 235–36.
91 Barkham, 'Fisheries of the Spanish Basques', p. 240.

discussed above. It is more difficult to estimate how these numbers may have fluctuated year on year, but it seems evident that Spanish activity in Ireland diminished over the century. The Spanish fleet was probably smaller than the English but would still have significantly impacted the local economy. Furthermore, the Spanish seem to have almost exclusively visited Gaelic Irish regions. The Irish had a complex system of taxation in place to exploit this and profited greatly from the Spanish visitors. That profit would increasingly frustrate the English authorities as the century progressed, and would eventually lead to the violent takeover of the most important Gaelic Irish fishing sites.

Compared to the English and Spanish, we know considerably less about the French fishery in Ireland during the medieval and early modern periods, in part because we lack published studies in English on the relationship between France and Ireland during this period. However, some material has been published that illuminates this topic, and new archival research in France holds great potential to tell us much more.

Notarial records from Bordeaux in the sixteenth century show that many French shipowners combined both trading and fishing in their voyages to Ireland. In 1534, the *Isabelle de Losquirec* was chartered in Bordeaux by merchants from Dingle, with a cargo of wine. The merchants agreed that the ship would stop at Loctudy in Brittany en route to Dingle, to pick up a crew of seven, who were meant to unload the wine in Dingle. Then the ship would 'set out fishing off the Irish coast before sailing back to La Rochelle with fish'.[92] While it is only one example, this voyage corroborates the idea that visiting fishers traded wine for the right to fish in Ireland. It was also more profitable for the fishing vessels to arrive in Ireland with a full cargo for sale or exchange rather than with an empty hold. Similarly, Breton fishers from Cap Sizun would travel to Bordeaux each year to exchange their hake for wine.[93] It is not clear how much was caught in Ireland, but given the proximity of Brittany to Ireland, it is feasible that some portion was.

The letter from Sir Anthony St Leger in April 1543 also discusses the presence of foreign ships on Irish shores. St Leger writes that the east coast was 'frequented mostly by English, and by Bretons and Spaniards in time of peace'; the north coast 'by Bretons and Scots in both peace and war, and those of the west, to which no Englishmen come save to Galway and Limerick, by Spaniards and Bretons at all times'. St Leger also said that the Gaelic Irish of Munster traded hides for 'salt, iron, guns and powder', but made no mention of fish.[94] His letter reveals

92 Jacques Bernard, 'The Maritime Intercourse between Bordeaux and Ireland, c. 1450–1520', *Irish Economic and Social History*, 7 (1980), pp. 7–21, at p. 15.

93 Bernard, 'Maritime Intercourse', p. 15.

94 *CSPI* (IMC), 1509–47, pp. 349–50, no. 627, p. 349.

that those areas furthest away from English control were more open to Spanish, French, or Breton ships, especially during times of war with England.

Further evidence for Breton fisheries in Ireland comes from a letter written by Sir Thomas Phillips in 1611 during the planning for the Ulster Plantation. 'There was long since at Port Rush a fishing used by Bretons in France who came every season thither for dogfish and rays.' According to Phillips, the fishers sold their catch in Spain where the fish were a 'great commodity'.[95] The discussion of dogfish[96] and ray[97] fishing is unusual, as the species are not mentioned in many other accounts from this period.

Ships from the French Basque country also took part in the Irish fishery. A notarial record from La Rochelle in 1540 mentions a bottomry loan[98] to the *Trinity* of Saint Jean de Luz near the Spanish border, stipulating that the ship would go to Ireland to fish hake and herring, without giving an exact location.[99]

Most visiting fishers, including the French, probably did not come to Ireland every year but decided on an annual basis what voyage would provide them with the most profit. Hake spawning also shows substantial interannual variability, so in some years the hake would have been particularly abundant in the Bay of Biscay, and in other years there would have been better catches in the Celtic Sea and off the Irish coast.[100] Fishers presumably knew of these fluctuations and shared information as to the location of the best fishing prospects.

While we cannot be sure how much French hake fishing took place in Ireland, we can be confident that it declined over the sixteenth century, as shown by evidence in French local records. The money received by the governor of Penthièvre on the Brittany coast from the hake fishery there declined sharply over 100 years from 1,200 livres in 1536 to only 80 in 1637; Figure 2.1 shows the complete trend.[101] This decline follows a similar pattern to that of the qualitative evidence that indicates that French fishing in Ireland was always rare and became non-existent after 1568. The decline happened for a variety of reasons, but

95 *Cal. Car.* MSS, 1603–25, no. 95, p. 150.

96 This could be *Scyliorhinus canicular* or *Scyliorhinus stellaris*, it is not clear from this reference alone.

97 While we cannot be certain, this was likely *Raja clavata*.

98 Bottomry is a transaction where a shipowner borrows money using the ship itself as collateral and will forfeit the ship if the money is not paid back before the ship returns.

99 Bottomry loan on the Trinity of Saint Jean de Luz to go fish Hake and herring on the coast of Ireland, 17 April 1540 (Archives départementales de Charente-Maritime [La Rochelle], 3E-9). This material was kindly provided by Dr Bernard Allaire from the NorFish project team.

100 ICES, *Report of the Working Group on the Assessment of Southern Shelf Stocks of Hake*, p. 53.

101 Camille Vallaux, *Penmarch, aux XVI et XVII siècles* (Société Nouvelle de Libraire et d'Edition, Paris, 1906), p. 23.

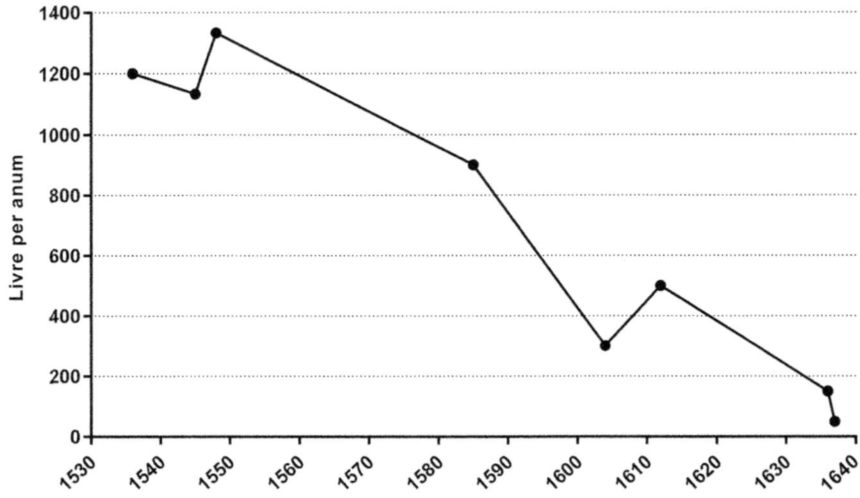

Figure 2.1. Governor's income from Penthièvre hake fishery, 1536–1637. Source: Vallaux, *Penmarch, aux XVI et XVII siècles*, p. 23.

especially because from the late sixteenth century onwards the wars of religion in France severely disrupted normal fishing operations in Brittany.[102]

While the evidence for French fishing in Ireland is absent in the last few decades of the sixteenth century, there are some indications that French merchants still traded in Ireland. In 1601 an English ship's captain, Charles Plessington, on patrol on the Irish west coast, sent detailed reports which provide information on the French trade. 'Every year, till now', he wrote, 'this coast was much frequented with French-Men.' He claimed that French merchants commonly carried cargoes of 'wines, salt and other commodities', 'which they did commonly discharge in the harbour of Killybegs, Donegal, Ballyshannon and Sligo: and from here they lade hides, tallow and fish'. In the summer of 1600 one French merchant brought 60 tons of wine to 'Donegal Abbey'.[103] His letter illustrates that the French trade with the west of Ireland continued at least until the end of the sixteenth century, although it provides no evidence that the French were still conducting fishing expeditions. The mention of fish does suggest that some exchange was still taking place, involving goods like wine for fish, but we have no way of knowing the scale.

102 Vallaux, *Penmarch, aux XVI et XVII siècles*; and personal correspondence with Dr Bernard Allaire.

103 Plessington to the earl of Nottingham, and Secretary Cecil, 2 September 1601 (TNA, SP 63/209/1, f.148).

This problem extends to all the French fishing activity in Ireland. Unlike English and Spanish fishing, there is not enough data to estimate the number of French ships visiting Ireland in any year. There is also scant archaeological evidence for the presence of French, or any foreign, fishers in Ireland. Unfortunately, by its nature, this type of transient fishing activity left little physical evidence.[104]

However, material evidence suggests a strong connection between parts of Ireland and continental Europe. Archaeologists have uncovered stockpiles of European pottery at Dunboy Castle near Berehaven. Some of this pottery originated from Saintonge on the west coast of France and dated from the early sixteenth century. Other potsherds came from northern France, the Netherlands and as far away as Ming China.[105] Overall, pre-1601 pottery discoveries at this site were dominated by continental finds, while the seventeenth-century assemblages were 'almost exclusively English', which speaks to the radical change in ownership occurring in the early seventeenth century.[106] It is of course difficult to date pottery accurately, and later English settlers may also have used pottery from France and beyond, but despite these limitations, the discoveries provide further supporting evidence for the idea that Gaelic Irish families, such as the O'Sullivans of Berehaven, had robust connections to continental Europe that were centred around their fishing resources.[107]

The sixteenth century also saw a European-wide shift of fishers going to Newfoundland instead of domestic waters. The distant waters fisheries in Newfoundland had many commonalities with those in Ireland, and Irish fisheries may have been a forerunner to operations across the Atlantic. One of the primary targets of visiting fishers in Ireland was hake, and this relates directly to the later fishery in Newfoundland, which focused almost exclusively on cod. Cod and hake are closely related; they are white fish in the same taxonomic order of Gadiformes. The process of drying fish on rocks at Baltimore was probably related to hake fishing, and similar methods would be used in Newfoundland to dry cod.[108] A promissory note dating from 1506 in San Sebastian promised 'twenty cargas of cured hake from Ireland, half of which dried in Ireland and the other half dried in this land',[109] backing up the idea that fishers dried hake on rocks, but also showing that 'wet' fish could be brought back and dried later. 'Wet fish' refers to fish that was not wholly dried out but salted and packed together.[110] These same

[104] O'Sullivan and Breen, *Maritime Ireland*, p. 171.

[105] O'Sullivan and Breen, *Maritime Ireland*, p. 172.

[106] Breen, *O'Sullivan Beare*, p. 116.

[107] O'Sullivan and Breen, *Maritime Ireland*, pp. 171–72.

[108] Locker, 'Stored Fish', p. 53.

[109] Barkham, 'Fisheries of the Spanish Basques', p. 234.

[110] Locker, 'Stored Fish', p. 46.

methods of drying fish and transporting 'wet' fish would be employed in the Newfoundland fishery by the Spanish, French, and English.

The Basques were some of the first European fishers to exploit these waters. Barkham writes that Basque 'fishing and whaling voyages' to Ireland and elsewhere were clearly 'precursors of the transatlantic cod fishing and whaling expeditions to Terra Nova'.[111] Salted and dried cod from Newfoundland, or *bacalao* as the Spanish call it, would later play an important cultural and culinary role in Spain and Portugal.[112] In addition, similar systems of organisation and finance used by fisheries on voyages to Ireland were subsequently employed in Newfoundland.[113]

By the 1540s, French vessels were also commonly visiting the waters of Newfoundland. In 1540, Spanish cartographer Alonzo de Santa Cruz noted a 'great abundance of fish' in Newfoundland 'for which reason many Bretons and Irish come there every year to fish'. Cod caught in Newfoundland was commonly shipped to France and Galicia to be sold.[114] The mention of Irish fishers in Newfoundland is intriguing, as it would place them there at an early date. Overall, the lack of other evidence makes it implausible that Irish ships were going in large numbers to Newfoundland in the 1540s, but it is possible that Irishmen were serving in the crews of Breton and Spanish ships.

Even though in the late sixteenth and early seventeenth centuries Basque and French fishers were pushed off the shores of Newfoundland by the English, there remained a demand for Newfoundland cod in Spain, which English fishers and merchants fulfilled.[115] A large portion of the cod brought to Spain by the English was exchanged for wine that was in turn transported back to England.[116] A similar trade may have been at play in Ireland before the rise of the Newfoundland fishery. There are several examples of wine being imported to Ireland by fishers. It is hard to know if this trade was substantial, but it does seem that the trade of wine for fish existed in Ireland before the English exploited Newfoundland and it became a much larger exchange.

The English fishery in Newfoundland was part of a broader westward shift that first began in Ireland and Iceland.[117] Again, the methods developed for processing, transporting and fishing in Ireland were employed in Newfoundland.[118]

111 Barkham, 'Fisheries of the Spanish Basques', p. 232.

112 Regina Grafe, *Distant Tyranny: Markets, Power, and Backwardness in Spain, 1650–1800* (Princeton, NJ, 2012), pp. 52–79.

113 Peter Pope, *Fish into Wine: The Newfoundland Plantation in the Seventeenth Century* (Chapel Hill, NC, 2004), p. 16.

114 D. B. Quinn, *New American World. A Documentary History of North America to 1612* (London, 1979), pp. 276–77.

115 Barkham, 'Fisheries of the Spanish Basques', p. 243.

116 Pope, *Fish into Wine.*

117 Kowaleski, 'South-Western Fisheries', p. 442.

118 Kowaleski, 'South-Western Fisheries', p. 443.

The same is true of French fishers, who entered into the Newfoundland trade before the English.[119] Ireland was not the only location these fishers developed their skills, but it was part of the process and played a vital role in the westward expansion of the European fisheries.

Overall, the history of foreign fisheries cannot yet be written with confidence. These fishers only visited for a short time each year, and left little by way of an archaeological footprint.[120] Furthermore, the documentary evidence outside Ireland is scattered among several European countries and languages. Nevertheless, it is clear that English, Spanish and French fishers did visit the Irish coast regularly in the fifteenth and sixteenth centuries. The Gaelic Irish, and to a lesser extent some Old English, acted as a neutral third party facilitating fishing for the visitors. They provided access, protection, and space for the others to make and process their catch. It is also evident that distant water fisheries were a large part of the marine economy of fifteenth- and sixteenth-century Ireland, and the decline of these visiting fisheries in the later sixteenth century would have important consequences not just for the fishing industry but for Ireland's economy and society as a whole.

Organisation and Methods

How did fishers organise their operations in the sixteenth century, and what methods and gear did they use? We know that some fishing boats operated individually, with the organisational structure kept within the confines of a single ship. However, even these ships were subject to rules and regulations once on the Irish shore. The O'Driscoll inquisition from 1609 reported that Kinsale had previously elected a fishing admiral each year, who joined the lord of Baltimore 'in settling orders for the fishing' on the coast between the two towns. The fishing admiral held court every Monday, where fines were handed out for those who broke the rules, the revenues to be shared between Baltimore and Kinsale. He would, for example, extract a 11*s.* 6*d.* fine for every act of 'bloodshed'.[121] The rules stipulated that if Kinsale did not send an admiral, then the lord of Baltimore could appoint his own. This cooperation between the Old English town and Gaelic Irish is surprising but understandable: fishers had the freedom to travel anywhere along the coast, and the existence of a standard set of rules, and mechanisms to enforce them, clearly benefited everyone.

[119] Laurier Turgeon, *The Era of the Far-Distant Fisheries: Permanence and Transformation (circa 1500–1850)* (St. John's, Nfld, 2005).
[120] O'Sullivan and Breen, *Maritime Ireland*, p. 171.
[121] O'Donovan (ed.), *Genealogy of Corca Laidhe*, p. 105.

The mayor of Waterford also acted in the capacity of fishing admiral and was charged with 'keeping order among the fishers'.[122] A deposition from the Admiralty court, recorded in 1567, describes a similar system in Wales. Rice ap Edward was elected as the fishing admiral in Wales that year. His role was to oversee 'all contracts, conventions and punishment of fishermen using the seas of Ireland and certain places in Wales'. Like the admiral in Baltimore and Waterford, he would collect fines for breaches of the rules, and in 1567 he received a total of 23*s.* 8*d.* from the fishers.[123]

Edward informed the court that the Welsh admiral was elected yearly by the fishers themselves. He provided this deposition because some of the Welsh fishers who were 'fishing for herring at Aberdovey' got into a 'quarrel' with the local vice-admiral, Sir Nicholas Arnold.[124] In the sixteenth century the English Admiralty had a network of local admirals who would control and protect the coast in their regions. This particular dispute seems to have arisen because Arnold perceived that, as fishing admiral, Edward was a threat to his own authority and so accused Edward of extorting money from the fishers. In his defence, Edward stated that 'he has taken no money off any of the said fishermen, except such as has been accustomed to being taken for offences committed by them'.[125]

It is notable how similar the system described by Edward is to the one detailed for Kinsale, Baltimore and Waterford. These examples suggest that the election or appointment of a central fishing admiral was a common organisational method among fishers in Ireland and Britain in the sixteenth century. The dispute between the fishing admiral and naval admiral in Wales also suggests that, as time went on, these types of independent fisheries organisations ran into conflict with the Tudor state, which was seeking to expand its control in all economic activities. Historians have observed similar disputes between fishing towns and the central government in sixteenth-century England.[126] Ultimately English government officials felt fisheries were too important to be left in the hands of the fishers or local towns and sought to impose greater control. In the Irish case, the expansion of English authority late in the sixteenth century completely removed old

[122] *CSPI* (IMC), 1566–67, no. 595, p. 253.

[123] *Cal. HCA 1536–1641*, no. 145, p. 28.

[124] *Cal. HCA 1536–1641*, no. 146, p. 28. Dovey is an Anglicised spelling of the River Dyfi in central Wales.

[125] *Cal. HCA 1536–1641*, no. 145, p. 28.

[126] David M. Dean, 'Parliament, Privy Council, and Local Politics in Elizabethan England: The Yarmouth-Lowestoft Fishing Dispute', *Albion*, 22 (1990), pp. 39–64; G. R. Elton, 'Piscatorial Politics in the Early Parliaments of Elizabeth I', in Neil McKendrick and R. B. Outhwaite (eds), *Business Life and Public Policy* (Cambridge, 1986), pp. 1–20; R. C. L. Sgroi, 'Piscatorial Politics Revisited: The Language of Economic Debate and the Evolution of Fishing Policy in Elizabethan England', *Albion*, 35 (2003), pp. 1–24.

structures of Gaelic and Old English cooperation and governance that supported the appointment of admirals in places like Baltimore and Kinsale.

Fishers in Ireland also employed several different methods and types of fishing gear. Generally, the target species and local seascape determined the type of method and gear used. Local coastal conditions and the level of access to different technologies were also factors. Arthur Went's work focused on the methods used for riverine fish, particularly salmon, but many of the methods used for species like salmon were also used for marine fish like herring and hake.[127]

One of the most fundamental technologies used to fish at the time was the hook and line. The purest form of this method was the handline, which held a single hook. The longline was a more intensive form of fishing that used lines with multiple hooks. The longest of these could hold hundreds of hooks and even in the early modern period could be kilometres in length.[128] There is scant documentary evidence for the use of the hook and line in Ireland, and what is available has been summarised by Arthur Went.[129] However, despite the lack of direct evidence, we have to assume the method was used in Ireland because it was such a pervasive and fundamental technology across Europe.

Generally, fishers used hook and line on larger fish, such as hake and cod.[130] Targeting a smaller species like herring with this method would have been very inefficient. We have no direct description of hake fishing methods in Ireland, but fishers in the later Newfoundland cod fishery used hook and line commonly.[131] Thus it is likely that fishers had employed similar methods in the earlier Irish hake fishery. The line portion of the hook and line could also be suspended at different levels in the water column to bring it to the required depth to target a particular species.[132] Fishers presumably knew the daily cycles of their target species and would set their lines accordingly. For example, hake would stay lower in the water column during the day and rise to the surface in the evening and at

[127] A. E. J. Went, 'The Galway Fishery. An Account of the Modes of Fishing Together with Notes on Other Matters Connected with the Fishery', *Proceedings of the Royal Irish Academy*, 49C (1943), pp. 187–219; 'Eel Fishing at Athlone: Past and Present', *JRSAI*, 80 (1950), pp. 146–54; 'Fishing Scenes from Irish Topographical Prints', *JRSAI*, 81 (1951), pp. 156–60; 'The Pursuit of Salmon in Ireland', *Proceedings of the Royal Irish Academy*, 63C (1964), pp. 191–244; '"Sweeper" Nets for Salmon in County Kerry', *JRSAI*, 102 (1972), pp. 178–80; 'Historical Notes on the Fisheries of the Estuary of the River Shannon', *JRSAI*, 111 (1981), pp. 107–18.

[128] Peter Jones, 'Technological Innovation and Resource Management in the Fisheries of the British Isles, ca.1400–1900' (PhD thesis, University of Strathclyde, 2016), p. 195.

[129] Went, 'Pursuit of Salmon', p. 199.

[130] C. F. Hickling, *The Hake and the Hake Fishery* (London, 1934), pp. 32–34.

[131] Cutting, *Fish Saving*, p. 255.

[132] B. A. Bjarnason and M. Carlesi, *Handlining and Squid Jigging* (FAO, Rome, 1992).

night.[133] If fishers weighted their lines correctly, they could target the fish during the day or night.

So, while hook and line was a simple technology, fishers could employ it in ingenious ways that exploited the traits and behaviours of particular species. Furthermore, with enough fishing effort, hook and line could make significant catches.[134] According to Laurier Turgeon, in a single season fishers could 'catch 300–400 hake on the coasts of Brittany, using a hand-held line'.[135] We can assume similar catches were possible in Irish waters. Comparatively, Turgeon claims that a good fisher could catch the same number of cod in a single day in the more abundant waters of Newfoundland.[136] Other studies have further demonstrated that primitive technology such as the handline could have a surprising and potent impact on fish stocks.[137]

Just as important as fishing methods were the processes for preserving fish. Without preservation, no commercial fishery could exist. Fish simply rots too quickly to be brought to distant markets without processing. Maryanne Kowaleski attributes improvements in curing methods to the growth of the south-west English fisheries in the late medieval period, when Dutch fishers employed the most technically advanced methods for curing herring, gutting and salting their catch on board ships before repackaging it on land for transport to distant markets. There is some evidence that English fishers adopted these methods, but they also set up curing stations on land, including in Ireland.[138]

The records of imports to England from Kowaleski's sources, and from the Bristol port books, speak to the wide variety of cures used in Ireland. The trade records indicate that both brined and smoked herring was imported.[139] Cod was imported salted, probably after being dried on rocks and racks, like those at Baltimore. Salmon, eel, and other fish were pickled in brine.[140] Records show 'wet hakes' exported from Ireland too; most likely, fish that had been 'gutted and salted in the Dutch style'.[141] The similarity between Irish and Dutch cures suggests that methods pioneered by the Dutch made their way to Ireland via English fishers visiting the coast. It is also possible that Spanish and French fishers

133 Henk J. L. Heessen et al. (eds), *Fish Atlas of the Celtic Sea, North Sea, and Baltic Sea* (Wageningen, 2015), pp. 184–85.

134 Bjarnason and Carlesi, *Handlining and Squid Jigging.*

135 Turgeon, *Far-Distant Fisheries*, p. 65.

136 Turgeon, *Far-Distant Fisheries*, p. 65.

137 Andrew A. Rosenberg et al., 'The History of Ocean Resources: Modeling Cod Biomass Using Historical Records', *Frontiers in Ecology and the Environment*, 3 (2005), pp. 78–84.

138 Kowaleski, 'South-Western Fisheries', pp. 449–50.

139 Evan Jones and Susan Flavin (eds), *Bristol's Trade with Ireland and the Continent, 1503–1601: The Evidence of the Exchequer Customs Accounts* (Dublin, 2009).

140 Locker, 'Stored Fish', p. 50.

141 Kowaleski, 'South-Western Fisheries', p. 450.

employed similar curing techniques in Ireland, although we have no specific evidence of this.

The development of new curing methods in Europe in the later medieval period and the subsequent permeation of those methods to Ireland allowed fisheries to develop. Preservation was necessary for commercial fisheries, and these techniques allowed Irish fish to be transported long distances and to reach markets in England, Spain, or elsewhere on the continent. Additionally, fish processing onshore created additional work for coastal communities and added to the fisheries' economic impact.

Most fishing in sixteenth-century Ireland was conducted close to shore in small vessels such as pinnaces, whose shallow drafts allowed for safe operation in coastal waters. They had a crew of four or five while fishing, could be equipped with one or two masts, and often had oars as well.[142] We know the O'Driscolls had a fleet of ninety pinnaces at Baltimore in 1537, which could have been used for fishing and transportation in the area.[143] Such small craft could have made short sea voyages around the coast of Ireland or from England to Ireland on their own. They were also sometimes carried on the decks of larger ships.[144] Plate 2.3 shows a typical single-masted pinnace with holes for oars, with a smaller rowing boat off its side, which provides a sense of scale.

The deposition of Rice ap Edward, the fishing admiral in Wales, also reveals that three types of boats were commonly employed in fishing there, and we can assume fishers used similar craft in Ireland. We can tell the relative size of these because a different fine was imposed on each type of boat for breaking the rules. Edward recounts that a drover boat would pay 4*d.* and a skiff 6*d.* per offence. Both were small craft, similar to the pinnace. Larger boats called 'barks' would pay 8*d.*[145] The bark was a fully rigged ship with a forecastle and quarterdeck.[146]

Several larger multi-sail vessels were in common usage during the sixteenth century as well. The development of the carvel technique of shipbuilding, where planks were set end-to-end, allowed for the construction of larger, ocean-going vessels, in particular the carrack and the galleon.[147] There is no shortage of contemporary illustrations of larger ships on sixteenth-century maps. Plate 2.4 depicts an English galleon from a map of the coast of County Antrim in 1580.

[142] Kenneth R. Andrews, *Trade, Plunder and Settlement: Maritime Enterprise and the Genesis of the British Empire, 1480–1630* (Cambridge, 1984), pp. 22–23.

[143] O'Donovan (ed.), *Genealogy of Corca Laidhe*, pp. 94–95.

[144] *Cal. Car. MSS*, 1515–74, no. 70, p. 85.

[145] *Cal. HCA 1536–1641*, no. 145, p. 28.

[146] A. J. F. Duffy, 'The Early History of the Rye Fishing Industry', *Sussex Archaeological Collections*, 107 (1969), pp. 37–64, at p. 46.

[147] O'Sullivan and Breen, *Maritime Ireland*, p. 197.

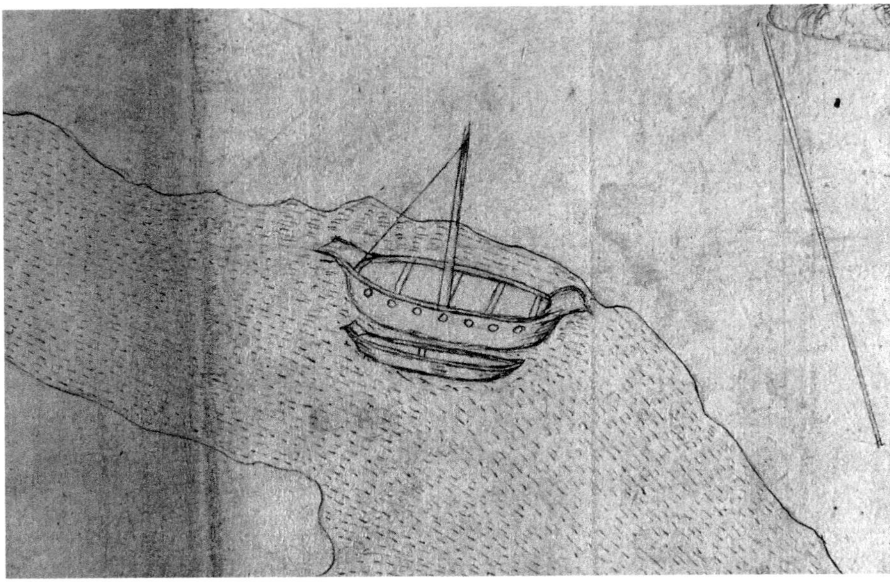

Plate 2.3. Image of a pinnace from 'Plat of the fort of Enniskillen', 1594.

Plate 2.4. Image of a galleon from 'Map of the Coast of County Antrim', 1580.

This image is an English naval vessel, but galleons used for trading or fishing would have had a very similar hull and sail configuration.

An illustrated map of Carlingford from 1568 (Plate 2.5) shows a larger vessel sailing into the lough towing a smaller rowing boat, probably a rescue craft, which shows that larger ships could have brought smaller boats for fishing in this manner. While this map does not depict fishing taking place, it does indicate a high level of marine traffic going into Carlingford, one of the most prominent fishing locations in the sixteenth century.

Galley ships were also widely used among coastal Gaelic Irish families, some of whom, like the O'Malleys of Mayo, were renowned for their expertise in using these vessels.[148] Irish galleys were in part derived from Viking longships and shared many similarities with similar vessels used in the Scottish islands.[149] The galleys remained in use in Ireland and Scotland owing to their ability to manoeuvre around islands and the craggy coastlines of the Irish south and west coasts. We have little documentary or archaeological evidence for the use of galleys in Gaelic Ireland, but several medieval and early modern maps and drawings depict these ships.[150] An etching on the gatehouse of Dunluce Castle in Antrim depicts a galley with a single mast, while the early sixteenth-century municipal seal of the town of Youghal (Plate 2.6) also displays a galley, implying that Old English towns used these craft as well.

The Genealogy of Corca Laidhe describes how in 1537 the 'great galley' of Waterford city joined in the attack on Baltimore and helped capture the Gaelic lord's 'chief galley of thirty oars',[151] indicating that Old English towns and Gaelic Irish lords used these vessels for combat. There is no evidence that galleys were directly utilised in fishing, but as previously mentioned, they played a critical role in the maritime strength of the Gaelic Irish and were used to protect visiting fishing fleets and deter pirates.[152]

The vessel types covered here are some of the most discussed in the sixteenth-century sources, but many other sailing craft were used by fishers and merchants in the sixteenth century. Several historians have written about these vessels at length, and the details are not necessary to repeat here.[153]

[148] Colin Breen and John Raven, 'Maritime Lordship in Late-Medieval Gaelic Ireland', *Medieval Archaeology*, 61 (2017), pp. 149–82, at p. 157; O'Sullivan and Breen, *Maritime Ireland*, p. 197.

[149] Denis Rixson, *The West Highland Galley* (Edinburgh, 1998), pp. 167–68.

[150] O'Sullivan and Breen, *Maritime Ireland*, p. 196.

[151] O'Donovan (ed.), *Genealogy of Corca Laidhe*, pp. 94–95.

[152] Breen, *O'Sullivan Beare*, p. 116.

[153] See, for example, O'Sullivan and Breen, *Maritime Ireland*, pp. 193–98; John de Courcy, 'A Survey of Early Irish Maritime Trade and Ships', in Michael McCaughan and John C. Appleby (eds), *The Irish Sea: Aspects of Maritime History* (Belfast, 1989), pp. 21–26.

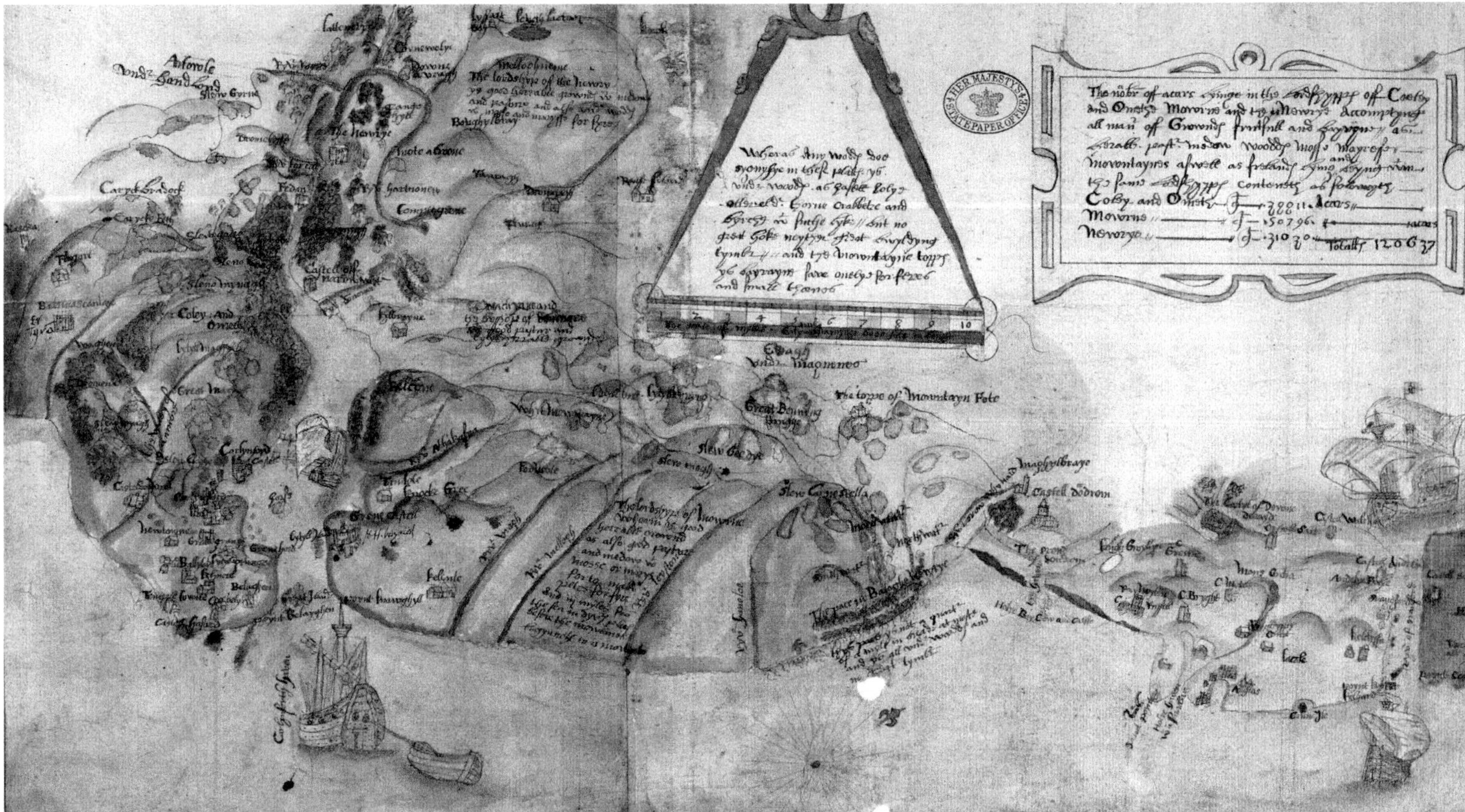

Plate 2.5. Coast of County Down from Lough Strangford to Carlingford, 1568.

Plate 2.6. Single-masted galley on the municipal seal of Youghal, c. 1527.

While most fishing for demersal fish seems to have been done by hook and line, a range of gear and boats were used for pelagic fish like herring. Because herring are relatively small shoaling fish (compared to hake or cod) and gather in large numbers near the coast during spawning periods, large nets were much more effective, and several types were used in early modern Ireland. A drift net was floated by buoys and attached to a vessel, which was commonly called a drifter.[154] These vessels had masts that could be lowered while the net was out so the ship could drift freely.[155] Some of the nets could be kilometres long and would entrap entire shoals of herring. In the sixteenth century, the Dutch herring fishers in the North Sea made the best use of this technology, but it was also in use in Ireland,

[154] D. H. Cushing, *The Provident Sea* (Cambridge, 1988), p. 1.
[155] Cushing, *Provident Sea*, p. 86.

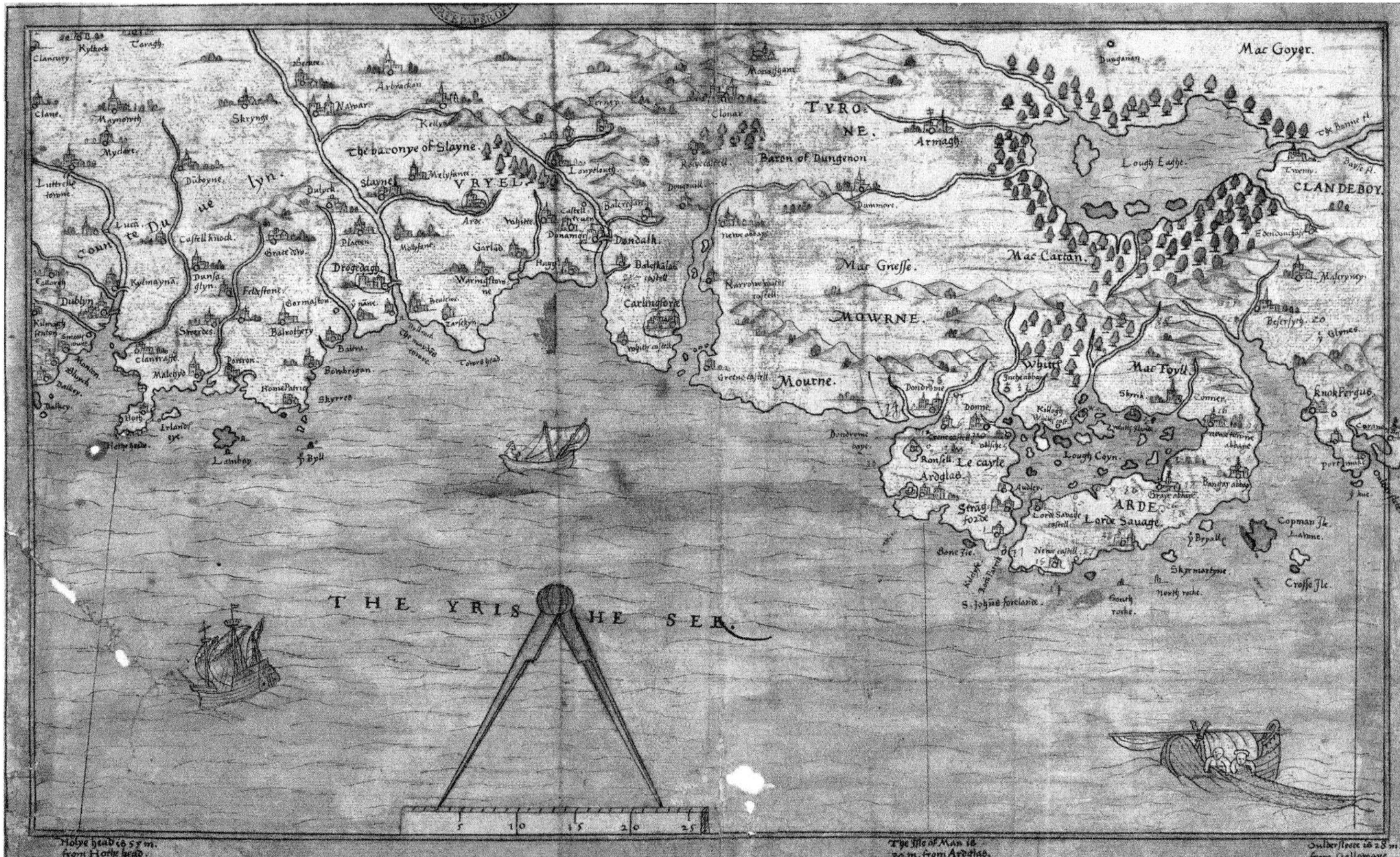

Plate 2.7. 'Map of the coast of Ireland from Dublin to Carrickfergus', 1580.

Plate 2.8. Image of drifter from 'Map of the coast of Ireland from Dublin to Carrickfergus', 1580.

albeit on a smaller scale.[156] In 1512 a merchant from Youghal, Patrick Walsh, ordered a 'herring-net / 120 fathoms long and 60 stitches broad',[157] a net similar to those used by drifting vessels in the North Sea.[158] Plate 2.7 is a map showing a collection of vessels off the coast of Ulster, including a small single-masted boat, a larger sailing ship and a drifter in the bottom right. Plate 2.8 shows a closer image of the drifter in action; it can be seen with its net in the water and mast lowered in characteristic fashion.

A more standard method for catching herring and pilchards in early modern Ireland was the seine net, a long net used to surround fish. The seine was usually dragged between two boats: the seine boat and the follower boat.[159] The seine boat was the larger craft and had a crew of about ten; the smaller follower boat had a crew of four or five.[160] Another crucial element was the spotter, or 'huer', who would stand in a prominent position on the coast and direct the boats towards

[156] Bo Poulsen, *Dutch Herring: An Environmental History, c. 1600–1860* (Amsterdam, 2008), p. 137.

[157] Bernard, 'Maritime Intercourse', p. 16.

[158] Poulsen, *Dutch Herring*, p. 60.

[159] A. E. J. Went, 'Pilchards in the South of Ireland', *Journal of the Cork Historical and Archaeological Society*, 51 (1946), pp. 137–57, at p. 137.

[160] Liam Downey and Ingelise Stuijts, 'Historical Fisheries: Irish Pilchard Fishery and Historical Fish Harvesting and Processing Methods, Fish Products and Consumption', *Kerry Archaeological & Historical Journal*, 2 (2014), pp. 1–35, at p. 17.

shoals. At night, the fish could sometimes be identified by the phosphorescence they caused in the water. Once fishers had located the shoal, the seine boat would be rowed as fast as possible around the fish to trap them. The net would then be closed and slowly tightened into a purse, making the fish easy to haul on board in baskets.[161]

A variation of the seine was the draft net. One end was anchored to the shore while the other was latched to a boat that would be rowed around the fish. Once the fish were trapped, fishers on land would drag the whole net onshore. The draft net could only be used if the coastline was suitable and in later centuries was commonly used to catch salmon.[162] These methods remained mostly unchanged from their medieval origins and were only superseded by the invention of modern trawlers.[163]

The best visual illustration of these methods comes from a map of Baltimore drawn around 1620 by an unknown illustrator. A close-up view of the fishing scene from the map can be seen in Plate 2.9.[164] The left side of the image shows a net anchored close to shore while another boat rows around a shoal of fish with the other end. On the right side, two boats can be seen surrounding another shoal of fish. The mapmaker was unlikely to have been a fisher and may have misrepresented some details, but for the most part, his image gives an accurate picture of seine nets in action. Some historians have suggested that seventeenth-century English planters introduced the seine net to Ireland, but the documentary evidence does not support this.[165] The settlers may have made more use of it on pilchards, but records from Baltimore's Gaelic period discussed earlier clearly indicate the net was in use then. The O'Driscoll inquisition taken in 1609 explicitly required 'That no man ought to draw a seine in the said Harbour, nor in any part of the said country without a licence first obtained from the Lord.'[166] It is feasible that the licences were required at Baltimore because these nets were extremely large and would have caused disruption to other shipping in the harbour.[167]

In 1620 Baltimore was an English colony, controlled by Thomas Crook.[168] The main catch at the time was pilchards, which is the fish depicted in the map, but

161 Went, 'Pilchards in the South of Ireland', p. 137.

162 Went, 'Pursuit of Salmon', p. 224.

163 Cutting, *Fish Saving*, p. 203.

164 The map is undated, but David Dickson posits it was drawn around 1620: David Dickson, *Old World Colony: Cork and South Munster 1630–1830* (Cork, 2005), p. 17.

165 *NHI, vol. III*, p. 181.

166 O'Donovan (ed.), *Genealogy of Corca Laidhe*, p. 103.

167 The same methods were also used by fishers to catch mullet in Stokenham in the fourteenth century; Kowaleski, 'South-Western Fisheries', p. 441.

168 Michael MacCarthy-Morrogh, *The Munster Plantation: English Migration to Southern Ireland 1583–1641* (Oxford, 1986), p. 151. The following chapter will explore the rise of the English colony in more detail.

Plate 2.9. Fishing scene from 'Baltimore, Ireland', c. 1620.

the methods used for pilchards and herring would have been the same in the sixteenth and seventeenth centuries.[169] Elsewhere in Munster, Richard Boyle was operating a pilchard fishery at Ardmore, County Waterford, in 1616, and gave his servant money 'toward the buying of a seine to take pilchards with', along with further funds to pay another man 'to be our hewer there the next season'.[170]

Fishers in Ireland also used cast nets to catch herring, a simpler technology, where nets were individually cast on shoaling fish by fishers from the shore or from small boats. A statute from the Irish Parliament in 1470 indicates how the herring fishery operated around Dublin and how cast nets were used there. It orders that any 'person or persons who interrupts or disturbs any of the said fishermen in throwing or drawing their nets in the light or in the dark' must

[169] A. E. J. Went, 'The Irish Pilchard Fishery', *Proceedings of the Royal Irish Academy*, 51B (1946), pp. 81–120, at p. 89.
[170] Went, 'Pilchards in the South of Ireland', p. 139.

pay £10.[171] The fishers probably operated at night because herring move toward the surface of the water column at dusk and disperse over a larger area at night.[172] Their movement closer to the surface would have made the herring easier to catch using a cast net. In 1590, there were multiple reports from English officials of an oversupply of herring in Ireland. Several merchants wrote to officials in Dublin asking permission to export the fish they had in storage. One describes how fishers had caught upwards of 10,000 barrels of herring that year using 'cast fishing'.[173] No other details are provided about the method, so it is hard to verify this claim, but given the earlier description of fisheries around Dublin, it does seem plausible that these catches were made with cast nets.

Of course, other fishing methods were used, including a whole variety for river fishing, but this chapter is concerned with the methods adopted in Ireland's commercial fisheries. Some of the most basic technologies, such as handlines, were used alongside newer technologies, like the herring drifters, and in this respect, Ireland was on a par with other European nations. Ultimately, the types of gear and methods used by fishers were determined by local conditions, as well as the behaviour and biology of the species they targeted. As the next chapter will show, through the use of these technologies, fish became the single largest export commodity in sixteenth-century Ireland.

[171] *Statute Rolls, Ireland*, vol. III, 10 Edward IV, A.D. 1470, p. 667.

[172] Henk J. L. Heessen, 'ICES FishMap Species Factsheet: Herring' (ICES, 2005), p. 2.

[173] Lord deputy to Burghley, 26 January 1589–90 (TNA, SP 63/150, f.40).

3

The Fish Trade

I caste to speke of Irelonde but a lytelle.
Commoditees of it I woll entitell
Hydes and fish, samon, hake and heryng

Unknown Author, *The Libelle of Englyshe Polycye*, 1436/7

This chapter attempts systematically to collect, standardise, and analyse the available data for the fish trade in fifteenth- and sixteenth-century Ireland, to explore the importance of fish to the export economy. It offers the most complete picture yet provided of the extent and nature of the exportation trade of fish. Appendix 2 contains a full outline of the methods used to generate the data.[1]

The Trade to Great Britain

As we lack any quantitative data detailing the export of fish from sixteenth-century Ireland, we have to rely on records of Irish imports to other locations. This section will examine trade data from England, Wales, and Scotland to see what it can tell us about the export of fish from Ireland. Most of the data will be taken from customs accounts and port books of the fifteenth and sixteenth centuries. It will also focus on several pivotal ports and regions which traded heavily with Ireland and for which there is also data available. The fishers of the English West Country often drew their catch from Irish waters in the sixteenth century. Their most important base was Bristol, one of the major trading ports in the south of England, which had an active partnership with several ports in Ireland, particularly those in Munster.[2] Bristol's customs accounts are some of the most extensive that remain extant for England,[3] but we also have data from other ports in the region, especially Bridgwater.[4] In the north-west of England, Chester was the primary port, trading heavily with the east and south-east of

[1] The FTD is available online at, Patrick Hayes, 'Fish Trade Dataset' (2022) (10.6084/m9.figshare.20367882) (25 July 2022).

[2] Ada K. Longfield, *Anglo-Irish Trade in the Sixteenth Century* (London, 1929).

[3] Wendy R. Childs, 'Ireland's Trade with England in the Later Middle Ages', *Irish Economic and Social History*, 9 (1982), pp. 5–33, at p. 12.

[4] Duncan Taylor, 'The Maritime Trade of the Smaller Bristol Channel Ports in the Sixteenth Century' (PhD thesis, University of Bristol, 2009).

Ireland.[5] Analysis of Chester's trade is facilitated by publication of several of the town's fifteenth- and sixteenth-century customs accounts.[6] Welsh ports were some of the closest geographically to the east coast of Ireland, and available trade data will be examined to see if there was an active fish trade between the two regions.[7] Finally, the fish trade between Ireland and Scotland will be explored. Other ports, like Southampton and London, were considered, but the volume of Irish shipping to these ports was negligible compared to the ports on the west coast.[8] The data recovered represents what is currently the best available quantitative information about Ireland's export trade to England, Scotland, and Wales in our period. It cannot provide anything like a complete inventory of all exports, but nonetheless enables trends in Ireland's fish exports to be examined in far greater detail than before.

The Bristol trade data comes from three main sources; the fifteenth-century data were compiled by Margaret M. Condon and Evan Jones and published on the University of Bristol's Research Data Repository.[9] This data comprises fourteen 'particular' accounts which all cover varying periods from 1461 to 1483.[10] The distributed nature of the months covered by this fifteenth-century data makes it far less reliable than the sixteenth-century accounts, which all cover the same period (September to September). As we shall see, the importation of fish was highly seasonal, so unfortunately, many of these early accounts miss the crucial season for fish imports. Wendy Childs has also published summarised data for several more Bristol accounts covering the years 1404/5, 1479/80, 1485/6, 1486/7 and 1492/3.[11] These accounts cover whole accounting years but contain only summarised values; they show the overall picture of trade each year, but do not provide the same level of detail as the other accounts.

Sixteenth-century data was sourced from the *Bristol's Trade with Ireland and the Continent* project. The project team, which included Susan Flavin and Evan Jones,

[5] C. Armour, 'The Trade of Chester and the State of the Dee Navigation, 1600–1800' (PhD thesis, University of London, 1956); Donald Woodward, 'The Trade of Elizabethan Chester' (PhD thesis, University of Hull, 1970), p. 1.

[6] K. P. Wilson (ed.), *Chester Customs Accounts, 1301–1566* (Record Society of Lancashire and Cheshire, vol. CXI, Liverpool, 1969).

[7] Arthur Edward Lewis (ed.), *The Welsh Port Books, 1550–1603* (Cymmrodorion Record Series, xii, London, 1927).

[8] Childs, 'Ireland's Trade with England', p. 31.

[9] M. Condon and Evan Jones (eds), *Bristol 'Particular' Customs Account Transcriptions* (http://www.bristol.ac.uk/history/research/cabot/publications/) (accessed 9 September 2020). A full list of the individual accounts used in this research can be found in the FTD.

[10] Jones has also published tonnage estimates for several other late fifteenth-century accounts: Evan Jones, *Tonnage Estimates of Goods Recorded in the Bristol 'Particular' Customs Accounts: 1465/6, 1485/6, 1503/4* (Bristol, 2017).

[11] Childs, 'Ireland's Trade with England'.

worked to computerise eleven of the extant 'particular' accounts and port books for Bristol covering the years 1503/4, 1516/17, 1525/6, 1541/2, 1545/6, 1550/1, 1563/4, 1575/6, 1594/5 and 1600/1.[12]

Combined, these sources can tell us which species of fish were exported most from Ireland during the fifteenth and sixteenth centuries. The simplest way to measure the relative importance of species is to compare the total value of their exports. Customs accounts and port books do not contain market values but instead list fixed 'nominal values' set by a centrally controlled book of rates.[13] The rates charged for fish did not change over the fifteenth century. However, in 1558, the duties for most commodities roughly doubled to align with rising market prices.[14] To adjust for this inflation, customs values after 1558 were pegged to pre-1558 levels to provide comparable figures across all records.[15] Instances where this reduction has been applied are marked by footnotes.

The first extant Bristol customs account is from 1404/5. This is not as detailed as the later records, but figures published by Wendy Childs show that the total value of Irish trade in that year was £2,091, with fish products accounting for £1,754 of the total and herring making up half the total value of all fish.[16] While these figures do not provide much detail, they show that fish constituted a large portion of the overall trade between Ireland and Bristol and was thus very economically significant at this early point. There is then a long gap in records until 1461, when accounts capturing all international trade once again became available.

Table 3.1 shows the values of different species of fish in available accounts from 1461 to 1500. Almost all currency values, other than those quoted from the sources, are listed in pre-decimal pounds (with a pound containing 240 pence).[17] White herring (salted herring) was the most valuable species, followed closely by salmon, with hake in third place. We can also observe red herring on this list; this was not a separate species of fish, but rather herring cured with salt and then smoked, giving it a distinctive red colour.[18] Charles L. Cutting describes

[12] Accounts in this period used the Christian festival of Michaelmas in September to divide years.

[13] R. J. Hunter and Brendan Scott (eds), *The Ulster Port Books, 1612–15* (Belfast, 2012), p. xii.

[14] Susan Flavin, *Consumption and Culture in Sixteenth-Century Ireland: Saffron, Stockings and Silk* (Woodbridge, 2014), pp. 13–14.

[15] This was done by taking the custom rate charged for each individual unit of fish before 1558 and applying that rate after 1558. In a few cases there were no matching units and in these cases the total was reduced by half, which was the method used by Susan Flavin in 'Consumption and Material Culture in Sixteenth-Century Ireland' (PhD thesis, University of Bristol, 2011), p. 11.

[16] Childs, 'Ireland's Trade with England', p. 15.

[17] The pre-decimal pound figures are calculated by converting everything to pence and dividing by 240.

[18] Alison Mary Locker, 'The Role of Stored Fish in England 900–1750AD; the Evidence from

the process of making red herring in nineteenth-century Yarmouth and claims that it had not changed significantly since the early modern period. The herring were first washed and then put in salt for fourteen days, after which they were put on spits and smoked for a further fourteen days in large smoking houses.[19] The fourth most valuable fish is a product marked generically as 'salted fish'. There is no way to know for certain to what species this refers, but, based on its value, it was probably lower quality fish including species like whiting, ling, and inferior quality hake.[20]

Table 3.1. Nominal value of fish imported to Bristol from Ireland, 1461–93.[21]

Fish product	Nominal value (£)
Herring white	2978.51
Salmon	2479.63
Hake	2,143.08
Salted fish	778.61
Herring red	465.55
Pollock	50.10
Eels	34.75

Figure 3.1 shows the balance of imports between species over the fifteenth-century accounts. The graph sets the share of trade held by the top three species, with all other species grouped together.[22] Imports in the last four years far outweigh exports in any previous years. However, we must remember that the earlier accounts are mostly fragments of full years, so a good deal of information is missing from them. Exports peaked in 1479/80 at £2,843, the highest total nominal value of any year in the fifteenth and sixteenth centuries. This was helped by a substantial import of salmon in 1479/80, a fish demanding a high customs value. There were also strong imports in the 1465/6 accounts, which cover a more extended period than most other accounts before 1479/80. Across the whole century, herring was the most valuable import.[23]

Historical and Archaeological Data' (PhD thesis, University of Southampton, 2000), pp. 43–44.

19 Charles L. Cutting, *Fish Saving* (London, 1955), p. 76.

20 Longfield, *Anglo-Irish Trade*, p. 49.

21 Several other fish and marine mammals were also found in the accounts in small amounts, these included 'Harburdan', cod, whiting, seals, 'Scroff', dogfish and porpoise.

22 Overall, the 'other' category accounts for a tiny amount of the total in both the fifteenth- and sixteenth-century data.

23 There were small shipments of herring in January 1483 as well, but the account is incomplete and does not list an origin for the fish, so it was not included in the totals; M. Condon and

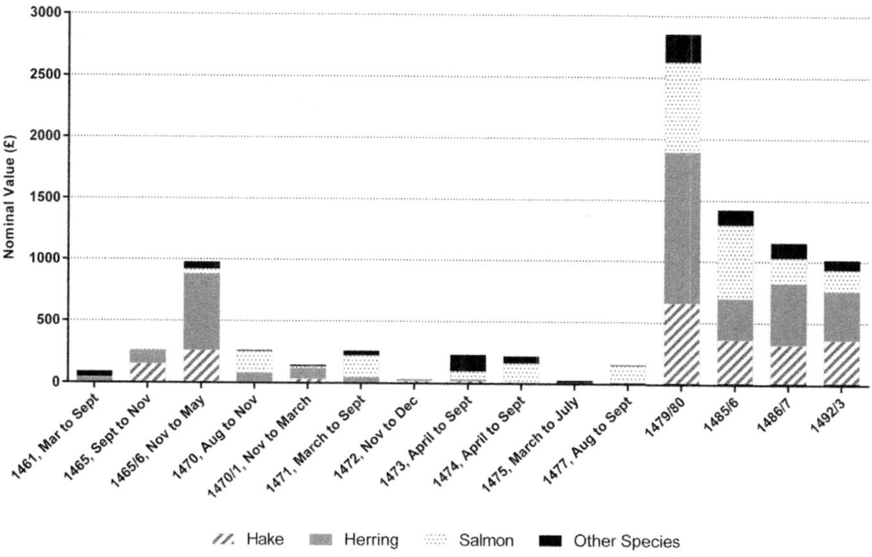

Figure 3.1. Nominal value of Irish fish imports to Bristol, 1461–77.

Moving on to the sixteenth-century data, Table 3.2 shows the six most valuable fish products imported in the eleven available accounts from 1503 to 1601. White herring was again the most valuable import, followed by hake and salmon. However, this relationship changed yearly, with hake the most valuable import in the 1516/17 accounts. After salted fish and red herring, eel is the next most costly species, followed by a fish called 'cole', which was probably *Salvelinus colii*, a fresh-water fish that exclusively inhabits Irish lakes. Compared to the fifteenth-century data, herring and hake imports were higher. Overall, however, salmon and salted fish imports had been higher in the fifteenth century.

Figure 3.2 shows the balance of imports from the sixteenth century. The sixteenth-century data only show figures up to 1595, while a further account for 1600/1 is complete but shows no fish imports from Ireland in that year. Imports were most robust in the first three accounts from 1503 to 1526. Subsequently, they began a sharp decline. For most years, herring were the most valuable, but this shifted in 1516/17, when there was a decline in herring and an all-time high import of hake. Salmon imports remained more consistent compared to herring and hake, but still experienced fluctuations, especially after 1541/2. A statute enacted in 1515 may explain some of the apparent drop after 1541. It stated

Evan Jones (eds), *Bristol 1483, January: Particulars of Account of John Walshe, Controller* (https://doi.org/10.5523/bris.1uimqzrjew48i21385ag2xzqsf).

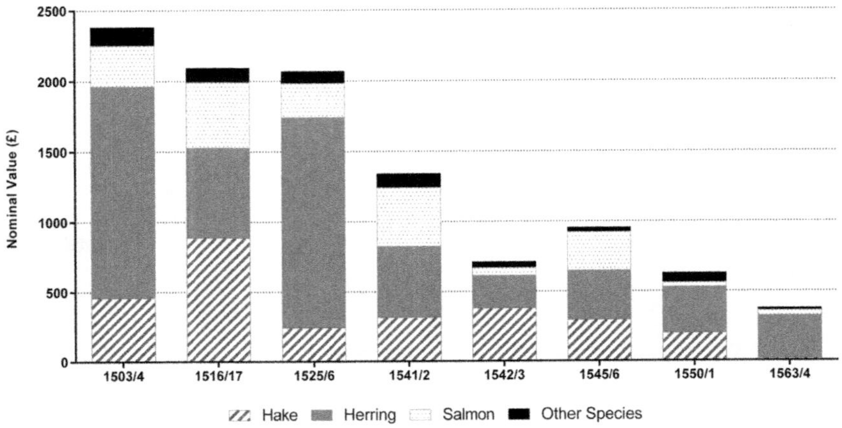

Figure 3.2. Nominal value of Irish fish imports to Bristol, 1503–95. Post-1558 customs values have been pegged to pre-1558 rates.

that 'merchants convey out of this land into France, Britain, and other strange parts, salmons, herrings, dry lings, hakes and other fish, so abundantly that they leave none within the land to vitall the King's subjects', and suggested that each merchant 'leave within the land the third part of his fish'.[24] Parliament enacted the statute in 1515, but there was no dramatic drop in imports until 1541, so perhaps the rule was not enforced effectively until 1525/6.

Table 3.2. Nominal value of fish imported to Bristol from Ireland, 1503–1601.[25]

Fish product	Nominal value (£)
Herring white	5,091.07
Hake	2828.36
Salmon	1,806.80
Salted fish	507.19
Herring red	291.98
Eel	46.25
Cole	12.5

24 Cited in Longfield, *Anglo-Irish Trade*, p. 42.

25 There were small amounts of sturgeon, seals, pollock, sardines, haddock, gurnard, mackerel, sprat, and whiting. Post-1558 customs values have been pegged to pre-1558 rates.

The main issue with the sixteenth-century data is the tax status of fish over the period. Fresh fish was never subject to import tax in English or Welsh ports, but this was not relevant to imports of processed fish from Ireland. However, in 1563, with the aim of growing the English merchant fleet, a law was enacted that suspended customs duties on imports of fish caught by English vessels in other waters and the export of sea fish. A note from 1588, written by Sir Francis Walsingham, indicates that this law was interrupted to include fish imported from Ireland. In 1591 the rules changed again, and some fish from Ireland were once more subject to customs.[26] Because of these rule changes, the customs accounts only haphazardly register any fish from Ireland between 1564 and 1591.[27]

After Parliament amended the law in 1591, small amounts of fish, mostly hake, were again recorded arriving from Ireland. However, a customs exemption was still widely applied that excluded fish personally caught by the merchant who brought it to port. A clerk in Gloucester estimated that, in his port, only one-third of the fish imported from Ireland was liable to duty after 1591.[28] Bristol's 1600/1 account contains no fish for Ireland. However, even if imports in 1594/5 were two-thirds higher, the amount would still be far lower than imports in 1563/4, the last year for which we have reliable information. That would suggest the export of fish from Ireland suffered a crash by the end of the sixteenth century. Other factors support this hypothesis. In 1581 the English Parliament passed a law banning the import of herring from Ireland to protect the English fishing industry.[29] This would have inflicted further damage to the Irish fish trade. That said, a small number of herring did arrive from Ireland in 1594, so it is possible that customs officials did not enforce the law entirely, or perhaps entry was permitted because the amount was so small (only two barrels and three hogsheads).

External factors also supported the idea that the export of fish virtually ceased by the end of the century. During the Nine Years War, from 1593 to 1603, there were widespread food shortages and confiscation of fish by the English authorities for the army. The war caused significant disruption to fishers, and the next chapter will explore how and why this happened. Thus, it would not be surprising to find exports declining as sharply as the data seems to show. Changes in the law over the century exaggerate and mask the decline of Irish imports; but while the real decline was not as dramatic as the data would suggest, a decline did occur.

It is possible that other factors, such as large-scale customs evasion and smuggling, also contributed to the apparent decline of fish imports from Ireland. Evan Jones has shown that the smuggling of restricted commodities, such as

[26] Taylor, 'Maritime Trade', p. 29.

[27] In certain accounts some fish is still recorded but it is not clear why some shipments were recorded when this was not legally required.

[28] Taylor, 'Maritime Trade', p. 29.

[29] Flavin, 'Consumption and Material Culture', p. 190.

leather and grain, was widespread in Bristol in the sixteenth century.[30] However, no evidence suggests that merchants smuggled fish to any great extent. The customs on fish in the early decades were not excessive, giving little incentive to smuggle.[31] Furthermore, between 1563 and 1591 fish was exempt from customs, so there was no need to hide the trade during this period.

The most valuable fish product imported to Bristol from outside Ireland was fish from Newfoundland, which was almost always cod. These imports help to explain the decline of Irish imports. Newfoundland fish was always exempt from customs duties, but the customs officials did record some shipments, even if they did not charge customs on them.[32] Around the time Irish imports first began to decline, in 1541/2, Newfoundland imports were peaking, with £414 worth imported that year. Later, in 1600/1, when no Irish imports were recorded at all, Newfoundland imports totalled £240. Administrative issues meant that the accurate scale of the Newfoundland imports was never documented; so, what we have are only fragments of the total trade. However, even these limited shipments illustrate that from the 1540s onwards Newfoundland fish began, in part, to replace Irish fish in the Bristol imports.

The customs value of fish imports to Ireland from Bristol in the sixteenth century was minute, with only one ship arriving from Dublin in October 1600 carrying £50 worth of Newfoundland cod.

The available data therefore suggests that the export of fish from Ireland reached its highest levels in the late fifteenth and earlier sixteenth centuries. However, customs values only allow us to compare the relative importance of goods and do not speak to the total weight of fish exported. To address this issue, the most important species in the accounts were converted to tonnes of live weight (TLW) to provide accurate figures on the weight of fish exports. TLW represents the weight of fish as it was alive before being caught and processed. The methods used to calculate these figures are described in Appendix 2. Overall, live weight figures far more accurately reflect the amount of fish extracted from the sea and allow us to better assess the potential impact of these extractions on marine ecosystems. While we need to modify the nominal value figures to account for rate changes, the same is not necessary for the TLW figures because they are based on the volume of fish transported.

30 Evan Jones, 'Illicit Business: Accounting for Smuggling in Mid-Sixteenth-Century Bristol', *Economic History Review*, 54 (2001), pp. 17–38.

31 Evan Jones and Susan Flavin (eds), *Bristol's Trade with Ireland and the Continent, 1503–1601: The Evidence of the Exchequer Customs Accounts* (Dublin, 2009), p. xiii.

32 Taylor, 'Maritime Trade', p. 29.

The main complication in calculating TLW is the wide variety of arcane units of measurement that were used in early modern ports. Each of these units has been converted to a modern equivalent in kilograms. Additionally, each cure of fish, such as white and red herring, had separate units but these were combined to give a total for each species. For these reasons, the calculations were only done for the three most important species that dominate imports: herring, hake, and salmon.[33]

The nominal value figures show that herring was the most important import to Bristol from Ireland in the fifteenth- and sixteenth-century accounts. Comparatively, Figure 3.3 shows the TLW of herring imported to Bristol from Ireland from 1461 to 1595.[34] In total, 2,076 TLW was imported during the fifteenth century and a further 3,338 TLW in the sixteenth.[35] Herring imports exhibit substantial variation; they peak first in 1479/80 before declining and rising again at the beginning of the sixteenth century. Imports reached their highest level in 1525/6 before dropping dramatically and remaining low for the rest of the century.

The trade records can also reveal the seasonality of imports, although not all accounts are suitable for this purpose. In the fifteenth century, only the accounts from 1461 to 1477 contain monthly details. These show that, on average, the most significant imports took place in February, with about 300 TLW imported that month, with October the next most frequent at only 64 TLW. On the other hand, all the sixteenth-century accounts contain monthly figures. Figure 3.4 shows the monthly trend for sixteenth-century herring imports. This graph shows a similar pattern to the fifteenth century, with no or very few imports during the summer months. However, the sixteenth-century data differs in that it indicates a more significant spike in the autumn. The trend year on year does vary slightly across the century, but in most cases, the peak of imports occurs in December and January. This pattern may be due to ships wishing to exploit the Lenten market, which usually began in February.[36]

[33] Salted fish was also impossible to accurately convert because it is not a single identifiable species. The full list of conversions and the sources used to find them are listed in the introduction to the FTD hosted online: Patrick Hayes, 'Fish Trade Dataset' (2022) (10.6084/m9.figshare.20367882) (25 July 2022).

[34] Again, there was one further account in 1600/1 with no fish imports from Ireland recorded.

[35] The TLW figures for the last four accounts provided by Wendy Childs in the fifteenth century were calculated by taking the value of each species and dividing it by the value of the most common unit used for that species. For example, in 1479/80 there was £1,059.55 worth of herring imported, the most common unit for herring was the barrel, which was worth £0.25 in customs. So, 1,059.55 divided by 0.25 means there were 4,251.86 barrels imported. The TLW figure for a barrel is 157.82kg. To reach a total TLW figure we multiply 4,251.86 by 157.82 and then divide by 1,000 to reach 671 TLW.

[36] Flavin, 'Consumption and Material Culture', p. 190.

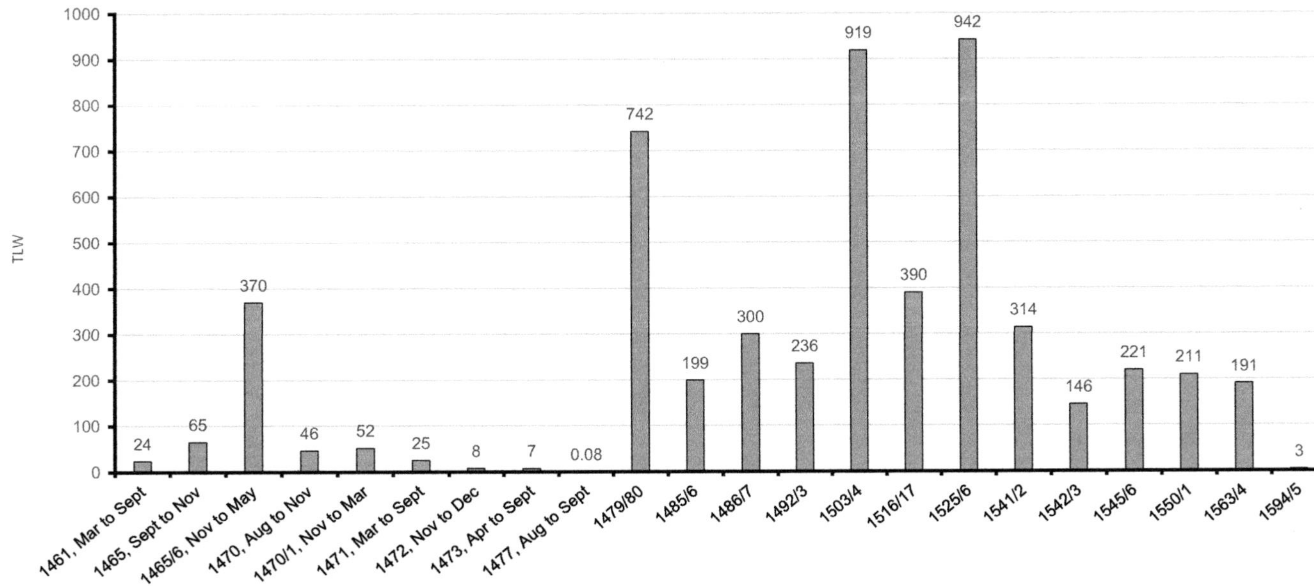

Figure 3.3. TLW of Irish herring imported to Bristol, 1461–95.

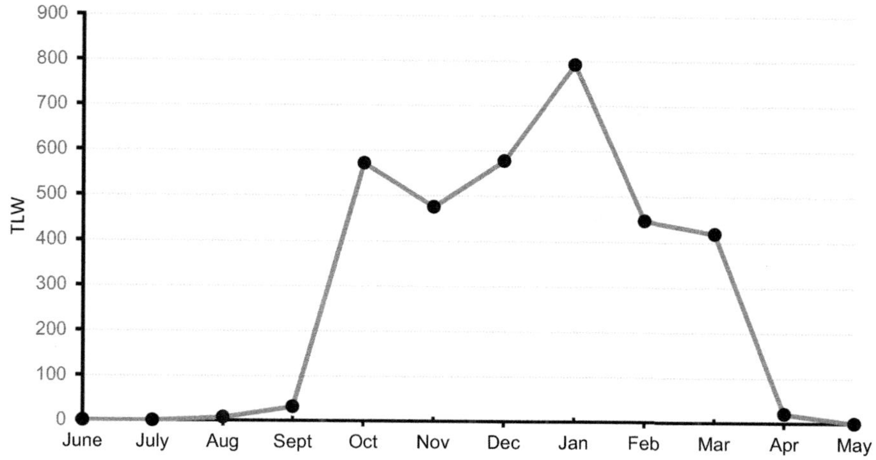

Figure 3.4. Irish herring imports to Bristol per month, 1503–95.

The spawning season of herring, which occurs during autumn and winter on different parts of the coast, also influenced the timing of imports. As mentioned, imports spiked in January/December and, based on modern observations, the most prominent spawning populations during these months are found around Waterford.[37] However, spawning takes place in different areas of south-east Ireland from October to March, so the pattern here suggests that merchants imported herring from a variety of locations in Ireland. That said, fishers could also preserve their catch to ready it for export at a later time. For that reason, we cannot assume in every case that the import timing of fish relates directly to its spawning period.

The two most popular preservation methods for herring in the sixteenth century were salting (white herring) and smoking (red herring). Figure 3.5 shows that white herring were overall the much larger import, but shipments began to decline from January onwards, and in contrast, red herring imports reached their peak later in the season during February and March. Ecologically this is the trend we should expect to see. The fat content of the catch determined the type of cure used on the herring. Generally, fattier herring were salted and barrelled, while herring with less fat were smoked to mask the poor taste.[38] Herring had higher fat content before they spawned than they did post-spawning when their energy stores were far more depleted.[39] It is therefore logical that we should see far more

[37] David O'Sullivan et al., 'An Inventory of Irish Herring Spawning Grounds', *Marine Institute, Irish Fisheries Bulletin*, 42 (2013), pp. 1–31, at p. 27.

[38] James Travis Jenkins, *The Herring and the Herring Fisheries* (London, 1927), p. 54.

[39] Locker, 'Stored Fish', p. 43.

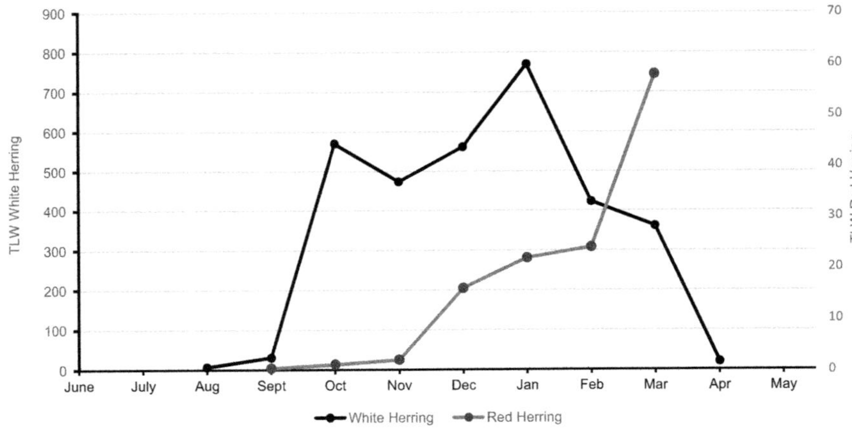

Figure 3.5. Comparison of white and red herring imports, 1503–95. Red herring is placed on a separate axis here as the total volume of red herring is far lower than white.

whites earlier in the season and more reds later into the new year. It also took far longer to produce the reds, meaning there was a greater delay between catching and export.

In terms of spatial data, the Bristol records do list origins of ships, but in most cases, these are simply recorded as 'Ireland', with no specific last port of call logged. However, the records do list a home port for almost every ship. Based on these home ports, the data shows that, in the fifteenth century, 41 per cent of the fish carried from Ireland to Bristol was done so by Irish-based vessels, another 41 per cent by English, and 12 per cent by Portuguese (with 6 per cent unknown).[40] The Portuguese involvement is intriguing and unique to the fifteenth-century Bristol accounts; by the sixteenth century, there were no Portuguese vessels involved in the fish trade with Ireland. Newfoundland's cod fishery may have drawn Portuguese fisheries and traders away from Ireland in the sixteenth century, but the evidence is limited.[41]

Conversely, the sixteenth-century accounts show that 27 per cent of the fish was carried by Irish vessels, 53 per cent by English and 19 by Welsh (with 1 per cent unknown). However, the share of the trade did fluctuate year by year. In 1550/1, 86 per cent of the herring was carried by English boats, while in 1563/4,

[40] These values are based on tonnes live weight of fish carried.

[41] For more information on the evidence for the Portuguese Newfoundland fishery, see 'Spanish and Basque Newfoundland Cod Fishery 1500–1790 Supporting Documentation', p. 4, which is part of a wider dataset: Poul Holm et al., 'Accelerated Extractions of North Atlantic Cod and Herring, 1520–1790', *Fish and Fisheries*, 23 (2022), pp. 54–72.

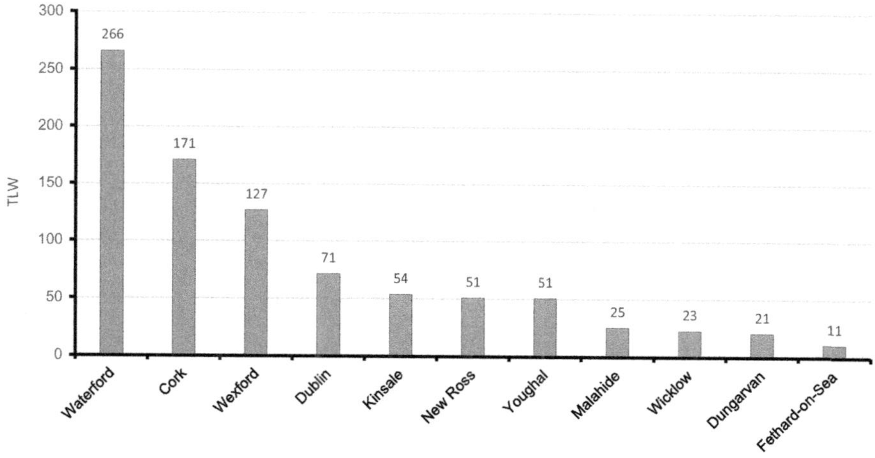

Figure 3.6. Import of herring to Bristol by Irish home ports, 1503–95. The towns of Waterford and Pilltown were combined in this total because Pilltown is up river from Waterford and the ships would have likely traded from Waterford.

86 per cent by Irish. It is not clear from the accounts alone why there was so much variation occurring year by year.

Boats with home ports outside Ireland reveal nothing about the ports from which the fish originated, but Irish home ports can inform us about the towns most involved in the trade. Because a ship has a particular home port does not guarantee that the cargo originated there, but given the lack of other information, the home ports are our best indication of where the fish originated. In the fifteenth century, Waterford and Cork were by far the most dominant exporters of herring, while Figure 3.6 shows the comparative breakdown of home ports for the sixteenth century. A higher number of towns were involved overall in the sixteenth-century trade, but Waterford and Cork remained the most prominent. Wexford was the next most frequent, followed by Dublin and Kinsale. Unsurprisingly, ports on the south-east coast that were geographically closest to Bristol dominated the trade. Many of these ports are also near herring spawning grounds, which means that catches were probably taken near the ports. We lack other evidence for the scale of fishing in Waterford in the fifteenth and sixteenth centuries, so these figures are our best indication that the town was one of the largest fishing ports in Ireland in the early modern period.

Hake was the second most imported fish into Bristol after herring, and as the previous section highlighted, it was one of the most sought-after fish by distant water and domestic fishers. Merchants would often count hake as individual fish

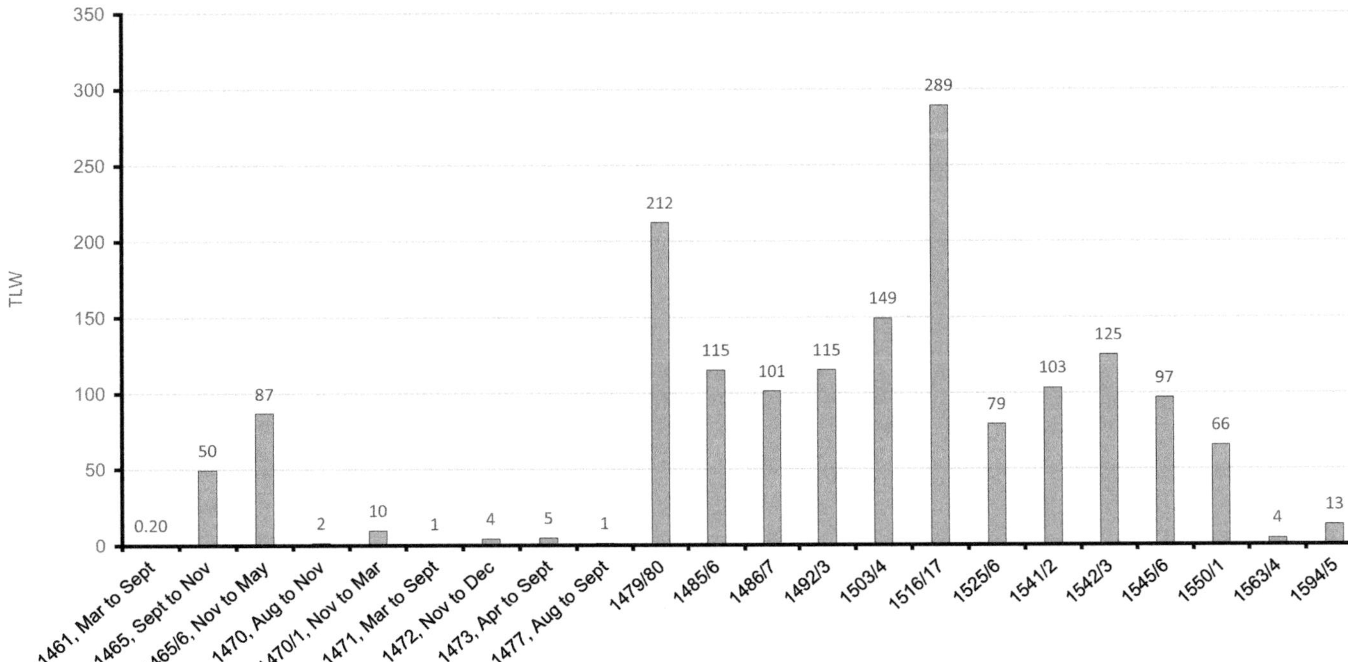

Figure 3.7. TLW of Irish hake imported to Bristol, 1461–1595.

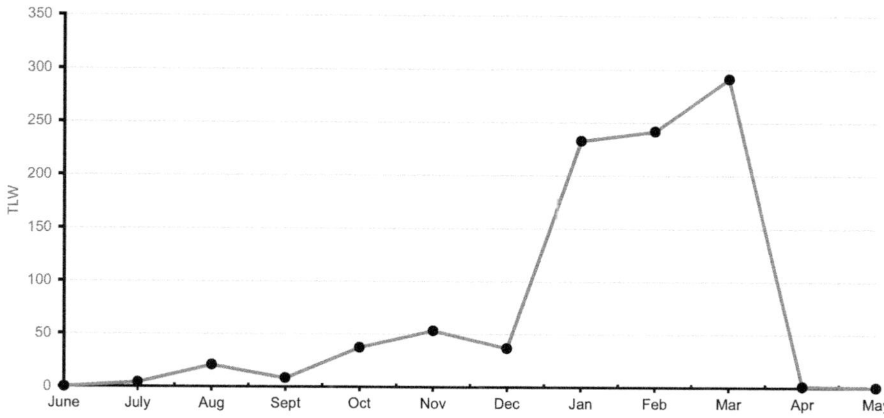

Figure 3.8. Hake imports to Bristol from Ireland per month, 1503–95.

or group them into a long hundred (120 fish).[42] Figure 3.7 shows the TLW of hake imported to Bristol from Ireland over the fifteenth and sixteenth centuries. In the fifteenth century, the most significant import took place in 1479/80. In the sixteenth century, hake imports peaked in 1516/17 as the most significant fish imported that year. However, in 1525/6, imports declined sharply and they only experienced a modest rebound before records ceased after 1563. When records reappear in 1594/5, imports are only a fraction of what they were in previous decades, suggesting that hake imports declined sooner compared to herring. The decline may be related to the qualities of hake itself, as it was a white fish comparable in taste and appearance to cod. The rise of the Newfoundland cod fisheries and the increase of cod imports to England in the middle of the sixteenth century may have reduced demand for hake. In the fifteenth century, hake imports took place mostly between October and February, with the most significant importation happening in the latter month. In contrast, Figure 3.8 shows the monthly import of hake in the sixteenth century. While the trend here is not entirely different from the fifteenth century, with similarly high levels of imports in February, the sixteenth-century data shows hake imports to be more concentrated in the first three months of the year.

This peak in the winter months seems contrary to the pattern of hake fisheries discussed earlier, which placed the English, French, and Spanish hake fisheries in Ireland firmly within the summer months. The summer is when hake spawning

[42] Longfield, *Anglo-Irish Trade*, p. 48; Hake live weight data, from 'Species Dashboard, Hake (Merluccius merluccius)' (https://shiny.marine.ie/speciesdash/) (accessed 30 September 2020).

off the Irish coast is at its peak, so this makes sense.[43] In contrast, the import figures from Bristol suggest the hake fishery was a winter activity. Hake caught in the summer may have been kept in storage until the winter, although there is no evidence to indicate this was a widespread practice, and it would have been logistically challenging for fishers and merchants. It is also possible that the hake recorded here was caught elsewhere, such as the Bay of Biscay, before being landed in Ireland and shipped onwards. The peak spawning period for the Bay of Biscay happens from January to March, which matches the peak of imports closely.[44] However, again, there is no indication that this was a common practice among fishers. This pattern of imports might also suggest that there were two hake seasons: one for the foreign fishers in summer and another for domestic fishers in the winter. Given the lack of contemporary sources discussing the hake fishery, we can only speculate.

In the fifteenth century, 38 per cent of hake from Ireland to Bristol was carried by Irish vessels, compared to 53 per cent by English over the same period. Over the whole sixteenth century, a much higher portion of the fish – 64 per cent – was taken by Irish vessels, compared to only 20 per cent by English. Once again, this does not mean that Irish fishers caught 64 per cent of the hake, but it shows that the sixteenth-century hake fishery had a greater involvement of Irish vessels compared to the herring trade. In the fifteenth century, only five Irish ports were involved in the hake trade, with Kinsale being the most frequent exporter, followed by Cork, Youghal, Waterford and Ballyhack. Figure 3.9 shows the most frequent home ports for the sixteenth century. In this case, ports on the southeast coast, such as Youghal, Kinsale and Dungarvan, showed the highest involvement in the trade. Modern observations show that Ireland's largest hake spawning and nursery grounds are located off the southeast coast, which may have given fishers in the region an advantage.[45]

Salmon was the second most valuable fish imported to Bristol from Ireland in the fifteenth century, and the third most valuable in the sixteenth century. Across all the accounts, it was by far the most valuable fish caught in freshwater, assuming most were captured in weirs and not in the open sea, which seems to have been

[43] Paula Alvarez et al., 'Distribution and Abundance of European Hake Merluccius merluccius (L.), Eggs and Larvae in the North East Atlantic Waters in 1995 and 1998 in Relation to Hydrographic Conditions', *Journal of Plankton Research*, 26 (2004), pp. 811–26, at p. 819.

[44] Maria Korta et al., 'European Hake (Merluccius merluccius) in the Northeast Atlantic Ocean', in Hugo Arancibia (ed.), *Hakes: Biology and Exploitation* (Chichester, 2015), pp. 1–37, at p. 8.

[45] Hake Nursery Grounds, from Marine Institute of Ireland, 'Ireland's Marine Atlas' (http://atlas.marine.ie/#?c=53.9108:-15.9082:6) (accessed 7 August 2019); 'Irish Groundfish Survey' (https://shiny.marine.ie/igfs/) (accessed 6 July 2020).

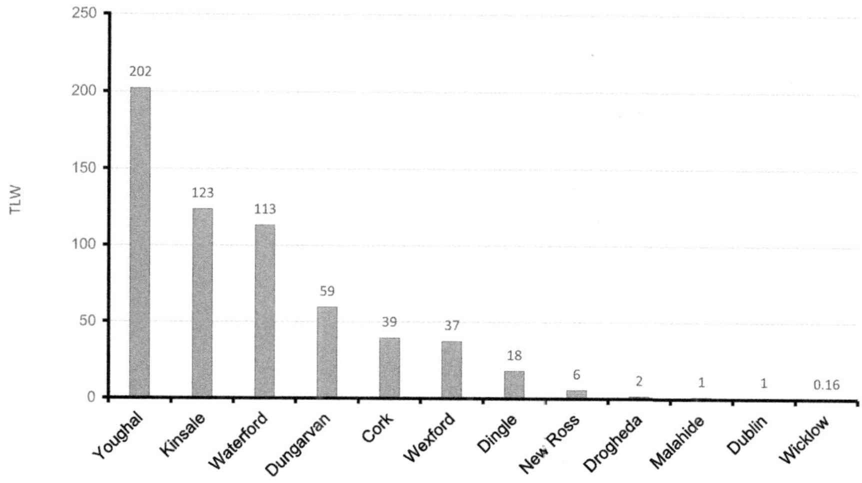

Figure 3.9. Import of hake to Bristol by Irish home ports, 1503–95.

the case.[46] The conversions used to convert salmon to TLW were derived mostly from the reported volume of a pipe – the most common unit for salmon and said to be 84 gallons in volume.[47] Figure 3.10 shows the TLW of salmon imported across all the Bristol accounts. The data shows that salmon imports peaked in the fifteenth century and would not reach the same levels in the sixteenth century. Salmon imports crashed dramatically in 1542/3 before recovering partially in 1545/6 and then declining again in 1550/1. Since the crash in salmon imports occurred before the change of law in 1563, the two cannot be related. When the customs accounts recorded fish again in 1594/5, only a minuscule amount of salmon was reported. This earlier decline of salmon compared to marine species such as herring may be related to environmental degradation and pollution. The documentary records tell us that excessive building of river infrastructure and pollution from industry and agriculture were of concern to people in the sixteenth century, and this could have negatively impacted salmon stocks.

Across the fifteenth century salmon was exported from Ireland from February to November, with a significant spike recorded in August. Salmon spawning in Ireland occurs in different rivers from around January to August. During this period, salmon return to the river in which they were born, to spawn and give rise to the next generation.[48] Salmon that have spent two or more years at sea tend to

46 A. E. J. Went, 'The Pursuit of Salmon in Ireland', *Proceedings of the Royal Irish Academy*, 63C (1964), pp. 191–244.

47 Jones, *Tonnage Estimates*, p. 15. For more on all the conversions see Appendix 1.

48 Marine Institute of Ireland, 'Salmon Life Cycle' (https://www.marine.ie/Home/site-area/areas-activity/fisheries-ecosystems/salmon-life-cycle) (accessed 7 August 2019).

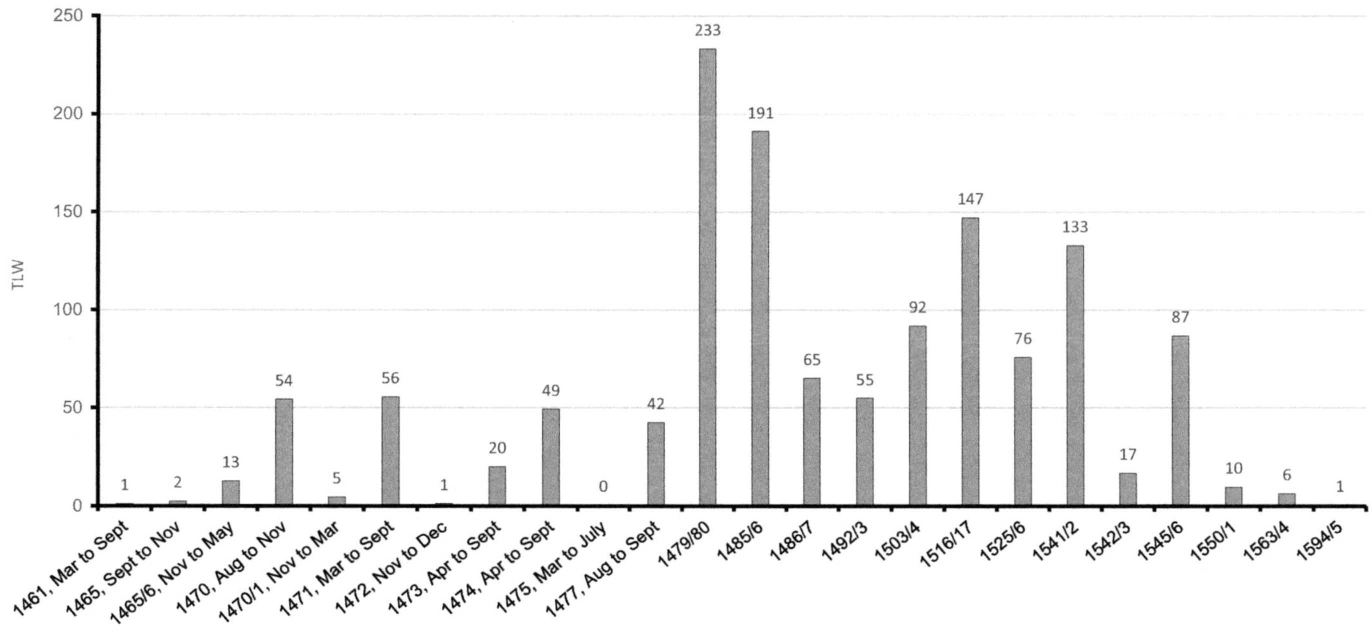

Figure 3.10. TLW of Irish salmon imported to Bristol, 1461–1595.

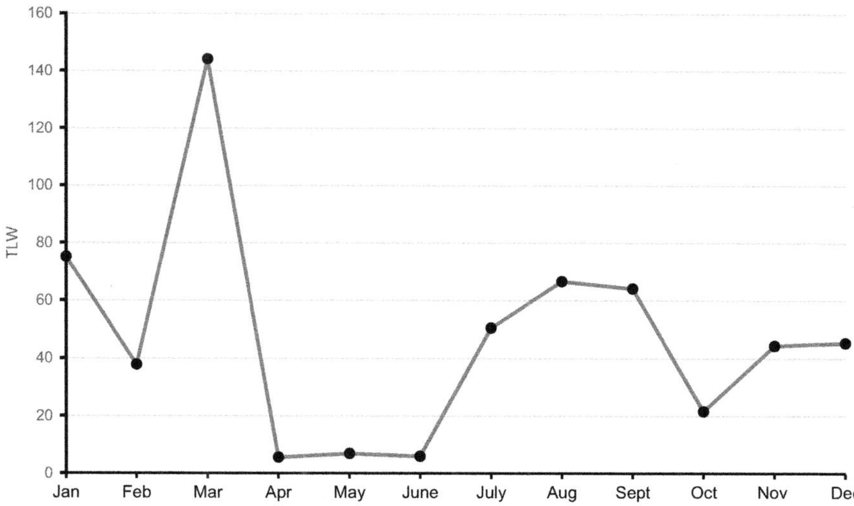

Figure 3.11. Irish salmon imports to Bristol per month, 1503–95.

return earlier in the season in the spring. These are larger fish, so anglers prefer them. Those that return later in the year are generally coming back after just one year at sea.[49] The high concentration of salmon in August would suggest most of the salmon caught in this period were returning after a single year.

Figure 3.11 shows the monthly trend for the sixteenth-century salmon data. These accounts show a different picture, with imports peaking in March and then declining again before picking up in the summer months. It is possible that the peak in March was related to the influx of fish for Lent but it might also indicate that fishers caught a higher number of salmon spawning in their second year. However, the trend does change year on year; for example, in 1541/2, there was a robust import in August and September, while in 1516/17, there was a massive import in March. That single month drives much of the trend seen in Figure 3.11. Overall, compared to herring and hake, salmon imports were spread more evenly throughout the whole year, a pattern which makes sense in light of salmon's lengthy spawning season.

In the fifteenth century only 9 per cent of the salmon coming from Ireland to Bristol was carried by Irish vessels, as compared to 78 per cent by vessels with English home ports. (Interestingly, French ships from Brittany carried 12 per cent of the total. The French involvement was the highest of any continental ships in the trade of fish from Ireland to England.) The sixteenth century witnessed a significant rise in the involvement of Irish vessels, with Irish boats carrying 53 per

49 Marine Institute of Ireland, 'Salmon Life Cycle'; Went, 'Pursuit of Salmon'.

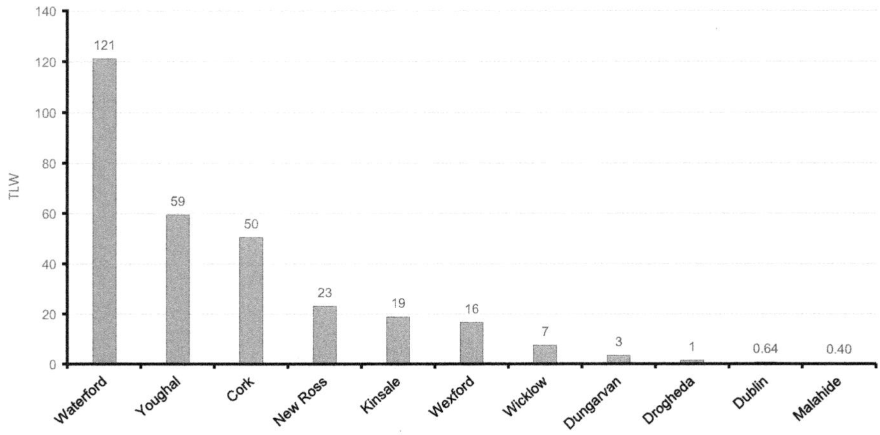

Figure 3.12. Import of salmon to Bristol by Irish home ports, 1503–95.

cent across the whole century, with a further 28 per cent carried in English vessels and 14 per cent in Welsh. In 1516/17 alone, Irish boats carried 82 per cent of the salmon, providing further evidence of the decline in the activities of continental and English merchants and fishers in Ireland.

Of the ships that had Irish home ports, ships from Cork and Drogheda carried the most salmon in the fifteenth century. Both towns are built on major fishing rivers, Cork located on the river Lee and Drogheda lying at the mouth of the Boyne.[50] Other smaller amounts were carried by boats from Waterford, New Ross, Youghal, Kinsale, and Ballyhack. Figure 3.12 shows the fish carried by vessels with Irish home ports in the sixteenth century. Waterford comes out as the highest carrier, followed by Youghal and Cork. Waterford lies at the meeting point of three salmon rivers – the Suir, Nore and Barrow – so it is not surprising to see it as the most significant trafficker of salmon.[51]

Salt was crucial for nearly all commercial fisheries in the early modern period, as almost all the fish transported overseas needed to be cured with salt to preserve it.[52] Little is known about salt production in fifteenth- or sixteenth-century Ireland, but it appears that there was no developed industry and little domestic production of salt. The first recorded salt works in Ireland only emerged in the seventeenth century at Ballycastle on the north coast.[53] In the absence of local

[50] Went, 'Pursuit of Salmon', p. 228; A. E. J. Went, 'Material for a History of the Fisheries of the River Boyne', *Journal of the County Louth Archaeological Society*, 13 (1953), pp. 18–33.

[51] Went, 'Fisheries of the River Boyne', p. 228.

[52] Only air-dried stockfish did not need salt and none of this was found exported from Ireland; Cutting, *Fish Saving*, pp. 173–74.

[53] Wes Forsythe et al., 'Persistence and Risk: Salt Production in Post-Medieval Ireland', *World*

supplies, salt came from France and Portugal, where it had been extracted from seawater by evaporation. It also came from England, most likely from natural springs in Droitwich in Worcestershire, some distance up the river Severn from Bristol.[54] This salt was said by contemporaries to be purer than the continental salt because it was not contaminated with sea bird droppings and was therefore considered better for curing fish.[55]

Figure 3.13 shows the total value of salt exported from Bristol to Ireland between 1461 and 1595. In the fifteenth century, the most significant importation of salt occurred in 1479/80, the year with the highest import of fish. There was also a substantial amount in 1465/6, another good year for the fish trade. During this year, the months in which salt was imported coincided with the months with the highest exportation of fish, suggesting that salt from England was meeting an immediate demand from fishers.

In the sixteenth-century data there is no clear link between salt imports and fish exports. Salt imports rose in 1542/3, while fish exports fell in the same year. There was also little salt imported in the year with the largest export of fish (1516/17). Salt imports fell off most dramatically in the middle of the sixteenth century, around the same time that fish exports from Ireland declined. By the end of the century the export of salt from Bristol to Ireland had all but ceased. Longfield has suggested that, in response to this, most Irish salt imports at this time began to come from the continent, filling the gap left by the decline of imports from England.[56] Unfortunately, we have little evidence to show whether this was the case or not. It is not easy to correlate the level of fish exported from Ireland directly with the corresponding salt imports early in the century. However, we can at least assume that the near collapse of Irish fish exports by the end of the century would have had an impact on the demand for salt in the country.

Moving away from Bristol, we can turn to the port of Chester, which lies far to the north of Bristol at the mouth of the river Dee, and almost directly across the Irish Sea from Dublin. Chester is not a major port today, but in the fifteenth and sixteenth centuries it was one of the principal trading ports of north-western England.[57] Several port books and customs accounts have also been printed for Chester, although these are far less detailed than those for Bristol.[58] Chester's first published custom account runs from October 1404 to September 1405. Another covers 1467 to 1468, but contains few details about cargo, rendering substantive analysis impossible. For the sixteenth century we have one local custom account

Archaeology, 50 (2018), pp. 603–19, at pp. 604–05.

54 Longfield, *Anglo-Irish Trade*, p. 164.

55 Jones, *Tonnage Estimates*, p. 24.

56 Longfield, *Anglo-Irish Trade*, p. 164.

57 Armour, 'Trade of Chester'; Woodward, 'Trade of Elizabethan Chester', p. 1.

58 *Chester Customs Accounts, 1301–1566*.

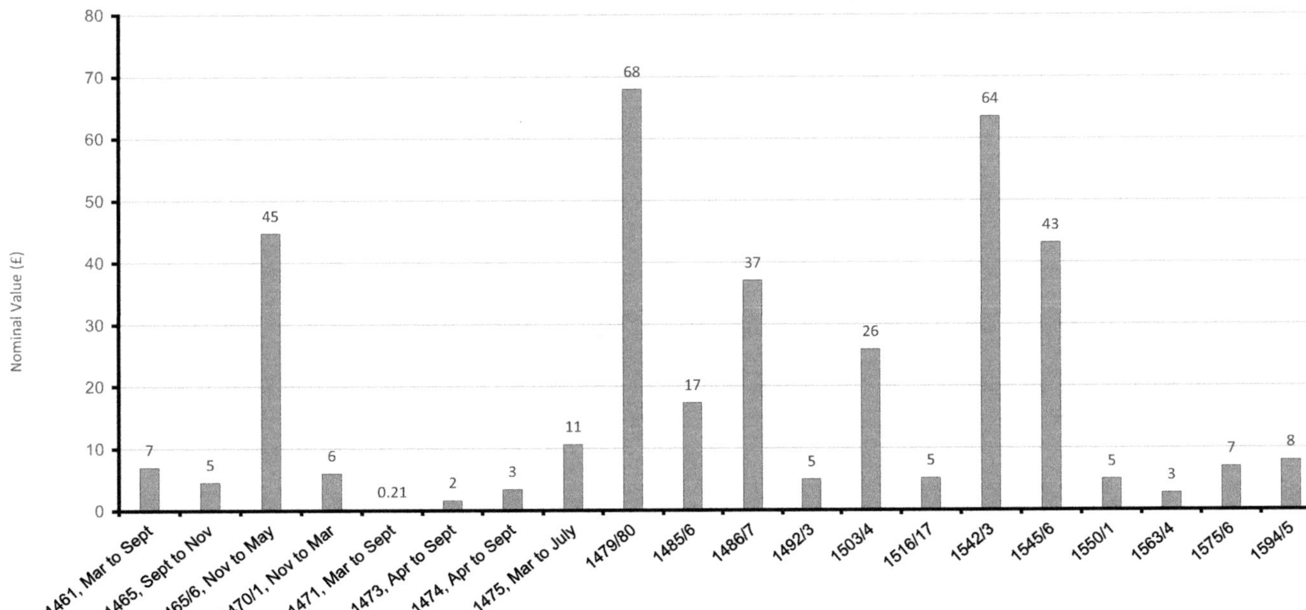

Figure 3.13. Nominal value of salt exports from Bristol to Ireland, 1461–1595.

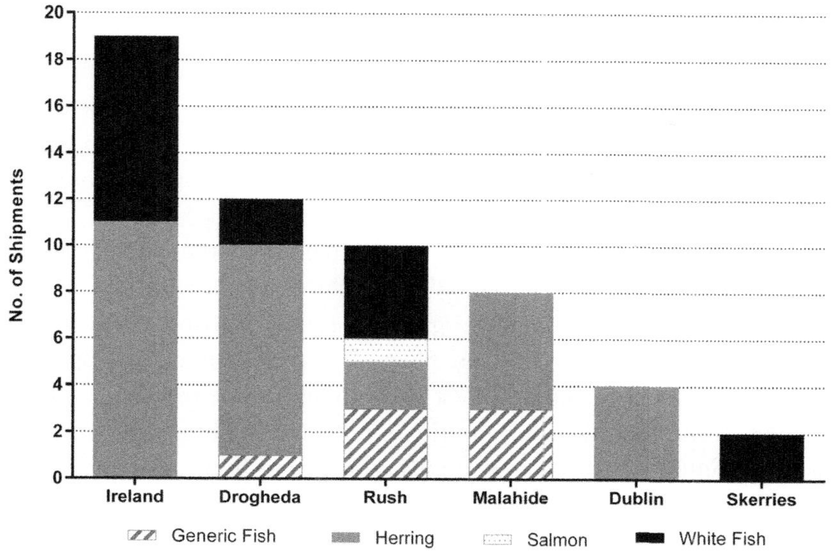

Figure 3.14. Shipment of fish from Ireland to Chester, 1404–05.

transcribed in detail for 1525/6, in addition to several more port books for Chester and Liverpool for the years 1565/6. In his doctoral thesis, Donald Woodward summarised Irish imports for many more of Chester's accounts in the sixteenth century.[59] However, his accounts were all taken after the law changed in 1563, so they recorded little fish from Ireland. In general the printed accounts are not sufficiently extensive to enable an estimate of the scale of the fish trade between Ireland and Chester, but they permit the identification of patterns in that trade and show which species and ports were most important.

The 1404/5 Chester account provides a useful comparison with the trade of fish a century later. Unlike the Bristol accounts, those for Chester do not contain nominal values to compare the volume of imports, so alternatively, Figure 3.14 shows the number of shipments made for each type of fish. Unlike many later records, the 1404/5 account does list the specific port of origin for most incoming ships. The first bar in Figure 3.14 shows shipments that had 'Ireland' listed as their origin, while the others show specific Irish ports. The data indicates that most loads of fish came to Chester from Drogheda, followed by Rush and Malahide. Dublin was the fourth most frequent shipper of fish. In terms of species, herring was the most popular, followed by undefined white fish. The Chester accounts also listed 'fish' as a commodity, which was adjusted to 'generic fish' for this study.

[59] Woodward, 'Trade of Elizabethan Chester'.

Predictably, the ports trading most with Chester were all within the Pale, since this area was politically and culturally connected to England and was geographically closest to Chester. While Dublin and its satellite ports were mostly absent from the Bristol data, the Chester records show these ports had a stable trading connection to northern England. Almost all the ships carrying fish into Chester in the early fifteenth century returned to Ireland with salt, presumably in part to process further catches.

Herring was by far the most traded species named in the 1404/5 account and the only species converted to TLW. Over the whole account, 38 TLW of herring was imported, with the most significant spike occurring in January and a secondary high the following September. Given modern herring spawning patterns, the peak in September is most likely to have come from the spawning populations on the Mourne coast and the Isle of Man, as these populations are most active in autumn.[60] As we have already seen, fishers from Ireland visited the Isle of Man from at least the thirteenth century.[61] The available information from Chester, combined with the data from Bristol in 1404/5, indicates an active and reasonably substantial trade in fish from Ireland to England from at least the start of the fifteenth century.

The next Chester account with any level of detail about fisheries relates to 1525/6. Table 3.3 shows the number of shipments of different fish products sent from Ireland to Chester in that account. The data illustrates that, once again, herring was the most frequently traded species, followed by salmon. Salmon is far more prominent in this account compared to a century earlier. Hake is noticeably absent. However, as mentioned earlier, the primary hake nursery areas are located in the south-east and west of Ireland, so it is not surprising to see an absence of hake traded from ports in the Irish Sea where the fish was not common. Some new species are listed, namely cod and ling, which were probably categorised into the more general 'white fish' classification in the earlier accounts.

In the 1525/6 accounts the sailing origin of the ship was only recorded in seven cases, with six ships coming from Dublin and one from Howth. Overall, twenty vessels were involved in the trade of fish from Ireland in 1525/6, and, of these, six had unidentified home ports, a further eleven came from England, and only three from Ireland.

60 D. Brophy et al., 'Temporal Trends in Stock Origin and Abundance of Juvenile Herring (Clupea harengus) in the Irish Sea', *ICES Journal of Marine Science*, 66 (2008), pp. 1749–53.

61 See p. 000.

Table 3.3. Number of shipments of fish from Ireland to Chester, 1525–26.

Fish product	No. of shipments
White herring	16
Salmon	14
Generic fish	6
Red herring	2
Cod	2
Ling	2
Eel	2

Herring and salmon dominated shipments to Chester, so these two species were converted to TLW to provide a clearer idea of the volume of fish shipped. Figure 3.15 shows the monthly profile of herring imports from September 1525 to February 1526. Once again, the monthly pattern is similar to the other accounts, and all lie within the herring spawning window. Overall, 150 TLW of herring was imported, not a substantial amount compared to some years in the Bristol accounts. However, fish represented a large portion of the goods imported from Ireland to Chester that year, and the trade was clearly of regional importance. It is also likely that much fish was consumed locally in urban centres like Dublin and Drogheda, and that any exports represented a surplus in good years.

Based on the evidence in the sixteenth-century accounts, Dublin was the primary supplier of herring to Chester, and a statute from 1470 provides further details on how the herring fishery operated around Dublin. The Act opens by

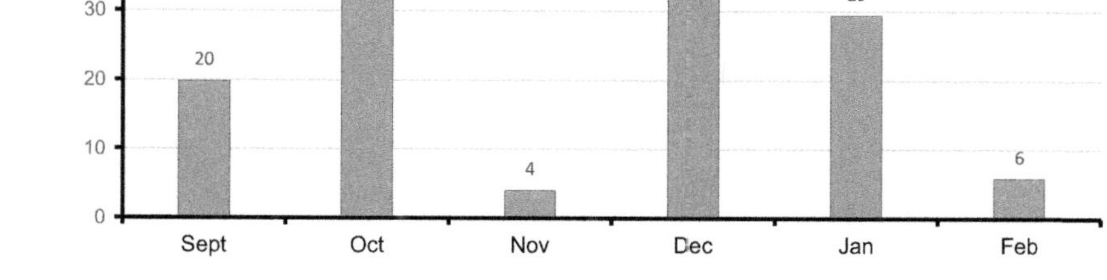

Figure 3.15. TLW of Irish herring imported to Chester per month, 1525–26.

saying, 'the herring fishery is always casual and removable by storms and winds from one place to another', and unless they 'be taken during the season they would disappear'.[62] This shows that early modern fishers targeted the herring during the spawning season and were conscious that the fish changed their migration patterns over time in response to the weather. However, they could not have understood the larger climatic and oceanographic forces driving these changes.[63] The statute ordered that any fishers who wanted to use the land of the Blessed Virgin Mary monastery near Dublin be allowed to do so without hindrance. Fishers were permitted to 'make their harbour' on the land, and 'throw their nets for herrings at all times according to their will'.[64] The abbey owned close to 17,000 acres of land in Dublin alone, including Bullock harbour, so it is difficult to know the exact spot to which the statute refers.[65] However, based on this description, it seems likely that herring were spawning in Dublin Bay in the fifteenth century, while none spawn there today.[66] The necessity and forcefulness of the statute also highlight the importance of herring fisheries to the food security of Dublin; it was a priority to keep the fisheries undisturbed and operating.

Salmon was the next most imported fish but was still far behind herring, with 24 TLW imported in the 1525/6 account. All this salmon was imported from November to February, with the most substantial spike happening in December. Like herring, the majority of this salmon came from the Dublin region, possibly originating in the salmon fishery on the river Liffey.[67]

The last transcribed account for Chester covers the accounting year 1565/6; it records ships entering and leaving both Chester and its satellite port of Liverpool. The customer's accounts for Chester in 1565/6 record no fish because of the statute of 1563 exempting Irish fish. However, the searcher's accounts for Liverpool do record a few shipments of fish. The reason is not apparent; it may have been in error, or perhaps the searcher was confirming that the goods carried by these ships matched their cockets.[68] The Liverpool accounts record three

[62] *Statute Rolls, Ireland*, vol. III, 10 Edward IV, A.D. 1470, p. 665.

[63] Adrianus Corten, 'Herring and Climate: Changes in the Distribution of North Sea Herring Due to Climate Fluctuations' (PhD thesis, University of Groningen, 1994).

[64] *Statute Rolls, Ireland*, vol. III, 10 Edward IV, A.D. 1470, p. 665.

[65] C. Ó. Conbhuí, 'The Lands of St. Mary's Abbey, Dublin', *Proceedings of the Royal Irish Academy*, 62C (1961–63), pp. 21–86; B. Phadraig, 'Speed's Plan of Dublin', *Dublin Historical Record*, 10 (1949), pp. 97–105.

[66] O'Sullivan et al., 'Irish Herring Spawning Grounds'.

[67] For more on the Liffey fisheries, see A. E. J. Went, 'Fisheries of the River Liffey. Notes on the Corporation Fishery up to the Dissolution of the Monasteries', *JRSAI*, 83 (1953), pp. 163–73; and 'Fisheries of the River Liffey: II. Notes on the Corporation Fishery from the Time of the Dissolution of the Monasteries', *JRSAI*, 84 (1954), pp. 41–58.

[68] A cocket was a type of certificate given to a merchant that verified he had paid customs on his cargo; Taylor, 'Maritime Trade', p. 2.

ships arriving with fish. The *Mary* arrived on 8 March 1566 carrying white and red herring and '40 small Newfoundland fish'; the *George* arrived on 25 February 1566 carrying herring and '1 C small Newfoundland fish'; the *John* arrived on 8 March with just red and white herring.[69] All three ships had sailed from Wexford. Two were carrying Newfoundland fish, another signal that the fish market was changing in this period, even though the amounts carried were small. We cannot be sure if the ships carrying the fish had caught it themselves or were re-exporting it. As we have seen, the Spanish cartographer Alonzo de Santa Cruz did claim that Irish ships were present in Newfoundland in 1540. At the very least, these imports suggest that Irish towns were receiving shipments of Newfoundland fish in this period. However, the real extent of the Newfoundland fish trade in this period is obscured by the tax-exempt status of fish.

Further evidence indicates that the herring trade to Chester continued after the change of law in 1563. Data compiled by Donald Woodward shows that 80 tons were imported in 1582/3 and 'over 90 tons' were brought from Ireland in 1584/5 and 1587.[70] Unfortunately, no further detailed information is provided about these imports. However, Woodward does state that one ton was equal to one barrel. From this we can calculate that in 1582/3, 101 TLW was imported, with a further 114 tons in 1584/5 and 1587. Overall, the statistics suggest imports of herring in the late sixteenth century declined by about a third from 1525/6 but did not collapse completely. However, after this point, Irish herring imports dropped further in the 1590s, and herring supplies in Chester 'were supplemented by landings of fish from other English and Welsh ports'.[71] Woodward's data also demonstrates that fish was surpassed by wool and skins as the most dominant imports from Ireland to Chester in the last decades of the century.[72]

Woodward also records large shipments of salt leaving Chester in the later sixteenth century, much of which was imported to Wexford, Waterford and New Ross to supply their fisheries. In addition, in the late sixteenth century, some English ships from around Chester were recorded fishing in Ireland. For example, the *Marten* sailed from the Dee River 'for the coast of Ireland to make herrings' in 1576.[73] These voyages may have been a response to the decline in imports from Ireland, as local fishers attempted to supplement supplies.

Returning to the south of England, the port of Bridgwater, on the south side of the Bristol Channel, was the head port for the coastline of the modern county of Somerset. In a PhD thesis from 2009, Duncan Taylor compiled extant

69 *Chester Customs Accounts*, pp. 88–89.

70 Woodward, 'Trade of Elizabethan Chester'.

71 Woodward, 'Trade of Elizabethan Chester', p. 10.

72 Woodward, 'Trade of Elizabethan Chester', p. 8.

73 Woodward, 'Trade of Elizabethan Chester', p. 20.

sixteenth-century customs data from Bridgwater.[74] His findings show that in the sixteenth century, fish from Ireland was the most valuable category of goods imported. Before 1550 all overseas fish brought into the port was carried by small Irish vessels with fish as their only cargo. They carried white and red herring, hake, and generic salted fish.[75] In the 1490s both Bridgwater and nearby Minehead imported Irish fish, with Minehead being the more prominent port.[76] This connection between south-west England and Ireland continued in the sixteenth century. Overall, Irish fish accounted for 36 per cent of all the goods imported to the Somerset ports. Comparatively, Irish fish accounted for 19 per cent of the total value of goods imported to Bristol. So, while the total imports to the smaller Somerset ports were lower than those to Bristol, fish from Ireland overall held a more significant share of the trade.[77] Many West Country fishers visited Ireland yearly. However, not only did English fishers travel to Ireland, but many Irish came to south-west England to trade and settle.[78] The connections between the two regions facilitated a commercial relationship that brought a high proportion of Irish fish into the area.

Figure 3.16 shows the value of fish imported from Ireland to Bridgwater over the sixteenth-century accounts. Unfortunately, the available data only distinguishes between Newfoundland cod and fish from Ireland, which included hake, herring, salmon, and salted fish, and it is not possible to analyse each species separately. The volume of fish carried is also not given, only the value. Thus, no live weight calculations are possible. Nevertheless, the data still illustrates the overall trend of imports across the century. The pattern is similar to that for Bristol; imports are highest in the first three decades of the century, after which they decline considerably. Fish imported after 1563 was exempt from tax, but when fish was recorded again in the 1597/8 accounts, the levels of imports had dropped markedly. In contrast, by 1597/8, the import of Newfoundland cod had risen significantly. Taylor believes that two factors led to the decline of Irish imports: first, the rise of the English involvement in the Newfoundland fishery from the middle of the sixteenth century; and second, the growth of a domestic English herring fishery.[79] In theory, the growth of both fisheries would have reduced the demand for Irish imports. However, the present book will argue that other factors, such as warfare, contributed to the decline of Irish fisheries, and fishing operations in England and Newfoundland came into the market to fill the gap.

[74] Taylor, 'Maritime Trade'.

[75] Taylor, 'Maritime Trade', p. 54.

[76] M. Kowaleski, 'The Expansion of the South-Western Fisheries in Late Medieval England', *Economic History Review*, 53 (2000), pp. 429–54, at p. 437.

[77] Taylor, 'Maritime Trade', p. 57.

[78] Taylor, 'Maritime Trade', pp. 55–56.

[79] Taylor, 'Maritime Trade', pp. 60–61.

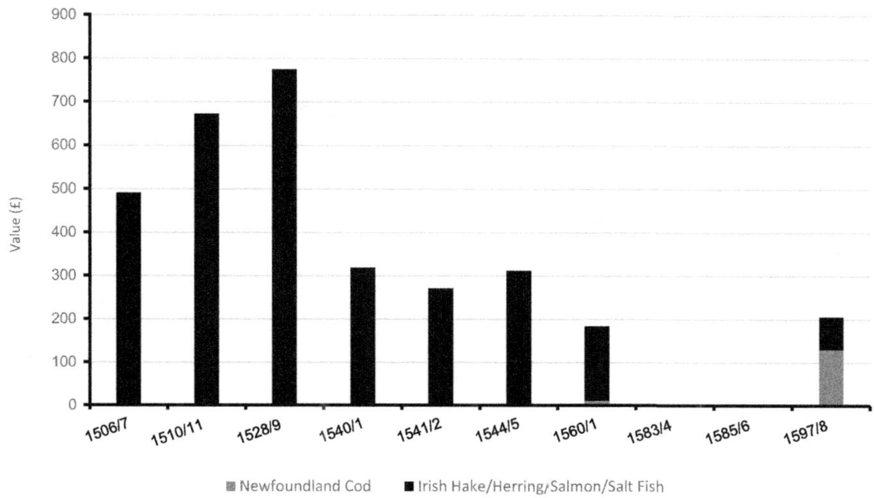

Figure 3.16. Nominal value of Irish fish imported to Bridgwater, 1506–98. Taylor, 'Maritime Trade', p. 59. Figures have been pegged to pre-1558 values by the original author.

We can also look at the evidence from Welsh ports trading with Ireland in the sixteenth century. Only in 1559 did the Welsh customs system come in line with the English. Before then, there are no quantitative records of trade between Wales and Ireland. The extant Welsh port books and customs accounts for the sixteenth century were published by E. A. Lewis in 1927.[80] The surviving accounts are scattered geographically and temporally. Furthermore, as the records only began in 1559, most were impacted by the change of law in 1563.

Table 3.4 details the shipments of different fish products from Ireland to Wales in all the Welsh accounts from 1559 to 1603. While the number of shipments does not reveal the size of individual loads, it shows which products were moved most. Only thirty-two shipments of fish were sent from Ireland over the period. Of these, herring was by far the most common, with ten loads of white herring and four of red herring. We also find shipments of 'shotten herring', herring that had already spawned and was therefore of lower quality.[81] Conversely, also present are 'full herring', herring that had not yet spawned and were of higher quality. These terms illustrate that not only did fishers target the herring around the spawning season, but they were also aware that the spawning itself had an impact on the quality of the fish caught. Almost all herring imports occurred

[80] Lewis (ed.), *The Welsh Port Books*, p. ix.

[81] Arthur Michael Samuel, *The Herring: Its Effect on the History of Britain* (London, 1918), p. 54; Cutting, *Fish Saving*, p. 27.

in February, matching the spawning season and import timings from the other accounts discussed previously.

Table 3.4. Shipments of fish from Ireland to Wales, 1559–1603.

Fish product	No. of shipments
White herring	10
Shotten herring	5
Red herring	4
Full herring	3
White fish	2
Irish cod	2
Newfoundland cod	1
Ling	1
Dry fish	1
Wet fish	1

Three Welsh ports were most involved in the trade of fish from Ireland: Beaumaris on Anglesey received eight shipments of fish, Milford nine and Carmarthen eight. The temporal trend of these shipments across all the ports is shown in Figure 3.17. Before 1563, no significant shipments of fish from Ireland to Wales were recorded. Likewise, after 1591, when the law changed again, there was no spike in imports. Instead, the most significant peak occurred in the 1580s when fish was meant to be exempt: the reason is unclear.[82]

Outside Irish imports, the most common fish imported to Wales was Newfoundland fish. Many of these shipments came via England, although some came directly from North America. Several more ships from Newfoundland are recorded carrying only train oil, which was an oil pressed out of fish; we can assume these ships also carried cod, but customs officials did not record it.[83] Again the documentation is not extensive enough for us to know the true extent of the Newfoundland trade, but we can assume that fish from Newfoundland took part of the market share away from both Irish imports and locally caught Welsh fish.[84]

Overall, shipments of fish to Wales from Ireland are too irregular to enable us to know definitively how the trade fluctuated over the century. The census of

[82] Without sufficient information, live weight calculations could not be attempted in the Welsh case.

[83] Cutting, *Fish Saving*, p. 80.

[84] For more on this in relation to Milford, see Taylor, 'Maritime Trade', pp. 195–98.

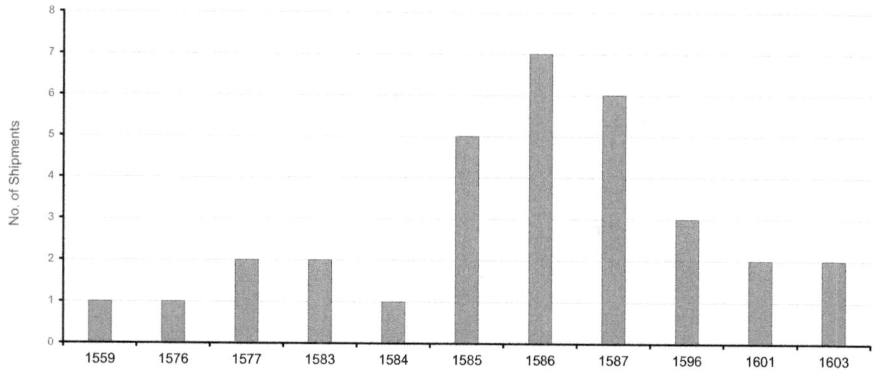

Figure 3.17. Shipments of fish from Ireland to Wales per year, 1559–1601.

ships in 1566 showed that Pembrokeshire had a fishing fleet of its own, and other Welsh towns may have had fleets and local supplies which meant there was little need for Irish imports.[85] Ultimately, the evidence suggests that Wales was not a significant importer of Irish fish in the second half of the sixteenth century.

Data regarding trade between Ireland and Scotland is even less extensive than it is for Wales. In a comprehensive survey of the available Scottish customs records, Martin Rorke reported that records of Scottish imports are scant, and many of those which do survive fail to distinguish between different types of goods or indicate where products originated.[86] However, he does suggest that, in the fourteenth century, Scotland may have imported more fish than it exported, and while we cannot know for sure, it is likely that a good deal of the importation came from Ireland.[87] From the beginning of the sixteenth century, the export of fish from Scotland increased rapidly. Exports, mainly consisting of herring, peaked in the 1540s and remained high for the rest of the sixteenth century.[88] These high domestic catches made Irish imports unnecessary. It is also notable that Scottish exports rose in the same period that Irish exports were declining, suggesting that Scottish fish took some of the market share previously held by the Irish.

However, there is one exception to this trend. In 1597/8, 4,832 barrels of salmon were imported to Scotland and then re-exported from the west coast ports

[85] Lewis (ed.), *The Welsh Port Books*, pp. 311–13.

[86] Martin Rorke, 'Scottish Overseas Trade, 1275/86–1597' (PhD thesis, 2 vols, University of Edinburgh, 2001); 'The Scottish Herring Trade, 1470–1600', *Scottish Historical Review*, 84 (2005), pp. 149–65, at p. 152.

[87] Rorke, 'The Scottish Herring Trade', p. 152.

[88] Rorke, 'The Scottish Herring Trade', p. 153.

of Ayr, Irvine and Glasgow.[89] Unfortunately, the customs officials did not record the origin of this fish, but there is a strong case to be made that it came from Ireland. First, the Scottish west coast is close to Ulster, which had strong political and economic ties to Scotland and many active salmon fisheries, including those on the Bann, Foyle and Erne rivers. In 1597 the Bann was controlled by Hugh O'Neill, and just the year before, in 1596, Nicholas Weston reported that O'Neill was at Castle Roe for the 'fishing salmon in the Bann'.[90] O'Neill was fighting the English in the Nine Years War from 1593 to 1603, but evidence suggests that the Bann fishery was still active and under Irish control in 1597/8. Given the demands the war would have placed on O'Neill's finances, these salmon exports may have been a way to generate additional income.

The Ulster port books, which run from 1612 to 1615, show significant exports of salmon from Ulster at a slightly later date, but little of this went to Scotland: instead, most headed to markets in England, Spain and France.[91] This seems to suggest that the export of salmon to Scotland was not a typical trade and was undertaken because other markets were cut off to O'Neill because of the war. Furthermore, all the salmon imported to Scotland was re-exported, so clearly there was little local demand for imported salmon. Instead, Scotland was used as a transit point to get the fish safely out of Ireland and on to better markets.

When we convert the salmon import to live weight, it equals 576 TLW, which would make it the highest single import of salmon across the fifteenth and sixteenth centuries. While these exports represent only one year, they do suggest that in previous years the salmon fisheries in Ulster had produced similar catches, which went to ports in England, France and Spain and have gone unrecorded. It appears that there was no regular trade in fish between Ireland and Scotland, mainly because Scotland had a domestic supply of salmon and other fish that made Irish imports unnecessary in most years.

Having surveyed the available evidence for the fish trade from Ireland to Great Britain, we can also ask what was the relative importance of fish compared to all other export commodities? The Bristol accounts, which are the most extensive and reliable, provide the best evidence. Figure 3.18 shows the nominal value of all fish products imported from Ireland to Bristol, compared to all other commodities from Ireland, starting in 1461 and running to 1601. In the fifteenth century,

[89] The original describes 402 lasts and 8 barrels (1 last containing 12 barrels). George Powell M'Neill (ed.), *Rotuli Scaccarii Regum Scotorum (The Exchequer Rolls of Scotland)*, vol. XXIII, *A.D. 1595–1600* (Edinburgh, 1908), p. 316; Rorke also partially quotes this figure: see Rorke, 'Scottish Overseas Trade', p. 294.

[90] *CSPI*, 1592–96, no. 27, p. 519. For more on Weston's connections to the earl, see Ruth A. Canning, 'Profits and Patriotism: Nicholas Weston, Old English Merchants, and Ireland's Nine Years' War, 1594–1603', *Irish Economic and Social History*, 43 (2016), pp. 85–112.

[91] R. J. Hunter and Brendan Scott (eds), *The Ulster Port Books, 1612–15* (Belfast, 2012).

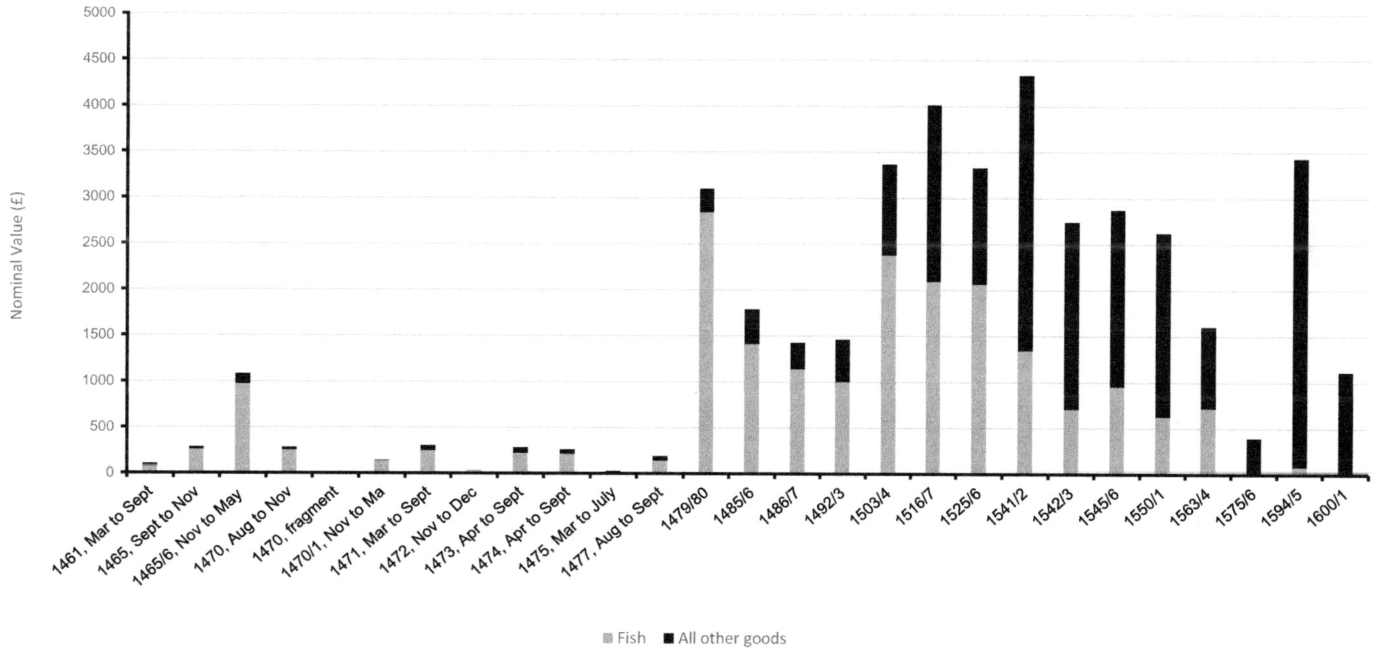

Figure 3.18. Irish fish imports compared to all other Irish imports, 1461–1601. Values after 1558 have not been adjusted in this data as the overall trend is not important compared to the relative importance of the series in each year.

fish accounted for 84 per cent of the nominal value of all goods imported from Ireland to Bristol. This shows that fish was by far the most dominant import over the century, despite many accounts not covering the crucial winter months when most fish was imported. Fish also accounted for 84 per cent of all imports at the very start of the fifteenth century: in 1404/5 fish constituted £1,754 of the £2,091 total value of Irish imports. The next most valuable commodities imported during the century were skins, hides and cloth, but these did not come close. However, fish declined in importance toward the end of the fifteenth century, accounting for 91 per cent of imports in 1479/80 and only 69 per cent in 1492/3.

Despite this decline, fish remained the dominant import at the beginning of the sixteenth century. In 1503/4 fish still accounted for 71 per cent of the total value of goods imported. However, even by 1516/17, this had further declined, to 52 per cent, and it dropped again to 24 per cent in 1550/1. In 1563/4, the last year before fish was exempt from tax, there was a slight recovery to 45 per cent, but this year also saw the general decline of all trade. By the time fish was once more recorded, in 1594/5, it only accounted for 2.5 per cent of imports. The export commodities witnessing the highest growth in the sixteenth century were cloth, followed by animal hides and re-exported wine. Additionally, from the 1540s onward, there was a general shift of exports towards more manufactured goods and away from animal products like fish and hides.[92]

Overall, this data shows that fish was a vital export commodity from at least 1404 and would reach a peak of importance in the 1480s. Subsequently, there was a gradual decline over the next several decades before fish was superseded by other commodities in the latter half of the sixteenth century. Over the whole period, fish was by far the most important export commodity Ireland produced. Understanding what drove this trade and ultimately led to its decline is crucial to the broader economic history of the country.

The Trade to Continental Europe

Evidence for the trade of fish from Ireland to continental Europe is far more limited. Once again, we are hampered by the lack of export records from Ireland, and there is little information available from continental ports. For the present, we have in the main only snippets of evidence.

From the Bristol accounts, we know that in the late fifteenth century, roughly ten to twenty Spanish ships were visiting Bristol every year, with many more going to London (forty-six in 1490/1). Wendy Childs surmises that the trade from Spain to Ireland in the same period could not have matched or exceeded these

[92] Flavin, *Consumption and Culture*, p. 27.

figures, which seems fair.[93] Other historians have shown that hides were probably the primary export from Ireland to continental Europe from the fourteenth to sixteenth centuries.[94] In 1479/80, when fish imports to Bristol were peaking, only 950 hides from Ireland were imported to the city. Tim O'Neill has suggested that Wales was supplying Bristol's leather industries, and therefore there was no market for Irish hides in England. He assumes, therefore, that Irish hides were destined for continental Europe.[95] If this was the case, then fish was not an essential part of the trade to continental Europe, as there was a ready market in England. O'Neill further claims that within Europe, Flanders was the primary market for Irish hides, but provides little proof to support his theory,[96] while other evidence would seem to contradict it directly. A pilotage account dating from 1486/7 from the city of Bruges shows no Irish ships; another from 1499 contains only three ships of Irish or English origin; and a third from 1500 again has no Irish ships. Three accounts alone do not prove there was no Irish trade to the Low Countries, but they suggest it was not significant.

While quantitative trade data is the backbone of this chapter, it is worth examining other sources, such as the State Papers and Admiralty court records, for what they might be able to reveal. Map 3.1 is a conglomeration of every mention of a trading voyage involving fish over the sixteenth century. Unfortunately, there were not enough records to do a similar analysis for the fifteenth century. In some cases, the sources do not list an exact origin or destination and only list the country of arrival or departure. In these instances, the lines connect to the centre of the countries in question.

This map includes voyages to all destinations, but overall, the number of fish trading voyages mentioned in these qualitative sources is not extensive, and their value is limited. The map indicates that south-east Ireland experienced the highest traffic in trade, and most of this was with ports in England and Wales. However, several voyages to Spain originated from Cork, Limerick, Dublin, and Waterford. In total, only seven recorded voyages went to England, five to France, another five to Spain and two to Wales. In contrast, no records mentioned voyages from the north coast of Ireland or heading to northern destinations like Scotland. These figures should not be interpreted as an exact representation of the balance of trade to different regions, as we probably lack the records accurately reflecting trade to places like Scotland. Moreover, there may have been external reasons why European voyages were mentioned more often in the later sixteenth century. The State Papers, in particular, record more Spanish voyages in the 1580s and 1590s.

93 Childs, 'Ireland's Trade with England', p. 9.

94 Timothy O'Neill, *Merchants and Mariners in Medieval Ireland* (Dublin, 1987), p. 77; Longfield, *Anglo-Irish Trade*, p. 61.

95 O'Neill, *Merchants and Mariners*, p. 77.

96 Longfield, *Anglo-Irish Trade*, p. 78.

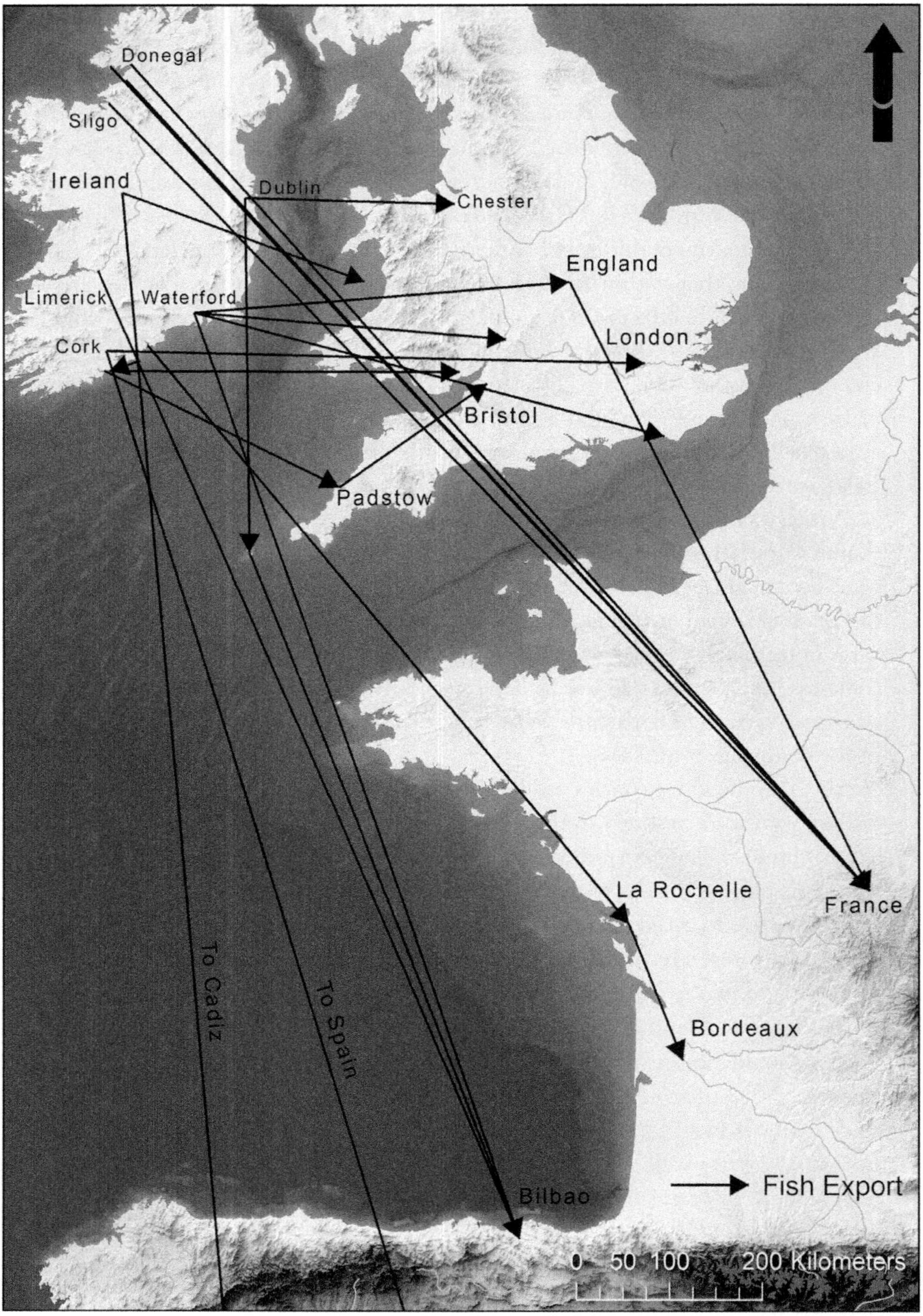

Map 3.1. Fish export routes from Ireland, 1500–1603.

But this increase was related to the war between England and Spain rather than better trading relations. The records of central government reflect periods of conflict or disruption more often than they do periods of calm. In peaceful times there was little reason for voyages to Spain to generate letters relevant to English government business. The number of records related to Spain increased rapidly during the war because merchants were interrogated on their return so that the English could gather intelligence on Spanish preparations for invasion.[97] A report from Waterford in 1587 recounts how the town had sent three smaller ships carrying 'dry hakes, herrings and hides' to Bilbao to 'learn some intelligences'.[98] In May 1590, another Waterford merchant, Walter Sherlock, reported from Bilbao that he had spotted several other Irish ships there, including the *Maurice* and *Thomas Fitz John* of Desmond. At the same time, he recounted what he believed to be the intended movements of the Spanish fleet.[99]

These reports indicate that trade between Irish ports and Spain continued despite the war, but they do not record the true scale of trade as we have no earlier reference point from peacetime. These limited records also illustrate that fish was a small part of the trade to Spain and not the most important. Of those ships carrying fish, herring was by far the most traded species, with twenty-five voyages carrying herring domestically and internationally. Cod was the next most frequent with seven voyages, while hake and salmon each had one.

Karin Schüller has also analysed the State Papers but she focused on all voyages to Spain and Portugal from 1585 to 1604, not just those carrying fish. Her data showed that Waterford was the most frequent trading partner with the Iberian Peninsula, followed by Dublin. On the Spanish side, Bilbao was the most frequented port for Irish merchants, with 28 per cent of voyages. Lisbon was important, too, with 20 per cent of voyages heading there.[100] Schüller recorded few instances of fish being transported, with salt and wine the largest imports to Ireland from Iberia and hides the largest Irish exports.[101] As Schüller's data is also distorted by the Anglo-Spanish war, it is difficult to know what the trade looked like in the previous decades of the century. What seems clear is that by the 1580s, there was no substantial Irish fish trade to Spain or Portugal. It is also possible that no large Irish-operated fish trade ever existed to the Iberian Peninsula, as the demand for Irish fish could have been met by distant water fishing voyages from

[97] Karin Schüller, 'Special Conditions of the Irish-Iberian Trade during the Spanish-English War (1585–1604)', in Óscar Recio Morales et al. (eds), *Irlanda y la Monarquía Hispánica. Kinsale 1601–2001: Guerra, Política, Exilio y Religión* (Madrid, 2002), pp. 447–68, at p. 449.

[98] *CSPI*, 1586–88, no. 438, p. 438.

[99] *CSPI*, 1588–92, no. 7, p. 339.

[100] Schüller, 'Special Conditions', p. 452.

[101] Schüller, 'Special Conditions', p. 454.

the Basque country. Unfortunately for the Irish fishing industry, war (among other factors) also put an end to those voyages by the close of the sixteenth century.

The best French data we have comes from Bordeaux. In the early fifteenth century, the ports of New Ross, Cork and Waterford were the most frequent Irish trading partners with Bordeaux, mostly importing wine. However, Irish trade only accounted for 2 per cent of the shipping in the port in this period.[102] Notarial records from Bordeaux can tell us more about the late fifteenth- and early sixteenth-century trade. Of the seventy-one surviving notarial contracts for 1445–75, only three refer to cargoes from Ireland. Then out of a further 520 records up to 1500, Irish destinations are only quoted nineteen times.[103] Based on this evidence, it appears that Irish trade with Bordeaux was not particularly significant but was increasing toward the end of the fifteenth century. Elsewhere in France, port accounts for Dieppe from 1470 to 1480 show no Irish ships or merchants at all.

The French historian Jacques Bernard compiled further data from notarial records for the early sixteenth century, showing that in the first two decades, '150 Irish merchants engaged in the Bordeaux trade against some 75 Gascons engaged in the Irish trade'.[104] These records do not allow us to quantify the volumes of goods traded, but they do provide some indication of the goods traded most frequently. In the early sixteenth century, wine, salt, and honey dominated imports from France to Ireland. The main commodity exported from Ireland in the early sixteenth century, most commonly exchanged for wine, was hides. Other goods typically imported from Ireland included cloth, mantles, and fish. The main fish species imported to Bordeaux were herring and hake, while 'Bay salt' was the major export from Bordeaux to Ireland.[105] Notarial records also show that many merchants combined trading with fishing in Ireland. In 1601 Captain Plessington commented that French merchants in Donegal most often brought wine and salt which was exchanged for hides, tallow, and fish.[106]

The limited data we have suggests that Irish trade with France was on an upward trajectory from the late fifteenth century and into the sixteenth. However, accurately quantifying this trade with the data available is impossible. Bernard concludes that many aspects of the Irish/French trade – such as the currency used, insurance and the temporary nature of merchant partnerships – suggest that it was still largely medieval in character and did not undergo significant development in

[102] Childs, 'Ireland's Trade with England', p. 10.

[103] Childs, 'Ireland's Trade with England', p. 11.

[104] Jacques Bernard, 'The Maritime Intercourse between Bordeaux and Ireland, c. 1450–1520', *Irish Economic and Social History*, 7 (1980), pp. 7–21, at p. 13.

[105] Bernard, 'Maritime Intercourse', p. 13.

[106] Plessington to the earl of Nottingham and Secretary Cecil, 2 September 1601 (TNA, SP 63/209/1, f.148).

the sixteenth century. In her review of this trade, Mary Ann Lyons has written that overall trade with France may have increased, but the fish trade was in decline by 1600. This would be in line with export figures for Britain, but again, the data is too limited to understand how the trade changed over time.[107] The Irish fish trade to France seems to have been rather haphazard in the sixteenth century, with no large-scale operations in place, and was undoubtedly smaller than the trade of fish to England.

In 1563 the English Admiralty court took a deposition from a Hamburg merchant by the name of Peter Rutkins. He told the court that the merchants of 'the Hanse, as well as of Hamburg and other places haunt the realm of Ireland for traffic of hides to pass unto St Malo as they do all other nations'.[108] While this is only one anecdotal account, it does suggest that Hansa and German merchants engaged in Irish hide trade through French ports. The Hanseatic League was the most extensive commercial consortium in medieval and early modern Europe, so we would expect to see Hanseatic involvement in Ireland. The historian of the Hansa trade, Philippe Dollinger, makes no mention of a significant presence in Ireland but does note that the League's merchants preferred to sail along the west coast of Ireland to reach Spain in order to avoid the pirates infesting other routes.[109]

New archaeological evidence and analysis are shifting how we think about the economy of Gaelic Ireland.[110] The evidence presented here, especially relating to England, is mostly focused on the trade from the Old English ports in Ireland. There could have been a large, unregulated and unrecorded trade taking place in areas like Donegal, as described by Plessington. Ships would arrive from abroad and exchange goods through barter. As evidence of the growth of this exchange, Mark Gardiner and T. E. McNeill have written that the increase of conflict in Ulster in the second half of the sixteenth century required weapons that could only have been paid for 'by means of the commerce in fish and through the ports already serving it'.[111] The shipment of salmon to Scotland in 1597/8 may have been an example of such trade.

However, the Spanish and French both had fishing fleets which regularly visited Irish waters in the fifteenth and early sixteenth centuries. This activity facilitated

[107] Mary Ann Lyons, 'Maritime Relations between Ireland and France, c. 1480–c. 1630', *Irish Economic and Social History*, 27 (2000), pp. 1–24, at p. 7.

[108] *Cal. HCA 1536–1641*, no. 127, pp. 24–25.

[109] Philippe Dollinger, *The German Hansa* (London, 1970), p. 350.

[110] Máire Ní Loingsigh, 'An Assessment of Castles and Landownership in Late Medieval North Donegal', *Ulster Journal of Archaeology*, 57 (1994), pp. 145–58, at p. 148; Mark Gardiner and T. E. McNeill, 'Seaborne Trade and the Commercialisation of Fifteenth- and Sixteenth-Century Gaelic Ulster', *Proceedings of the Royal Irish Academy*, 116C (2016), pp. 229–62; Colin Breen and John Raven, 'Maritime Lordship in Late-Medieval Gaelic Ireland', *Medieval Archaeology*, 61 (2017), pp. 149–82.

[111] Gardiner and McNeill, 'Seaborne Trade', p. 262.

the exchange of fish for other goods such as wine, and brought money to coastal communities in Ireland through taxation. This activity would have reduced the demand for imported fish in the home markets of the fishing fleets. Additionally, traders from Spain and France were some of the first to become involved in the Newfoundland cod fishery, which returned massive catches to their home ports.[112] This new source of fish would have also reduced demand.

On the whole, we lack hard evidence to tell us the scale of trade between Ireland and continental Europe in the fifteenth and sixteenth centuries. Wendy Childs has written, 'unless a long series of customs accounts or similar records is unearthed, it will be impossible to assess orders of magnitude' for the continental trade.[113] Continental trade was undoubtedly important in relation to specific commodities, such as salt and wine, but there is no real indication that fish was a significant part of the trade. In contrast, the sum of all the evidence suggests that England was Ireland's largest trading partner in the fifteenth and sixteenth centuries, and fish was a vital part of that trade for many decades. England and Wales were the primary recipients of Irish fish in the fifteenth and sixteenth centuries. Some salmon also went to Scotland, but this was irregular. Likewise, some merchants carried Irish fish to continental Europe, but this trade was inferior to that with England and Wales. Regarding exports from Ireland to England and Wales, the most important fish species were herring, hake, and salmon. Significantly, for many years in the fifteenth and early sixteenth centuries, these fish were the largest exports from Ireland overall.

Figure 3.19 shows the combined TLW of the top three species imported from Ireland to England and Wales from 1450 to 1603.[114] There are significant gaps in the available data from England and Wales too, but we can observe a few general trends over the decades. Imports peaked in 1479/80 and then declined slightly before remaining relatively high and stable for the following four decades. Imports dropped considerably by the 1540s and continued to fall over the next two decades until 1563, when the law dictated fish would no longer be recorded in the accounts. Customs officials in Chester did record some Irish fish during the 1580s, but even this shows a decline from earlier highs. When records of fish resumed after 1591, only tiny amounts of Irish herring were recorded, suggesting that the drop observed earlier in the century continued and that ultimately a combination of factors meant the export trade from Ireland to England and

[112] Poul Holm et al., 'The North Atlantic Fish Revolution (ca. AD 1500)', *Quaternary Research*, 108 (2019), pp. 1–15.

[113] Childs, 'Ireland's Trade with England', p. 9.

[114] The accounts from Bristol and Chester for 1404/5 were not included here, because the Bristol data from that year was not detailed enough for live weight calculations, and only a further 38 TLW was recorded from Chester.

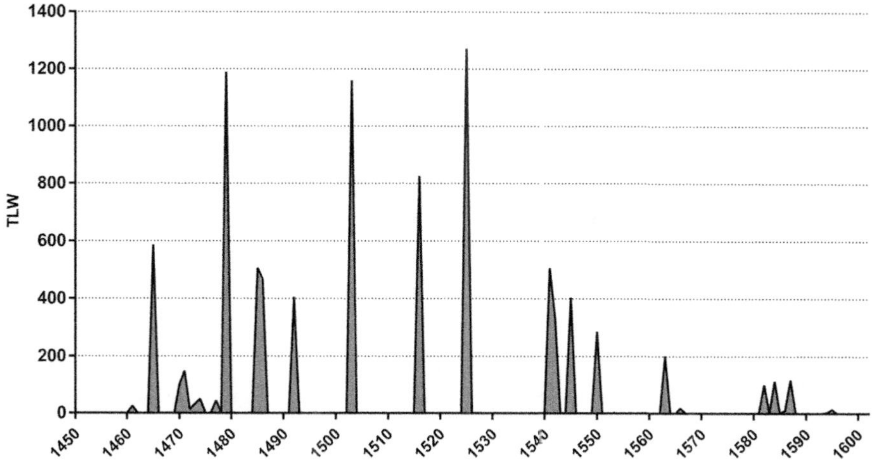

Figure 3.19. Combined TLW of Irish herring, hake and salmon imported to England and Wales, 1450–1603. Within this graph, data covering accounting years was grouped into the first year in that pair. For example, data for 1503/4 is shown as a single total for 1503. This provides a more realistic year on year comparison.

Wales had all but ceased by 1600. The data also highlights that herring dominated exports almost every year as the most important species.

The Scottish data from 1597/8 was left off Figure 3.19 because it was such an isolated event, and we have no longer-term series from Scotland from which to compare. The salmon exports in that year may have been an anomaly, but without a longer series of salmon exports to Scotland, we cannot know how it compared to previous years. But it is likely that in previous peaceful years before 1597/8, most salmon exported from Ulster was caught by visiting fishers. The Nine Years War may have disrupted this pattern, and O'Neill had to look for alternative export routes, with the result that the fish went to Scotland. If this was the case, even here traditional commercial relationships had been disrupted by the end of the century.

Figure 3.19 uses live weight, but owing to data limitations, we cannot undertake TLW calculations for every species and every trade account. To overcome this, Figure 3.20 indicates the nominal value of all fish imports per year. Where the nominal value is unavailable, the graph shows the number of shipments of fish made per year on a secondary axis. This graph contains data concerning every species of fish exported from Ireland, not just the big three. The main difference between the trend in Figure 3.20 and that in 3.19 is the increased shipments recorded during the period when fish was exempt from tax in the years after 1563. However, these shipments only strengthen the hypothesis that the trade in fish

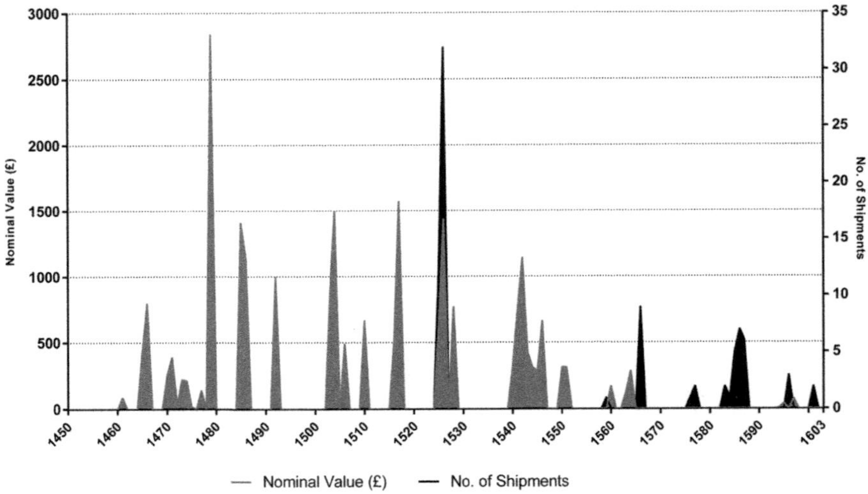

Figure 3.20. Nominal value and number of shipments of fish from Ireland to England and Wales, 1450–1603. Post-1558 customs values have been pegged to pre-1558 rates.

declined toward the end of the century, as they still show a dramatic fall from previous shipment levels.

These graphs represent a significant conglomeration of data and provide the most comprehensive picture yet compiled of the scale and trajectory of Irish fisheries in the fifteenth and sixteenth centuries. They are far from perfect, but the data presented here indicates broad trends. Irish fisheries experienced a boom period in the late fifteenth and early sixteenth centuries, when they were one of the most important industries on the island. In the later sixteenth century, they suffered a dramatic decline, which would have broader implications for the development of Irish society and culture.

In contrast to the declining trend of Irish imports, the sixteenth century saw a rapid increase in the importation of fish from Newfoundland to England and Wales. The data for Newfoundland imports is very patchy due to the tax-exempt status of the fish, but some shipments were still recorded. Figure 3.21 pulls together the total value and number of shipments of Newfoundland fish across all the accounts. The first peak in imports happened in the 1540s, just as Irish imports were sharply declining. By the end of the century, when Irish imports had ceased, we continue to see fish imported from Newfoundland. We cannot know for sure how much fish came into England and Wales from Newfoundland, but it seems that this new product took at least part of the market share previously held by Irish fish. Alongside other evidence, this demonstrates that many English

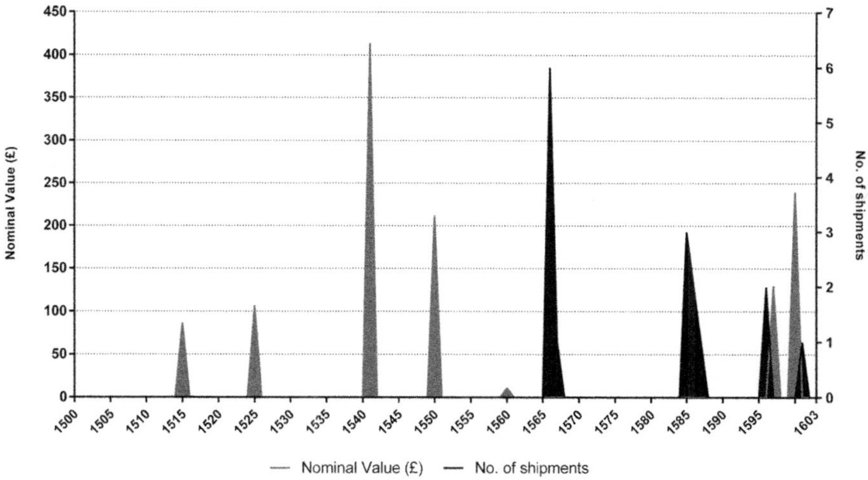

Figure 3.21. Nominal value and number of shipments of Newfoundland fish imported to England and Wales, 1500–1603. Post-1558 customs values have been pegged to pre-1558 rates.

and Irish fishers left behind old fishing grounds in Ireland in favour of the more profitable catches in the New World.

This trade data has also shown that the importation of most species was highly seasonal, and often those import seasons were dictated by the biology and spawning periods of the target species. Only the timing of hake imports seemed to contradict other historical information, suggesting that fishers targeted the species during the summer months when the hake spawned. However, in the case of salmon and herring, the level of imports month on month matched very well with modern observed spawning periods. Other factors, such as demand during Lent and trade fairs, impacted imports during some years. Ultimately, species' spawning periods determined when they were available in large numbers for fishers to target and consequently drove the timing of imports.

The data also highlights the Irish ports that exported the most fish to England and Wales. All the most dominant ports discussed here had either English or Old English roots. Some regional trends also emerged: Bristol in the south had a more reliable trading connection to ports like Cork, Youghal, Waterford and Wexford, compared to Chester, which had stronger ties with Dublin and Drogheda. The Welsh ports traded with ports in the south-east, like Waterford, and had a strong connection to eastern ports such as Dublin. Clearly there existed a robust mercantile community in many Old English ports. For the first half of the sixteenth century, fish was one of the most crucial commodities, if not the most crucial commodity, carried to England. Many of these towns, particularly those in the south-east, were also built close to vibrant marine habitats, such as herring spawning grounds, hake nursery grounds and salmon rivers, which provided the aquatic resources making the trade possible. It was this confluence of ecology, geography and strong trading towns that made the south-east the most prosperous region in Ireland for fisheries.

The accounting year with the highest combined export of herring, hake and salmon was September 1525 to September 1526, when exports totalled 1,271 TLW. The total of all fish extracted from Ireland in that year would be higher if we factored in domestic consumption, catches by visiting fishers and exports to other areas for which we have no records. In other research on the broader trends of North Atlantic fisheries, the trade figures presented here were combined with additional sources and estimates of domestic consumption to trace Irish herring catches from 1520 to 1790.[115] These figures still likely underestimate the true scale of herring catches, but when placed alongside estimates from eleven other European herring fisheries, they allow us to gauge how Ireland compared to its neighbours. The data indicated that the Dutch herring fishery was by far the largest in Europe over the early modern period, averaging around 30,000 TLW per year from 1520 to 1600 and achieving an estimated catch of almost 80,000 in 1602. However, when compared to smaller European nations, the Irish figures are not dissimilar. Our best estimates suggest that the Irish herring fisheries were more productive over the sixteenth century than those on the German North Sea coast, Norwegian west coast and the Limfjord and Kattegat fisheries in Denmark. In the early part of the sixteenth century, Irish herring catches were also not far behind those in Scotland and England, but while those fisheries grew over the century, fisheries in Ireland stagnated and declined. So, while these figures reveal that Irish fisheries were comparatively small compared to the big players in Europe, they also show that catches in Ireland were not insignificant. They also suggest that had the industry been allowed to grow uninterrupted, it may have been far more productive, as Irish fisheries even briefly became amongst the most productive in Europe toward the end of the eighteenth century.[116]

[115] Holm et al., 'Accelerated Extractions'. The full breakdown of methods used for the Irish figures can be found in the accompanying dataset to this paper: Poul Holm et al., 'Accelerated Extractions of North Atlantic Cod and Herring Dataset' (Dublin, 2021) (https://doi.org/10.6084/m9.figshare.13614452.v14) (accessed 25 July 2022).

[116] Holm et al., 'Accelerated Extractions', p. 62.

The Seventeenth-Century Perspective

None of the available fifteenth- or sixteenth-century data discussed thus far captures the total extent of Irish fish exports. However, data that does illustrate the complete picture of Irish trade is available only sixteen years into the seventeenth century. The data utilised in this section was compiled by David Dickson and Donald Woodward and published in 1999.[117] The first year covered is 1616/17, the next available year is 1621/2, then 1626 and finally 1640/1.[118] While this study is focused on the sixteenth century, this seventeenth-century data provides a useful comparison to the earlier figures and allows us to establish a quantitative estimate of what total exports looked like in the early modern period.

In the early seventeenth century the Irish customs system was reformed in an attempt to generate more income from foreign trade, resulting in the Irish customs being farmed out to a series of English syndicates.[119] To track how their investment was performing, these syndicates commissioned statistical series on the volume of trade in Ireland, which provide us with the figures under discussion here. Figure 3.22 shows the TLW of fish exports, while Map 3.2 shows the ports of departure. Unfortunately, the customs do not list destinations, so we do not know where this fish was going.[120]

Pilchards were the most exported species in this period, with a total of 14,722 TLW recorded in the data. Herring was the most dominant species throughout most of the fifteenth and sixteenth centuries, while pilchards appeared only in small amounts. Both are small oily, pelagic fish and roughly fill the same ecological and human dietary niche.[121] Pilchards are a mature form of sardines and are also in the same family as herring (Clupeidae).[122] They are generally found in warmer waters, but can flourish on the Irish south and west coasts when ocean conditions are right.[123] Herring and pilchards have also been shown to alternate

[117] The primary author is Woodward, but he acknowledges the help of David Dickson in compiling the data: Donald Woodward, 'Irish Trade and Customs Statistics, 1614–1641', *Irish Economic and Social History*, 26 (1999), pp. 54–80, at p. 54.

[118] Data for this section is taken from Woodward, 'Irish Trade and Customs Statistics', pp. 67–77.

[119] Woodward, 'Irish Trade and Customs Statistics', p. 54; Victor Treadwell, 'The Establishment of the Farm of the Irish Customs 1603–13', *English Historical Review*, 93 (1978), pp. 580–602.

[120] Any miscalculations present in the originals have been corrected here.

[121] Henk J. L. Heessen et al. (eds), *Fish Atlas of the Celtic Sea, North Sea, and Baltic Sea* (Wageningen, 2015), p. 139.

[122] Liam Downey and Ingelise Stuijts, 'Historical Fisheries: Irish Pilchard Fishery and Historical Fish Harvesting and Processing Methods, Fish Products and Consumption', *Kerry Archaeological & Historical Journal*, 2 (2014), pp. 1–35, at p. 9.

[123] Downey and Stuijts, 'Historical Fisheries'; A. J. Southward et al., 'Fluctuations in the Herring and Pilchard Fisheries of Devon and Cornwall Linked to Change in Climate since the

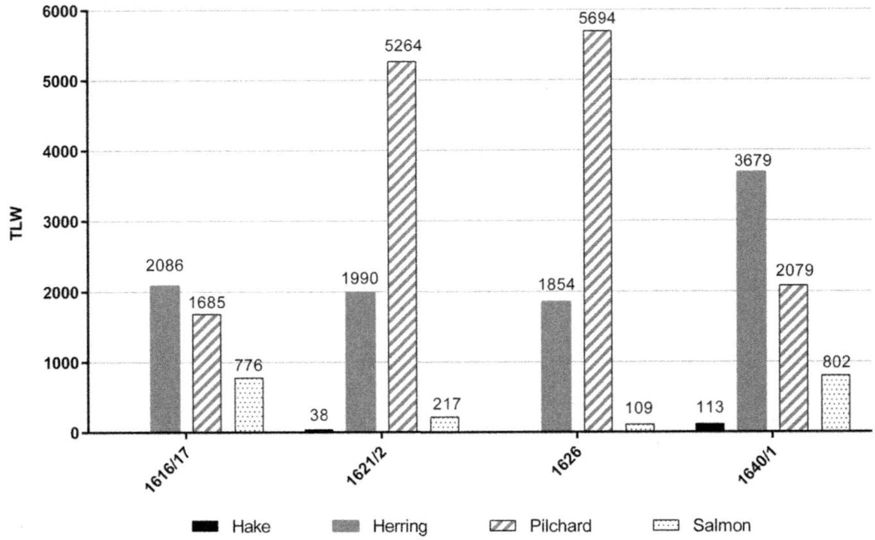

Figure 3.22. Irish fish exports, 1616–41.

in abundance on the south coast of England owing to longer-term cycles in the climate.[124] While no study has specifically studied this phenomenon in Irish waters, a similar pattern of alternation certainly does occur in Ireland.

The seventeenth-century data reveals that pilchards experienced periods of boom and then decline, with exports spiking in the 1620s before dropping again in 1640/1. Map 3.2 shows that pilchards were heavily exported from the south-east ports, with further exports coming from Galway and the Irish west coast.[125] Many of the most significant pilchard fishing sites in the seventeenth century, such as Baltimore, had previously been Gaelic Irish strongholds but had since been taken over and exploited by English settlers.[126] Pilchard catches best represent the changing pattern of Irish fisheries in the seventeenth century, both ecologically and politically.

16th Century', *Journal of the Marine Biological Association of the United Kingdom*, 68 (1988), pp. 423–45.

[124] N. C. Stenseth et al., *Marine Ecosystems and Climate Variation: The North Atlantic – A Comparative Perspective* (Oxford, 2005), p. 92.

[125] Most of the exports had an exact port described but some were generically described as the 'West of Ireland'; Woodward, 'Irish Trade and Customs Statistics'.

[126] For more, see A. E. J. Went, 'The Irish Pilchard Fishery', *Proceedings of the Royal Irish Academy*, 51B (1946), pp. 81–120; additional information on the slightly later period can also be found in David Dickson, *Old World Colony: Cork and South Munster 1630–1830* (Cork, 2005).

For the first three decades of the seventeenth century, herring was relegated to the second most important species, but by 1640/1, the boom in pilchards had declined, and herring was once again the largest export. Map 3.2 shows herring exports from places that did not feature heavily in the earlier data, such as Killybegs, Galway and Carrickfergus. The overall pattern remains broadly similar, with ports in the south-east, such as Wexford, remaining paramount in the seventeenth-century trade. In total, 9,609 TLW of herring was exported in the four years recorded, with the highest export – 3,679 TLW – in 1640/1. Comparably, the year in the sixteenth century with the highest export was 1525/6, with 1,092 TLW leaving Ireland. However, the earlier data was collected from only a small selection of English ports, so the actual total export would have been higher. Without consistent data across the centuries, it is not easy to establish whether the herring industry grew or declined over the period. At the very least, the figure from 1640/1 provides some indication of what the full scale of herring exports from the early modern period could have been.

In the seventeenth century the importance of hake seems to have significantly declined, with only minuscule amounts being exported. In the earlier data, hake was regularly the second most exported species and sometimes the most exported. Given the increasing significance of Newfoundland fish, it seems likely that cod from the New World largely replaced hake in the export market by this time.[127] Salmon also remained a small but crucial element in trade. Map 3.2 shows that salmon was exported from a wide variety of ports around the Irish coast. The strong export of salmon from Ulster was related to the newly imposed Ulster Plantation. Ulster port books from 1612 to 1615 further illustrate this trade.[128]

It is difficult to judge the relative importance of goods from quantity alone. Still, it appears evident from seventeenth-century data that fish was not among the most valuable Irish exports. Conversely, hundreds of millions of empty wooden barrels, hogsheads and pipe staves were exported from Ireland. In addition, hundreds of thousands of hides and pounds of wool were also shipped. Tallow was another major export, with 99,200 lbs exported across all the accounts.[129] Thus in the seventeenth century fish was superseded by timber, hides, wool and tallow as Ireland's most important exports. However, the data also indicates that the fishing industry had recovered from the near-total collapse experienced in the late sixteenth century. Traditional species like herring had a resurgence and continued to be extracted by fishers in large numbers, while other species, such as pilchards, were significant to local economies.[130]

[127] Holm et al., 'North Atlantic Fish Revolution'.

[128] Hunter and Scott, *The Ulster Port Books, 1612–15*.

[129] All the figures referenced here are calculated from data in the FTD.

[130] For more on the significance of the seventeenth-century pilchard fisheries in Munster, see Dickson, *Old World Colony*, pp. 19–20.

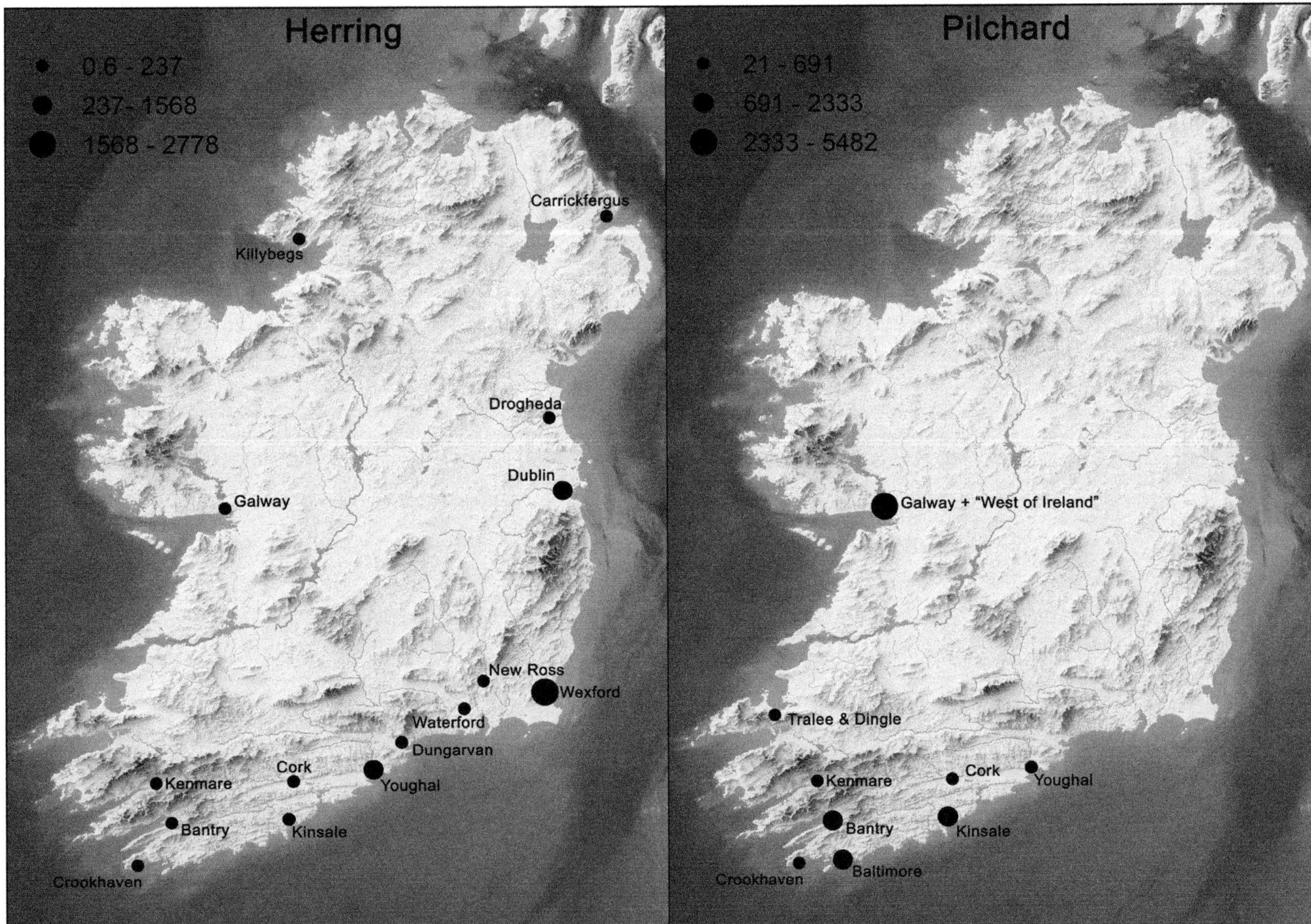
Herring
0.6 - 237
237- 1568
1568 - 2778
Carrickfergus
Killybegs
Drogheda
Dublin
Galway
New Ross
Wexford
Waterford
Dungarvan
Cork
Youghal
Kenmare
Bantry
Kinsale
Crookhaven
Pilchard
21 - 691
691 - 2333
2333 - 5482
Galway + "West of Ireland"
Tralee & Dingle
Kenmare
Cork
Youghal
Bantry
Kinsale
Crookhaven
Baltimore

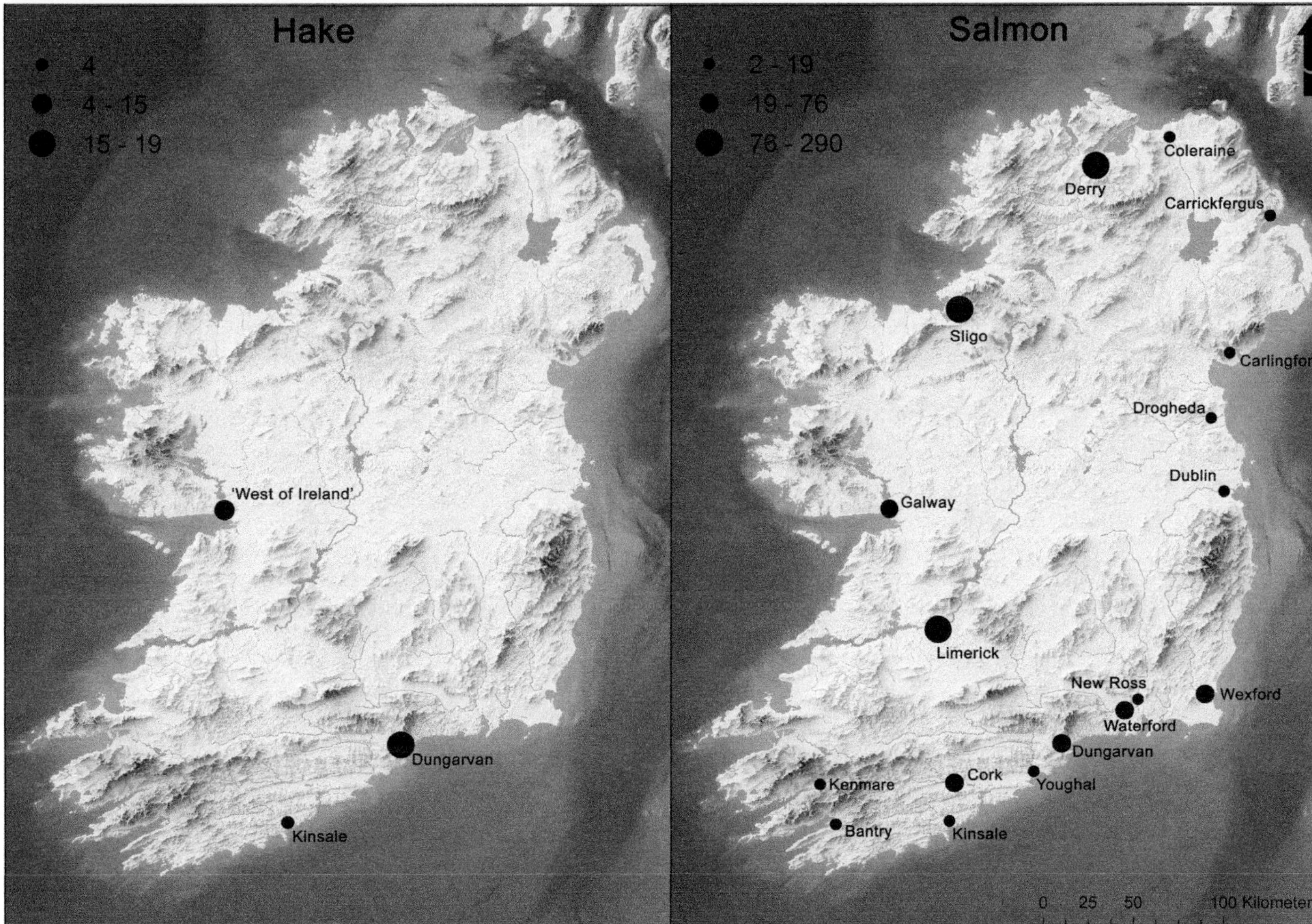

Map 3.2. Export of fish from Ireland by species and port, 1616–41.

The year with the total highest combined exports was 1625/6, in which 7,657 TLW of fish was exported. The full figure would have been higher if we were able to factor in domestic consumption. Even if we assume domestic consumption would increase this total twofold, these figures would still not place Ireland among the top fish producers in the North Atlantic.[131] Nevertheless, the numbers still represent a significant extraction, and fishing was evidently an essential economic activity in particular locations. The data also highlights that, in the seventeenth century, fish was no longer the primary export from Ireland, as many other products had superseded it.

Fish Prices

The nominal or customs values used hitherto were not market values, but fixed prices set by a central book of rates. To establish the actual monetary value of fish exports, we must rely on other sources, but also keep in mind that not all parts of early modern Ireland used currency or were integrated into the same market economy. Many Gaelic Irish communities continued to use elements of barter, so any price information would not reflect the value of fish across the whole of Ireland or act as a good measure of demand and supply. Nevertheless, price information can provide additional details on the workings of the fish trade.

As most fish exported from Ireland was brought to Bristol, it would be ideal to have an extensive price series from there. Unfortunately, none exists for this period, so we must use data from other locations around England. The best series for England for our period was compiled by Thorold Rogers and originally published between 1866 and 1892.[132] It has since been digitised and published online as part of the Allen-Unger global price dataset.[133] Rogers' data contains price series for several species, including herring, salmon, and cod, but this section focuses on the white herring, which, as mentioned, is a product and not a separate species.[134] The herring prices come from various towns and institutions around England, with most prices related to the south of the country.[135] The geographic dispersion of Rogers' data is not ideal, nor is the lack of price information on key species such as hake, but despite these drawbacks, it is still the best available.

[131] Holm et al., 'The North Atlantic Fish Revolution'.

[132] J. E. T. Rogers, *A History of Agriculture and Prices in England* (7 vols, Oxford, 1866–92).

[133] Robert C. Allen and Richard W. Unger (eds), *London Prices, Allen-Unger Global Commodity Prices Database* (http://www.gcpdb.info/) (accessed 29 July 2016).

[134] Rogers, *A History of Agriculture and Prices*, vol. IV, pp. 540–45.

[135] The full list of where prices relate to can be seen in Rogers, *A History of Agriculture and Prices*, vol. III, pp. 310–34.

Gregory Clark has also published a price series that includes fish, but his data is based on statistical interpolations and is not as reliable as Rogers'.[136]

The lowest sustained herring prices in England occurred from 1480 to 1520, which was also the period that saw some of the highest exports from Ireland.[137] This makes sense, as higher supply likely drove prices down. In this respect, it is encouraging that the trade and price data correspond.

To estimate the total value of herring exported from Ireland, all the various units used to carry the fish were first converted to the equivalent in barrels. The number of barrels per year was then multiplied by the price of a herring barrel in Rogers' series for the corresponding year. If a price was not available for that year, the price from the closest year was used. Figure 3.23 shows the value of one barrel of white herring per year (black line) and the total value of imports from Ireland for every year with import data. Only 1404/5 was left off the graph; in those years a barrel of white herring cost around £0.55 and £157 worth of Irish herring was imported. Overall, the price of a barrel of white herring remained relatively stable until the late 1550s, when the debasement of the pound caused rapid inflation.[138] However, the debasement does not significantly impact the total value of Irish herring imports across the century, as the largest herring imports took place before its occurrence.

The estimated total value of Irish white herring imports in the fifteenth century was £6,735, while £15,969 worth of fish was imported from Ireland in the sixteenth. Across both centuries the total was £22,704.[139] The accounting year with the highest single value overall was 1525/6, when £5,638 worth of white herring was imported. These value figures based on market prices are double the value determined using customs values alone. However, market value figures are far from perfect. They do not reflect the total economic value of all fish imports, only covering white herring for a limited number of ports and years. If we included every fish product and every export destination, the figure would be far higher. However, they are the best figures we can achieve and illustrate the inadequacy of the customs rates for calculating the real economic value of the

[136] Gregory Clark, 'The Price History of English Agriculture, 1209–1914', *Research in Economic History*, 22 (2003), pp. 41–123; 'The Long March of History: Farm Wages, Population, and Economic Growth, England 1209–1869', *Economic History Review*, 60 (2007), pp. 97–135; Clark's data is ranked as 'poor' in Robert C. Allen and Richard W. Unger (eds), 'Allen-Unger Global Commodity Prices Dataset Introduction' (https://iisg.nl/hpw/allen-unger_about.pdf) (accessed 28 January 2023).

[137] Rogers himself comments on the lower prices in this period: Rogers, *A History of Agriculture and Prices*, vol. IV, p. 527.

[138] C. E. Challis, 'The Debasement of the Coinage, 1542–1551', *Economic History Review*, 20 (1967), pp. 441–66; Y. S. Brenner, 'The Inflation of Prices in Early Sixteenth Century England', *Economic History Review*, 14 (1961), pp. 225–39.

[139] All figures have been rounded to the nearest pound. Based on a pound with 240 pence.

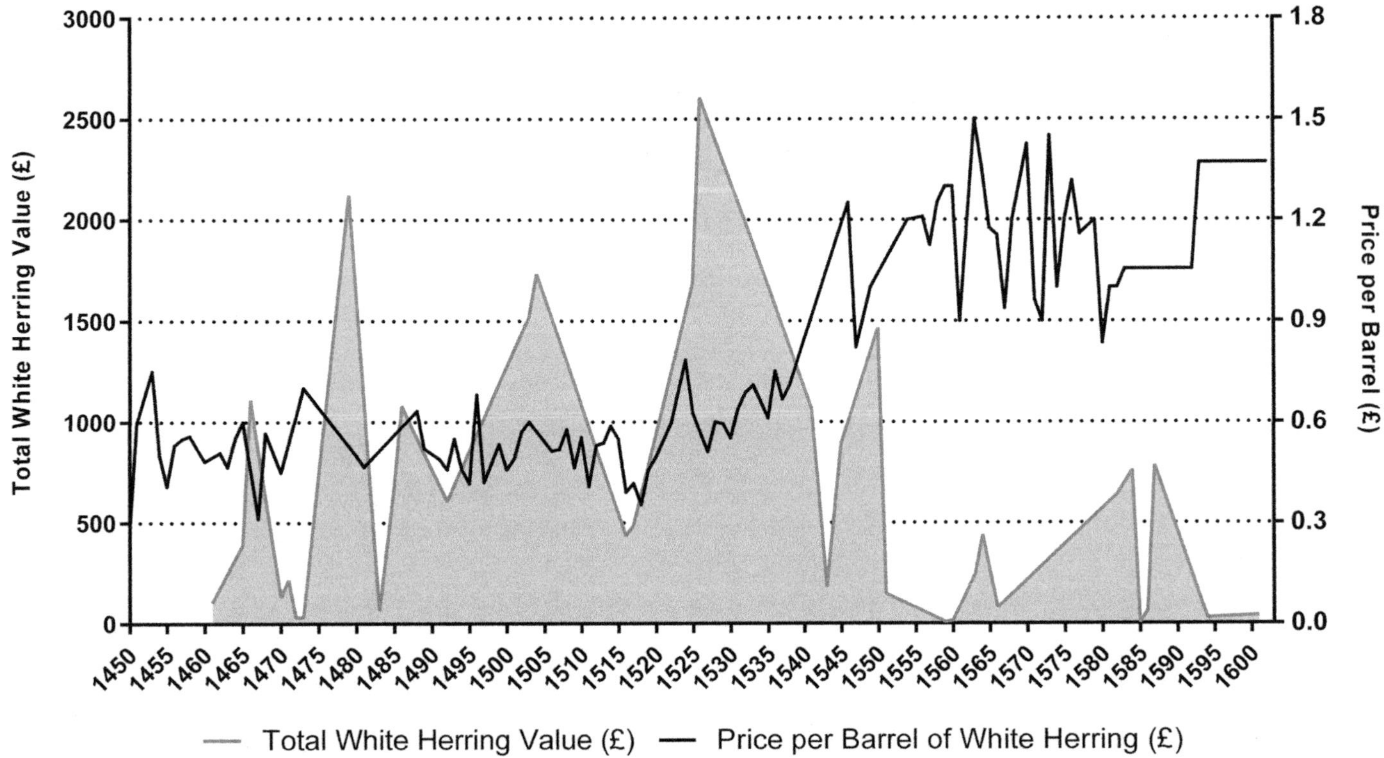

Figure 3.23. Market price and total value of white herring imports to England, 1450–1603.

fisheries. It is also difficult to compare the importance of this trade to others during the sixteenth century because of the lack of available figures for trade not dependent on customs charges. The inflation after 1558 also makes price comparison challenging.[140]

Since many of the ships carrying fish from Ireland were not Irish owned, it is difficult to determine what portion of revenue from the fish trade went back into Ireland. The question surrounding the ownership of Irish fisheries has been approached by other historians, often with disdain at the degree to which Ireland's medieval and early modern fisheries were in the hands of 'foreigners'.[141] However, the question of who profited and who controlled the trade is not simple to answer. As we have seen, a large portion of the fish trade from Ireland was carried out by vessels not based in Ireland. Unsurprisingly, English and Welsh boats were heavily involved in the transport of fish to England and Wales. This does not mean that resident Irish fishers, wholesalers, or merchants did not profit from the trade at different stages in the supply chain.[142] Furthermore, we know that many Irish families did not carry fish to market themselves or use currency but instead received goods in exchange for allowing fishing activity to take place in their waters. Ultimately, the complexity of the fishing industry in sixteenth-century Ireland makes it difficult to determine who profited most and what form that profit took. Nonetheless, we can certainly conclude that many coastal communities in Ireland would have benefited from the fish trade and would have been badly affected by its collapse.

While there is limited information on fish prices from England, there is no comparable price series for fish from sixteenth-century Ireland. Some historians have commented on the stability of Irish prices based on the scant evidence available, such as the corporation book of the Irishtown of Kilkenny.[143] However, that particular source does not assist the detailed study of fish prices as it only contains prices for salmon, which show no change at all. For this study, every mention of fish prices in a variety of qualitative sources was collected and collated in an attempt to construct a price series for fish products in Ireland.[144] Due to limited sources on the fifteenth century, prices were only generated for the sixteenth and early seventeenth.

[140] Raymond Gillespie, *The Transformation of the Irish Economy 1550–1700* (Dublin, 1991), p. 7.

[141] John de Courcy, *Ireland's Sea Fisheries: A History* (Dublin, 1981), p. 14; Lyons, 'Maritime Relations', p. 4.

[142] In this instance 'Irish' does not refer to the Gaelic Irish, but any residents of Ireland.

[143] Gillespie, *Transformation of the Irish Economy*, p. 9; John Ainsworth (ed.), 'Corporation Book of the Irishtown of Kilkenny, 1537–1628', *Analecta Hibernica*, 28 (1978), pp. 1, 3–78.

[144] See Appendix 1 for more details on the sources and methods used in this endeavour.

This exercise produced forty-two records relating to fish prices, twenty-one of which related to sixteenth-century herring prices, ranging from 1552 to 1602.[145] Most data points were collected from the Irish State Papers and Admiralty court records and because herring were the only extensive series available, they are the focus here.

Several issues limit the reliability of the series. First, the prices deal with different units, mostly either barrels or individual pieces of fish. To overcome this, all the price points were converted to a standard value of pounds per barrel.[146] Second, not all the prices are from the same area, with some from Dublin, Waterford, and Wexford, and others relating to Ireland as a whole. This regional variation is not ideal in terms of consistency, although it illustrates interesting regional differences. The most serious problem is that most of these price points are not market values but prices described within the English army purveyance records. The issue here is that the English army could purvey or purchase goods at rates below market value. Many aspects of the purveyance system were informal, and the rate of discount received was not fixed;[147] otherwise, we could simply increase the army prices by a fixed amount to reflect actual market value.

The same system of purveyance was used in England, and sellers there reported that purveyors often purchased goods at less than half the market value. It was also sometimes stated that purveyors would record a higher price in their records than actually secured, allowing them to pocket the difference. The informality of the system and abuses by some purveyors rendered the system very unpopular in England.[148] In Ireland, purveyance was part of a broader system of 'cess', or taxes, that supported the English army in Ireland and was also widely unpopular. The citizens of the Pale, in particular, felt that purveyance put an undue strain on them. Besides, soldiers' pay was often held up, meaning that they could not purchase supplies legitimately and would often resort to confiscating food by force.[149] The impact of confiscations on industries like fishing and agriculture is hard to quantify as this activity went unrecorded most of the time. However, it is crucial to keep these practices in mind. The increasing numbers of English troops in Ireland later in the sixteenth century put further pressure on food producers.

[145] The search for this analysis encompassed the period from 1500 to 1603, but no price points before 1552 were discovered.

[146] Pieces of fish were converted to a barrel price using the rate of 900 fish per barrel (see Wallop to Burghley, 13 January 1583 (TNA, SP 63/99, f.40)).

[147] C. S. L. Davies, 'Provisions for Armies, 1509–50; A Study in the Effectiveness of Early Tudor Goverment', *Economic History Review*, 16 (1964), pp. 234–48, at p. 237.

[148] Davies, 'Provisions for Armies', pp. 237–38.

[149] Steven G. Ellis, 'Thomas Cromwell and Ireland, 1532–1540', *Historical Journal*, 23 (1980), pp. 497–519, at p. 511.

Being forced to sell products far below market value would have adversely affected both fishers and merchants.

While the rate at which the army purchased goods was not fixed by law, a document in the Irish State Papers from 1597 describes the differences between the current market value and the value at which the army purchased supplies.[150] It describes the price of fourteen different commodities, including wheat, biscuit, beer, and herring. The market price is given first, and then the 'rates that the soldiers do answer for'. The percentage decrease between the market value and purveyance value differs between commodities.[151] In the case of wheat, the army price was 44 per cent lower than the market value, while in the case of beer, the difference was only 12 per cent. For a barrel of herring, the difference was 33 per cent, with the market value of one barrel being 28*s.* 8*d.*, and the army purchasing price 20*s.*[152] There is no way to establish if this difference in price applied to all army herring purchases in the sixteenth century.

Figure 3.24 shows assorted prices of herring barrels in Ireland per year, and the equivalent English prices – all of which have been converted to pre-decimal pounds.[153] In years when multiple Irish prices were available, the average price from that year was used. The lowest Irish price in the series is £0.55, and the highest is £1.17. This relatively narrow range strengthens the reliability of the data. Had prices fluctuated more widely, we would be forced to question the accuracy of the figures. There was one further price in Ireland from 1552 that is not displayed in Figure 3.24. In that year, lord deputy Croft reported 'the excessive prices of all things in Ireland', reporting that a measure of corn previously worth 3–6 shillings now cost 30, while six herring cost 4*d.*[154] If there were 900 herring per barrel, a whole barrel would have cost £2.50, more than twice the next highest price in the entire century. This would suggest that Ireland did not entirely escape the hyperinflation that impacted England over the same period. Because this 1552 price was so abnormally high, it was not included in the graph.

The Irish data indicates that herring prices decreased throughout the 1570s to early 1580s before peaking toward the end of the century, which could be related to a lack of supply. Since the data discussed in this chapter suggests a severe decrease in production from Irish fishers by the close of the century, in this instance the price rises make sense. Additionally, the arrival of new commodities in the Irish market, such as cod from Newfoundland, could have influenced the price of herring, as demand for domestic fish fell in the late sixteenth century. At

[150] The prices of victuals in Ireland, 4 February 1596–97 (TNA, SP 63/197, f.237).

[151] This is an incredibly detailed record and as well as different commodities the prices for different units are described too.

[152] The prices of victuals in Ireland, 4 February 1596–97 (TNA, SP 63/197, f.237).

[153] With 240 pence each.

[154] Lord deputy Croft to the marquis of Winchester, 22 March 1552 (TNA, SP 61/4, f.87).

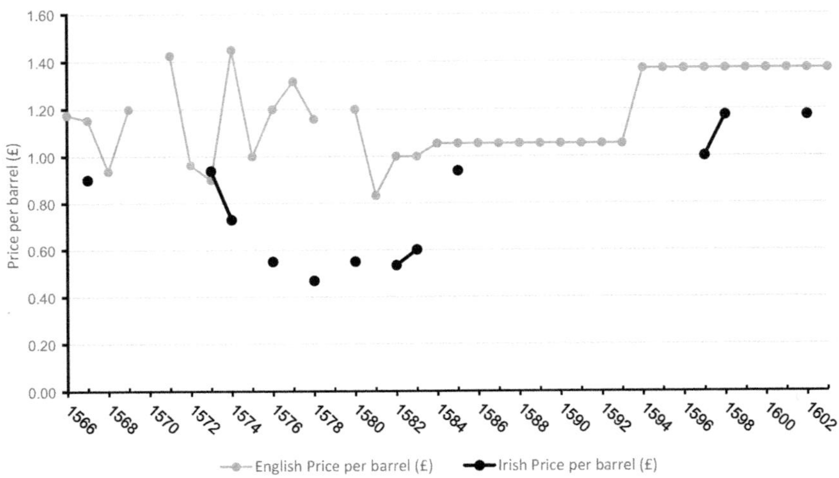

Figure 3.24. Irish and English herring prices compared, 1566–1603.

least in part, Newfoundland fish gradually replaced imports from Ireland in the English and Welsh markets, and the same was true to some extent in the Irish market. However, we must be careful about drawing too many conclusions from such limited data, especially considering that many parts of Ireland were not using cash but exchanging fish for other commodities. These prices best reflect the value of fish in the English parts of Ireland on the east and south coasts.

Even within this limited area, there were regional differences. The average cost of a herring barrel in Dublin across the whole series was £0.70, while in Waterford, it was £0.50, and in Wexford, £0.45. As the trade data showed, Waterford and Wexford were significant exporters and producers of herring. Dublin did have local fisheries, but these do not seem to have been as productive. We would expect prices to be lower in towns producing more fish, which is what the data indicates. A deposition in the Admiralty court records from 1577 highlights that quality impacted price. The deponent, William White, stated that one barrel of herring was worth 10*s*. in Waterford in that year, and 'the best sort of herring was worth not more than 12s'.[155] Clearly the production of herring in Ireland was not uniform, and quality did influence price.

Irish and English prices are not entirely comparable, as there was a difference in the inherent value of the currency. An expenses list for the English army in 1577 notes '£5,000 sterling makes £6,666 13*s*. 4*d*. Irish', meaning the exchange rate was £1.33 at that time.[156] While we do not have sufficient data to calculate the value

[155] *Cal. HCA 1536–1641*, no. 209, p. 41.
[156] *CSPI* (IMC), 1571–75, no. 851.2, p. 522.

difference across the whole series, comparison is still a useful exercise in relation to overall trends. Irish and English prices were similar at the start of the period, but diverged in the 1570s as Irish prices decreased and English prices remained stable. Possibly the turmoil in Munster during the Desmond rebellions of the 1570s shocked the herring market and caused this divergence in price.[157] Toward the end of the century, there were similar increases across both price series.

It is encouraging that both series show relatively similar trends across the period, confirming that the collection of scattered price points from a selection of sources is a legitimate method to generate a limited price series. In contrast, in his work on prices from the corporation book of the Irishtown of Kilkenny, Raymond Gillespie has written that Irish prices showed 'little movement during the late sixteenth century'.[158] The data collected here reveals that Irish herring prices did experience movement in the late sixteenth century and at a similar level to English prices. Even allowing for the lower value of the Irish currency, Irish prices were still lower in real terms than those in England, particularly in the 1570s and 1580s. This may have meant that it was in merchants' and fishers' interests to sell their products in England to get better prices, though we have no evidence to prove this.

Wendy Childs observed that 'there has been rather too pessimistic an assessment of the available quantitative evidence' concerning Irish trade.[159] This chapter has shown that there is in fact sufficient quantitative data to trace the significant ups and downs of Ireland's export of fish from the fifteenth to seventeenth centuries. While there are substantial gaps in the sources, the data used here captures the overall pattern of trade to some of Ireland's key partners. The strongest evidence comes from Ireland's trade with England, the principal importer of Irish fish, while the extent of the fish trade with Scotland and continental Europe remains mostly unknown.

As England was the primary importer, the available data still constitutes an excellent guide to the overall fluctuations in Ireland's fish exports. It shows that the fish export trade peaked in the late fifteenth and early sixteenth centuries, began to decline from the middle of the sixteenth onwards, and virtually ceased by 1603. Price data from England and Ireland further supports this overall trend. The early seventeenth century saw a resurgence in fish exports, but fish would never be Ireland's premier export commodity again. In the fifteenth and sixteenth centuries, fish exports were a vital part of the Irish economy and represented a significant total monetary value. The true value of the fisheries cannot

157 My thanks to Al Mathews from the NorFish team for his advice on the impact of conflict on fish prices. For more on this topic, see Al Matthews, 'European Fish Market Dynamics from 1500 to 1800' (PhD thesis, TCD, 2021).

158 Gillespie, *Transformation of the Irish Economy*, p. 9.

159 Childs, 'Ireland's Trade with England', p. 1.

be measured in money alone since fish was also exchanged for goods along the Irish coast. In these cases, the value of fish lay not only in the goods for which it was bartered, but in the relationships built with fishers and traders from across Europe. While the existing trade data cannot capture all those interactions, it demonstrates just how vital the industry was.

Economic historians have not fully grasped the significance of fish to the Irish economy in the early modern period. Fishing was an essential economic activity, and understanding the fisheries and the associated trade is necessary to answer broader questions about Ireland's medieval and early modern economy.

4

The Impact of War and Piracy

I wish that Ireland were a fishpool, for this age shall never see it settled in peace ...
George Carew to Robert Cecil, 28 June 1602 (*Cal. Car.* MSS, 1601–03, p. 258)

The Impact of War

In the late fifteenth century Ireland was a border region of English authority that saw a constant state of low-intensity conflict between and within the English and Gaelic Irish. Despite this, no large-scale wars occurred in Ireland during the fifteenth or early sixteenth centuries until the Kildare rebellion. One of the primary reasons was the tempering influence of the FitzGeralds, earls of Kildare from the 1460s onwards, who largely maintained stability by projecting their power into the Irish and English spheres of Irish politics. The earls of Kildare maintained a balance between the competing factions in Ireland, functioning as enforcers and mediators.[1] For this reason, before the 1530s there was no substantial alteration in the ownership or rights to fishing resources. The control of coastlines and fisheries remained stable. This period of relative peace allowed fisheries to grow and prosper, fish to become Ireland's most important export commodity, and foreign fishing and trade to flourish.

However, from 1534 onwards, this situation changed. In June of that year, Thomas FitzGerald, later the 10th earl of Kildare, popularly known as Silken Thomas, entered into open rebellion against Henry VIII. His revolt marked the end of the Kildare ascendancy and sparked the Tudor 'reconquest', culminating with the English taking complete control of the island in 1603. That reconquest was not a cohesive military campaign but rather the slow breakdown of the relationship between the English government and Ireland's native inhabitants.[2] The sixty-nine years from 1534 to 1603 witnessed successive conflicts and increasing violence on land and sea.[3] These conflicts are important to the story of fisheries; they had a severely disruptive and damaging impact on the trajectory of the fishing industry in Ireland.

1 *CHI vol. 1*, pp. 305–06.

2 Ciaran Brady, *The Chief Governors: The Rise and Fall of Reform Government in Tudor Ireland 1536–1588* (Cambridge, 1994), p. 1.

3 *NHI, vol. III*, p. 4.

It is not necessary or possible to discuss every conflict in the sixteenth century within this chapter, but the most significant conflicts regarding their impact on fisheries will be explored, how fisheries factored into English colonial policy and how English attitudes towards the exploitation of natural resources changed over the course of the century.

The defeat of the rebellion of 1534 spelled disaster for the FitzGeralds of Kildare. One of the immediate consequences was the confiscation and redistribution of their property by the English Crown. This included several valuable fishing towns, among them Ardglass, Strangford and Carlingford. While the Crown took control of all these towns, we know that the towns did not fare well during the war or its aftermath. Carlingford generated no income in 1534, one letter noting that the town 'remained waste during that time and no ship loaded or discharged at the port by reason of the war'.[4] The situation in Carlingford improved somewhat in 1535 when 600 English fishing boats were reported in the town. However, we only know about these fishers because William Brabazon reported that they had offered themselves up to fight with his forces if necessary.[5] This instance was part of the longer-term militarisation of Carlingford by the English, a trend that ultimately drove the fishing fleets away.[6]

Establishing the broader extent of damage inflicted on fisheries during and immediately after the 1534 rebellion is difficult due to a lack of detailed source material. In the broader context, the rebellion motivated Henry VIII to begin a series of reforms in Ireland, including making himself head of the Irish church and declaring himself king of Ireland in 1541.[7] In addition, Henry and his government began a campaign to dissolve Irish monastic houses.[8] A survey was conducted to facilitate this dissolution, from which we know that Irish religious houses controlled many fishing sites, the most common being salmon and eel weirs.[9] While these weirs were not damaged during the dissolution, their confiscation and redistribution must have caused disruption to the normal operations of the fisheries. We do not know how many weirs had been in monastic hands, but the trade data discussed shows a dramatic fall in salmon exports in 1542–43, suggesting that the dissolution did have an impact on catches.

4 D. B. Quinn, 'Anglo-Irish in Ulster in the Early Sixteenth Century', in Arthur Deane (ed.), *Proceedings and Reports of the Belfast Natural History and Philosophical Society* (Belfast, 1935), pp. 56–78, at p. 65.

5 *Cal. Car. MSS*, 1515–74, p. 85, no. 70.

6 Anngret Simms et al., 'Carlingford', *IHTA*, 23 (2011), pp. 1–15, at p. 3.

7 *CHI vol. 1*, p. 321.

8 Brendan Bradshaw, *The Dissolution of the Religious Orders in Ireland under Henry VIII* (Cambridge, 1974).

9 Newport Benjamin White, *Extents of Irish Monastic Possessions, 1540–1541, from Manuscripts in the Public Record Office* (London and Dublin, 1943).

There are not as many instances of monasteries controlling sea fisheries, but there are some examples. In the early 1540s, the Dominican monastery at Carlingford was dissolved, and the monastery's church used as 'accommodation of those who resort to the place in large numbers with the fleet of ships every year to catch herrings'.[10] Another survey of Crown lands taken from 1540 to 1541 reported that Carlingford's fishery was 'necessary in order for the defence of the country'.[11] The survey indicated that the town's castle was still in a state of disrepair, and the Gaelic Irish O'Hanlons and Magennis were threatening the inhabitants. The O'Hanlons were vassals of O'Neill but had previously had a close relationship with the earls of Kildare.[12] While fishing was still taking place in Carlingford, the town was now an isolated English military outpost, surrounded by hostile Irish. The power of Kildare had previously protected it, but those old relationships had been removed and replaced with suspicious and hostile attitudes between the English and Irish.

In the realm of ideological battles, from the 1540s onward, there was increasing interest among English politicians, writers, and adventurers regarding the reforming and settling of Ireland. The dissolution of the monasteries formed part of those reforms, but it was only the beginning. Plans were circulated for improving and settling Ireland before 1534, and after the rebellion, the number of such schemes increased markedly, with fisheries often forming a part.[13] In 1541 Robert Crowley's treatise 'for the reformation of Ireland' argued that Ireland should be shired in the fashion of Wales and England, which would make the Irish gallowglass and kerne 'forsake idleness, and take to tillage, mining and fishing'.[14] Tracts like these formed part of a broader cultural justification for plantation taken up by English writers and thinkers who espoused the same view that the Gaelic Irish were a barbarous and 'primitive' people that required civilising.[15]

The English desire to reform Ireland was partly driven by the presence and, as the English saw it, the potentially seditious influence, of foreign fishers on Irish shores. Visiting fishers were an essential part of the maritime economy in the early sixteenth century, and many Gaelic Irish and Old English areas had complex systems to tax and facilitate these operations. English fishers regularly visited Ireland too and largely in peace, but the English authorities were always

[10] White, *Extents*, p. 246.

[11] Translated from the original Latin; Gearóid Mac Niocaill (ed.), *Crown Survey of Lands 1540–41* (Dublin, 1992), pp. 75–76.

[12] *NHI, vol. III*, p. 16.

[13] For example, *CSPI* (IMC), 1509–47, no. 8, p. 2, no. 79, p. 31 and no. 80, p. 34.

[14] *CSPI* (IMC), 1509–47, no. 587.1, p. 325.

[15] This subject has a very extensive literature, a few examples of which include *CHI vol. 2*, pp. 559–61 and pp. 506–28; William Palmer, 'The Problem of Ideology in the Elizabethan Conquest of Ireland: A Comparative Approach?', *New Hibernia Review*, 19 (2015), pp. 128–42.

wary of the interaction between continental fishers and the Gaelic Irish. From at least 1465 attempts had been made to introduce licences to regulate foreign vessels.[16] Concern with these foreign fishers became even more significant after 1534. The following year, the Irish Parliament proposed an Act that stated, 'the fishing of Ireland is a great commodity which strangers haunt and carry away into Spain, France and Scotland'. The Act's articles specified that no ship could leave the country without first paying customs duties.[17]

There were several reasons why the English should have wanted to limit and control profits from foreign fishers in Ireland. The rebellion in 1534 had shown the English they were vulnerable in Ireland and increased their mistrust of not only the Gaelic Irish but also the Old English. The earl of Kildare himself had previously controlled several foreign fishing locations, so legislation like this would limit the profits other potential English enemies could gain from fisheries. Again in 1541, the reforming Parliament in Dublin passed legislation that ordered foreign fishers to pay tithes of fish in the parishes in which they made catches.[18] At this point, the main aim of the Dublin legislators was to capture lost revenue from the Crown and further deny it to the Irish, but there was as yet no talk of forcefully conquering territory for the purpose of exploiting the fisheries. That would soon change.

Conflict with other European powers also motivated the desire to limit foreign fishing operations in Ireland. In 1543 lord deputy St Leger wrote a letter to King Henry, which stated that 'There has long been a bruit of war with the Scots and Frenchmen', which made it imperative to protect Ireland against invasion.[19] In response to the threat of war, St Leger dispatched a ship to O'Donnell territory to report on any French vessels there. There was clearly a fear among the English authorities of a French-sponsored invasion of Ireland to restore the earls of Kildare to power. St Leger went on to say that the Irish in Munster were receiving supplies and arms by means of the trade of hides with foreign merchants. He did not explicitly say that foreign fishers were bringing in arms but informed the king that possession of the fisheries at Baltimore would be beneficial to the Crown, although this could not happen 'without some war on those who possess it'.[20]

The French invasion never materialised, but in January 1551, specific plans for capturing Gaelic fisheries at Baltimore and Berehaven were proposed.[21] The idea was conceived with the threat of a French or Scottish invasion in mind, but

16 *CSPI* (IMC), 1566–67, no. 117, p. 62.

17 *CSPI* (IMC), 1509–47, no. 119.3, p. 56.

18 *Cal. Car. MSS*, 1515–74, no. 157, p. 182.

19 *CSPI* (IMC), 1509–47, no. 627, pp. 349–50.

20 *CSPI* (IMC), 1509–47, no. 627, pp. 349–50. At that time the O'Driscolls were lords of Baltimore and vassals of the MacCarthy Reagh.

21 *CSPI* (IMC), 1547–53, no. 250, p. 99.

was also clearly motivated by greed. A scheme, penned by an unknown author, noted that Baltimore and Berehaven were the principal harbours in the south-west of Ireland, and proposed that six ships with mariners, soldiers, and settlers should reach the towns in April that year and take control. In order to exploit the fisheries there, 'fishermen with fishing nets should be set up'.[22] Settlers should be recruited from Cornwall and Devon, an obvious suggestion given the well-established relationship between West Country fishers and Ireland.[23]

The author concluded by stating that the lord deputy should agree with the earl of Desmond and MacCarthy Mór that their ports should come under royal control 'to prevent the Spanish plundering the fisheries of the King's subjects there and in England'.[24] Letters from May 1551 mention several times that the Spanish were 'mistreating the royal subjects at Baltimore'.[25] By August, King Edward VI wrote to deputy Croft to admonish him for not having finished the fortifications at Baltimore. Croft was instructed 'to have the site ready for trench-digging and building before the Spaniards arrive for the next fishing season'.[26] It is clear that the English were as much concerned with disrupting Spanish operations as they were with protecting their own fishing interests.[27] The scheme also included similar strategies for capturing Olderfleet and the Bann 'where there is much fishing'.[28] In all cases, the author hoped that the fishing at these locations would help fund the troops and forts necessary to keep hold of them.

This memorandum highlights the change in the English attitude toward Gaelic Irish fisheries. The relationships between the English, Gaelic Irish and Old English that had facilitated foreign fishing operations were breaking down from the middle of the sixteenth century. Further evidence comes from the attack on Baltimore by the Waterford fleet in 1537, in which they destroyed the O'Driscoll castle and wrecked the clan's fleet of boats.[29] Baltimore and Kinsale had jointly elected an admiral to administer the fisheries earlier in the century. By 1537 Waterford had sacked Baltimore, and in 1551 the English were seriously considering a full invasion.

This breakdown of old relationships accelerated after Elizabeth I ascended to the English throne in 1558. Most of Ireland was still at peace in 1558, but the

[22] *CSPI* (IMC), 1547–53, no. 250, p. 99.

[23] See too M. Kowaleski, 'The Expansion of the South-Western Fisheries in Late Medieval England', *Economic History Review*, 53 (2000), pp. 429–54.

[24] *CSPI* (IMC), 1547–53, no. 250, p. 99.

[25] *CSPI* (IMC), 1547–53, no. 271, p. 108 and no. 275, p. 111.

[26] *CSPI* (IMC), 1547–53, no. 296, p. 126.

[27] *CSPI* (IMC), 1547–53, no. 286, p. 118.

[28] *CSPI* (IMC), 1547–53, no. 250, p. 99.

[29] This raid was discussed in more detail in Chapter 2; *CSPI* (IMC), 1509–47, no. 627, pp. 349–50.

question of what should ultimately be done with Ireland remained an open one within the English court. Of all parts of Ireland, Ulster was the most problematic.[30] In Ulster, Shane O'Neill stood as an independent, powerful and locally ambitious Irish lord, who also took an active role in Scottish politics, and who was suspected by the English of working with the French.[31] The English Crown and ministers saw Ulster as a potential avenue for their enemies to interfere in Ireland, something that warranted action. A messy war was fought in Ulster in 1561 that ultimately ended in failure. Other conflicts followed and efforts to end the dispute between Shane O'Neill and the Crown came to nought. The saga eventually ended following O'Neill's defeat in May 1567, not by the English but by the O'Donnells. Soon afterwards, Shane fled to the Scots at Cushendun and was murdered by them in June 1567.[32] The events of the 1560s would have caused considerable immediate disruption to fisheries, but their longer-term importance lies in the events they set in motion. In subsequent decades, the English would seek to mould Ulster in their own image through the plantation of settlers, which would ultimately transform the nature of fisheries in the province.

While the plantations in Ulster would become the largest of their kind in Ireland, they were not the first. Already in the 1550s schemes were unfolding in Leix and Offaly.[33] However, these had little impact on fisheries as they were both inland counties and had no major fishing rivers. Faced with more financial constraints than her predecessors, Elizabeth turned to private enterprises to establish new plantations in Ulster. Under these new circumstances, profit became a key motivation and goal for plantations in Ireland, and fisheries were often a vital part of the bottom line.[34]

Many of these new plantation schemes focused on the river Bann, which had long been the site of conflict between rival Gaelic and Scottish clans. In 1566 Valentine Browne and merchants from Bristol were competing for the right to plant settlers in Ulster and take over 'the Bann fisheries and other waters'.[35] Lord deputy Sidney favoured Browne but these proposals were never implemented.[36] In July 1567 Sir Francis Knolly gave his opinion that English 'artisans and sea fishers' should be planted on the Bann and at Strangford, 'so that after they may

[30] *NHI, vol. III*, p. 83.
[31] *NHI, vol. III*, p. 80.
[32] *NHI, vol. III*, p. 86.
[33] *NHI, vol. III*, p. 78.
[34] *NHI, vol. III*, p. 95.
[35] *CSPI* (IMC), 1566–67, no. 292, p. 132.
[36] A letter in April 1567 gave Sidney permission to lease the fishery: see *CSPI* (IMC), 1566–67, no. 394, p. 169; and in June 1567 it was reported two offers had been made for the fishery: no. 444, p. 188.

grow to be haven towns'.[37] In the same month Sidney wrote that forts should be constructed in 'Coleraine, Castle Toome, Massereene, Newry and Strangford' and 'those who receive the fishing of the Bann are to maintain these forts'.[38] This plan was very similar to those proposed for Baltimore and Berehaven; the fisheries would fund the defences required to hold the newly acquired land.

Later, in July 1568, Captain George Thorton informed William Cecil of the capture of a pirate near the Isle of Man. At the end of the letter, he reminded Cecil about his application to be granted Islandmagee on the Antrim coast. He had 'already fortified the harbour and plans to have a fishing town there inhabited by Englishmen, ready to serve the queen at sea'.[39] In the 1570s further schemes were proposed to plant Ulster's east coast; these also listed coastal fisheries and the fisheries on Lough Neagh as potential income sources.[40]

In the end, two major plantation schemes went ahead in Ulster, the first led by Thomas Smith, who secured a grant in 1571 for a plantation around the Ards and Clandeboye on the coast of Belfast Lough.[41] In his promotional pamphlet, *A letter sent by I.B.*, Smith outlined the grounds for the scheme. He praised the Irish countryside as 'full of excellent fishe and foule'.[42] We should not read too much into this description, as Smith applauded all aspects of the Irish environment and did not specify what fisheries he intended to exploit. However, fisheries were certainly part of the plan: during the first year, fishing boats would be brought 'for our better victualing', in other words, to keep the plantation fed.[43] In letters to his son, Smith outlined how he thought the plantation would turn a profit. Naturally, a large portion of the revenue would come from rents, but Smith also hoped that a great deal would also be generated by exploiting forests and letting out the fisheries.[44]

Walter Devereux, second earl of Essex, spearheaded a second, much larger, plantation from 1573 onwards. It included the lands of Clandeboye in County Down and large parts of County Antrim. Essex's grant included the 'moiety of the fishing of the Bann and free fishing in Lough Neagh'.[45] Soon after arriving, Essex requested 'fishermen and their boats to be sent hither for fishing' to help

[37] *CSPI* (IMC), 1566–67, no. 490, p. 202.

[38] *CSPI* (IMC), 1547–53, no. 483, p. 483.

[39] *CSPI* (IMC), 1568–71, no. 149, p. 67.

[40] *CSPI* (IMC), 1568–71, no. 553, p. 218.

[41] Hiram Morgan, 'The Colonial Venture of Sir Thomas Smith in Ulster, 1571–75', *Historical Journal*, 28 (1985), pp. 261–78, at p. 263.

[42] Sir Thomas Smith, 'A Letter sent by I. B. Gentleman vnto his very frende Mayster R. C. Esquire, 1572', ed. Beatrix Färber (*CELT*, 2019), p. 409.

[43] Smith, 'A Letter sent by I. B. Gentleman', p. 412.

[44] D. B. Quinn, 'Sir Thomas Smith (1513–1577) and the Beginnings of English Colonial Theory', *Proceedings of the American Philosophical Society*, 89 (1945), pp. 543–60, at pp. 556–57.

[45] *CSPI* (IMC), 1571–75, no. 664, p. 393.

support this plantation.[46] Later in 1573, the queen inquired about the fishery on the Bann and asked who was profiting from it.[47] In response, Sidney said he had let the fishery out for a third of all the fish caught in the river that year. He also told the queen that 'if it prove well' the fishing on the river was worth £2,000 a year.[48] To put that figure in context, in the same year Essex projected that the total revenue from the plantations in Clandeboye and the Ards peninsula would be £5,500 'besides the fishing of the Bann and of the Lough'.[49] So, by Essex's estimate, the fishing on the Bann alone added another 40 per cent to the total revenue of the plantation. Evidently, the income from the fisheries was integral to the financial success of the scheme.

One of the fiercest Irish opponents of these Ulster plantations was Brian McPhelim O'Neill, lord of Clandeboye, who had fought Smith and Essex politically and physically since 1572.[50] In 1573 a sympathetic English priest, Roland Turner, wrote to Brian, declaring that the English 'rob you continuously, they take your livestock without paying, they take your fish for nothing, and they have seized your lands by force of arms and lies'.[51] Brian was undoubtedly aware of this, but Roland's letter does signal that fisheries were part of the economic makeup of Ulster and were being exploited by English planters. In October 1574, Brian was captured by English forces and taken to Dublin where he and his wife were executed. The Irish Annals recounted that this 'was a sufficient cause of hatred and disgust [of the English] to the Irish'.[52]

Despite the removal of opponents like Brian McPhelim O'Neill, both plantation schemes came to nothing, with Smith's project falling apart and Essex only succeeding in establishing a small sphere of control around Carrickfergus.[53] Fishing was taking place around the town in 1578 when it was reported that 'above 40 fishermen daily frequent the sea', but this activity did not rival the scale of fisheries in other parts of Ulster.[54] The important fishery on the Bann even reverted back to Irish control in 1575 when Essex signed an agreement with Turlough O'Neill, Shane O'Neill's successor, by which O'Neill received the customs of the Bann in exchange for allowing English fishers to operate there peacefully.[55]

46 *CSPI* (IMC), 1571–75, no. 756, p. 447.
47 *CSPI* (IMC), 1571–75, no. 790, p. 480.
48 *CSPI* (IMC), 1571–75, no. 930.1, p. 566.
49 *CSPI* (IMC), 1571–75, no. 1174, p. 725.
50 Morgan, 'Colonial Venture of Sir Thomas Smith', p. 271.
51 *CSPI* (IMC), 1571–75, no. 567, p. 342.
52 *NHI, vol. III*, p. 98.
53 *NHI, vol. III*, p. 98; *CHI vol.* 2, p. 567.
54 *Cal. Car. MSS*, 1575–88, no. 114, p. 148.
55 *CSPI* (IMC), 1571–75, no. 1463, p. 877.

Smith and Essex had both relied on the fisheries to bring viability and profit to their plantation schemes, and their failure to control and successfully exploit Ulster's fishing resources contributed to and hastened their demise. These plantations also highlight the value of fisheries for the Gaelic Irish; these were economic assets worth fighting to keep. The Bann's return to Turlough O'Neill demonstrates that when the English could not dominate fisheries, they had to revert to sharing them. The value of the Bann to English fishers was too great to miss out on, and a compromise was preferable to losing access altogether. Ultimately the English planters of the 1570s failed to dominate Ulster, which then returned to relative peace. This situation would not last, however, and the 1590s would see even greater upheaval in the province, with much longer-lasting consequences.

Violence and upheaval were not limited to Ulster. From the mid-century, the province of Munster would again become the arena for violent rebellion and widespread English plantation.[56] Munster was the most frequented province for foreign fishers in Ireland and was home to some of the most bountiful fishing grounds in the country. The province was also the centre of the fish export trade from Ireland and the main source of fish coming into ports like Bristol. For these reasons, any disruption in Munster would have an enormous impact on the whole fishing economy of Ireland and the broader export trade of fish.

Gerald FitzGerald, 15th earl of Desmond, was one of the most powerful lords in Munster in the 1560s. His family held sway over large parts of the province, their territory stretching from Dungarvan in the east to Limerick in the west. The Desmonds also controlled key fishing towns, including Kinsale and Youghal, and effectively dominated Gaelic Irish chiefs like the O'Sullivans. Towns like Youghal may have asserted their independence from the Desmonds but ultimately relied on them for protection.[57] In the 1550s and 1560s the earl of Desmond became embroiled in a bitter dispute with Black Tom Butler, the 10th earl of Ormand, whose family were the traditional enemies of the Desmonds.[58] The dispute climaxed in 1565 when the two sides fought a pitched battle at Affane. Desmond lost and was captured but was not removed from power.[59]

Following this conflict, Sir Henry Sidney marched through Munster in 1567 to restore English authority. On the way, he described the condition of the country. Tipperary was 'depopulated by war and famine'; while Waterford was in good condition but troubled by the followers of Lord Power, a local Gaelicised landlord. Dungarvan was 'a place worthy of repair' and Youghal a 'good town', although 'its trade' was 'much damaged by piracy'. Of Limerick, he wrote that the town

[56] *CHI vol. 2*, p. 567.

[57] *NHI, vol. III*, p. 87.

[58] Anthony M. McCormack, 'Fitzgerald, Gerald fitz James 15th earl of Desmond' (*DIB*, 2009) (https://doi.org/10.3318/dib.003149.v1) (accessed 29 July 2022).

[59] *NHI, vol. III*, p. 87.

was 'much impaired'. Sidney also reported that 'large numbers of Spanish traders and fishermen' visited Munster each year, and their actions 'lose the queen much revenue'.[60] In another letter, he wrote that the Spanish fishing in Munster and Connacht 'decays the city of Cork and the towns of Youghal and Kinsale'.[61]

Additionally, Sidney warned that these continental connections could lead the Spanish king to 'invade and conquer Munster'. The feared invasion would eventually come to pass in 1601 with the Spanish landing at Kinsale, but in 1567 Sidney's claims were little more than scaremongering aimed at increasing support for plantation and reform in Munster. As highlighting the Spanish presence in Munster could only help his cause, he made sure to emphasise lost revenue and make it very clear that the Irish were too well connected to their fellow Catholics in Europe for English comfort.

Of Munster, Sidney concluded that 'the constant wars and disorders have left this part of Ireland more waste than the battlefields of Europe', which speaks to the impact of the fighting already occurring in the province. During the journey, Sidney and Desmond met but could not reach a settlement, and Desmond was subsequently arrested and brought back to Dublin. After eleven weeks, Sidney reported to the queen that 'he found no one in the land sufficient to govern it properly or reform it'.[62] That lack of leadership would create a power vacuum which would lead to even more conflict in Munster. It may have been in Sidney's interest to exaggerate the destruction in the province, as he was attempting to build a case against Desmond, but, even allowing for this, it does seem that Munster was already in a state of general disorder by 1567.

Similar arguments relating to the threat of a French invasion had been used to justify building defences and fishing settlements at Baltimore and Berehaven in 1551.[63] Those proposed settlements had never come about, but following the unrest in the 1560s, there were far more English adventurers and politicians advocating and proposing plantation schemes for Munster. By 1565, the English had already moved to reduce the income Gaelic Irish lords were obtaining from fisheries. In June 1565 Sir Owen O'Sullivan signed a surrender and regrant agreement with the English Crown, in which he promised to give the earl of Clanricarde, an English ally, 2*s.* 6*d.* from every fishing boat that came fishing in O'Sullivan's country, which included Berehaven.[64]

Before any invasions could take place or the foundations of any forts could be laid, English adventurers first turned to the pen to outline their plans and drum

60 *CSPI* (IMC), 1566–67, no. 400, p. 172.
61 *CSPI* (IMC), 1566–67, no. 421, p. 179.
62 *CSPI* (IMC), 1566–67, no. 400, p. 172.
63 *CSPI* (IMC), 1547–53, no. 250, p. 99.
64 *Cal. Car. MSS*, 1515–74, no. 247, p. 366.

up support in the queen's court. Fisheries featured heavily in their arguments to justify and motivate plantations in Munster.

By 1568 the Crown was receiving a flood of proposals for planting Munster,[65] among them a proposal from a group of English adventurers including Humphrey Gilbert and Sir Warham St Leger, a grandson of the former lord deputy. They petitioned the queen for the fishing rights of almost the whole south coast, from New Ross to the Blasket Islands, and also to establish incorporated English towns at Baltimore and Berehaven, and have the right to trade and operate from all the islands along the coast.[66] This area covered some of the most productive fisheries in Ireland, most of which were still controlled by the Gaelic Irish. To sell their scheme to the Crown, Gilbert, St Leger, and their fellow planters wrote that havens which were 'now enjoyed by the Spanish, Frenchmen and other strangers' would be brought under the Crown's control and taxation. Furthermore, they claimed their plantation would stop the flow of weapons and armour from foreign fishers to the Irish and thus help prevent rebellion. The 'abundance of fish' on the Munster coast would now be 'enjoyed by natural Englishmen', and foreign fishers would have to pay for their catch. The fisheries to be established would increase the number of skilled mariners in Ireland, who could join the English navy if required. Plantation would also reduce the number of European pirates operating in Ireland.[67]

Others added their voices to those lobbying for action against the Irish and Spanish in Munster. In a letter to William Cecil in 1569, John Corbine claimed that two Spanish ships would arrive with armaments, armour and coin for the Irish by Easter that year, and that 200 Spanish vessels fished on the coast annually, operating mostly out of Gaelic Irish areas while paying no dues to the queen. He ended by offering his life in the execution of his duty.[68] In 1572, Humphrey Gilbert argued for the 'trade with Spain to be destroyed', and for the subjugation of Munster by force. Both actions would bring 'great profits to English merchants'.[69] He also requested to be made admiral of the seas in Munster, presumably to replace the system of local fishing admirals.

The arguments of these English planters reveal their attitudes toward the system underpinning fisheries in Munster in the sixteenth century. To them, the presence of foreign fishers and traders was an argument to increase support for intervention and plantation. They were sure to emphasise the seditious influence of the Spanish and the danger posed to English order. They also stressed the

65 For example, see *CSPI* (IMC), 1568–71, no. 273, p. 111.

66 *CSPI* (IMC), 1568–71, no. 302, p. 121; *CSPI*, 1509–73, no. 81, p. 399.

67 *CSPI* (IMC), 1568–71, no. 273, p. 112, and no. 327, p. 131.

68 *CSPI* (IMC), 1568–71, no. 362, p. 145.

69 *Cal. Car. MSS*, 1515–74, no. 290, pp. 422–23.

revenue lost each year and the profits that could be had from imposing their own system of taxation and authority over the Munster fisheries.

In 1569 the legal groundwork necessary for the takeover of fisheries in Munster was enacted. The privy council established that the queen held the rights to all lands in County Cork and could grant them at her pleasure. Furthermore, she legally held the rights to all the fisheries between 'Rosscarbery and Blasket Sound', virtually the whole south Munster coast,[70] an area encompassing the Gaelic Irish fisheries at Baltimore and Berehaven. The rights to fisheries in this area were subsequently granted to English petitioners, including Warham St Leger, along with the power to seize the lands of any Irish subjects the Crown deemed rebels.[71]

However, plantation plans in Munster did not proceed without interruption or resistance. In 1569 the Butlers of Ormond rebelled against the encroachment of English planters. They were later joined by other prominent families in Munster, including the Desmonds and McCarthys. G. A. Hayes-McCoy has written that English plans to colonise Munster and take over the province's fisheries were major motivating factors in the decision of the Irish lords to rebel.[72] Religious strife was also a key motivation, but the English plantations and the material loss they represented for the Irish were evidently on the minds of the Butlers and Desmonds when they joined forces against the English. Soon after the rebellion began, the situation in Munster spiralled out of control, and the whole province was engulfed in violence.

In a document written at the beginning of the rebellion in 1574, the earl of Desmond threatened to destroy Kilmallock, which was achieved in 1571. He also made a promise to besiege and capture Cork, which never came to fruition. Interestingly, he also pledged to seize 'seven English ships that are fishing in Baltimore, about Berehaven and Crookhaven', 'when he thinks good to his own possession'.[73] The threat indicates that not only did the English have a stake in the province's fisheries, one they were looking to expand, but the earl was asserting a right to decide if these vessels operated there or not. The war in Munster would drag on for several years, with English expeditions headed by Humphrey Gilbert and Sir John Perrot using brutal tactics to suppress the rebellion. By 1572, James FitzMaurice FitzGerald, who had commanded the rebel forces, gave himself up to the English.[74]

Our understanding of conflict in Ulster and Munster is hampered by the lack of sources from the Irish and Old English perspectives. Much of the violence and suffering inflicted on ordinary people goes unmentioned in the English

70 *CSPI* (IMC), 1568–71, no. 372, p. 150.
71 *CSPI* (IMC), 1568–71, no. 430, p. 170.
72 *NHI, vol. III*, p. 89.
73 *CSPI* (IMC), 1571–75, no. 988, pp. 605–06.
74 *NHI, vol. III*, pp. 91–92.

State Papers. This unknown broader impact on citizens' daily lives would have severely disrupted and damaged the regular economic activities of both provinces. However, in some instances, fisheries were also a saving grace for Munster towns whose terrestrial agriculture was disrupted by war. In 1571 the mayor of Youghal reported to Queen Elizabeth that the townspeople were sheltering inside the walls as their fields were destroyed. 'Our poverty is such that if our accustomed fishing of herring does not answer our expectation this season, few will live in the town for lack of food.'[75] Evidently, in times of war, towns like Youghal relied on their fisheries for subsistence when other sources of food failed.

As an English victory in Munster became apparent, officials reported on the improvement of fisheries in the province. In May 1572, Sir John Perrot told lord deputy Fitzwilliam that 'the English fishermen traffic safely on the west coast, which before they could not'.[76] This letter was part of a renewed propaganda campaign to convince the Crown of the benefits of removing the Irish and Spanish from the Munster fisheries. In 1573 programmes for planting Munster and controlling the province's marine resources were once again being proposed in London.[77] These plans featured all the same arguments put forward before the rebellion began.

During the first Desmond rebellion, the English sought to take more direct control of Ireland's provinces by installing regional presidents. Sir Edward Fitton was installed as president of Connacht in 1569,[78] followed two years later by Sir John Perrot in Munster.[79] However, the presidents were often absent from Ireland, and it was the earl of Desmond who most profited from this absenteeism. Despite his rebellious tendencies, Desmond was allowed to return to Ireland in 1573 on a promise to keep the peace. Soon after his return, he began to reassert his authority in Munster.[80]

In 1576 William Drury replaced Perrot as president and took a far more active and harsher approach to controlling the province. Drury used terror and violence to impose himself, murdering 'malefactors' who took a stand against him and obstructed his ambitions.[81] The ultimate aim of Drury, and Nicholas Malby, the new president of Connacht, was to increase tax revenue in Ireland to support a larger English army on the island. The move to increase taxation upset all facets

75 *CSPI* (IMC), 1571–75, no. 131, p. 75.

76 *CSPI* (IMC), 1571–75, no. 275.3, p. 179.

77 *CSPI* (IMC), 1571–75, no. 506, p. 301, and no. 507, p. 302.

78 Terry Clavin, 'Fitton (Fytton), Sir Edward' (*DIB*, 2009) (https://doi.org/10.3318/dib.003115.v1) (accessed 29 January 2023).

79 *NHI, vol. III*, p. 92.

80 *NHI, vol. III*, pp. 99–100.

81 Benignus Millett, 'A Report from the President of Munster, Sir William Drury, 24 Mar. 1577 (1578 new style)', *Collectanea Hibernica*, 46/47 (2004/2005), pp. 7–15, at p. 7.

of Irish society, from the citizens of the Pale to the Old English and Irish lords in Ulster and Munster.[82] The growing military presence only added to discontent: as more soldiers arrived in Ireland, the practice of 'cess' increased, which saw soldiers purchase goods at a discounted rate.[83] The combined pressure of violence, taxation and cess created the environment for another rebellion.

In 1579 FitzMaurice FitzGerald, the military leader of the first Desmond rebellion, returned to Munster and resumed the conflict he had abandoned years earlier, this time with renewed religious zeal.[84] Many of the earl of Desmond's supporters went over to FitzMaurice, and the earl was faced with a difficult choice: to defeat the rebellion himself, ignore it and allow the English to take the initiative, or lead it.[85] Later that year William Drury died and was replaced by Nicholas Malby, who acted immediately to quell FitzMaurice's rebellion. Malby's actions had the unintended consequence of forcing Desmond's hand, and under pressure, the earl chose to join the rebellion.

In response, the English marked Desmond as a traitor and stripped him of the legal right to his land, meaning that the Crown could confiscate his lands after the rebellion was put down. The authorities in Dublin hoped these confiscations would cover the costs of the war.[86] When put into effect they would have far-reaching consequences for the fisheries in Munster, as would the destruction and famine caused by the conflict.[87]

During the 1570s new arguments were put forward to limit foreign fishing in Munster. In 1575, a proposal was put forward to require foreign fishers to purchase licences.[88] When the second Desmond rebellion broke out, the desire to limit the activities of continental fishers in Munster acquired a new impetus. In a paper from 1580 titled 'A Plot for Munster', Pelham argued that five galleys should be placed in Munster 'to inhibit and withstand fishing of strangers'.[89] Pelham also advocated building coastal forts, to prevent 'superstitious priests and runagate friars' from entering. These forts would also deter the 'Portingalls and Spaniards, that yearly come to fish' and prevent the Irish from 'entertaining' pirates and receiving weapons.[90]

82 *NHI, vol. III*, p. 102.

83 Steven G. Ellis, 'Thomas Cromwell and Ireland, 1532–1540', *Historical Journal*, 23 (1980), pp. 497–519, at p. 511.

84 Ciaran Brady, 'Faction and the Origins of the Desmond Rebellion of 1579', *Irish Historical Studies*, 22 (1980–81), pp. 289–312, p. 309.

85 Brady, 'Origins of the Desmond Rebellion', p. 310.

86 Michael MacCarthy-Morrogh, *The Munster Plantation: English Migration to Southern Ireland 1583–1641* (Oxford, 1986), pp. 3–4.

87 See Brady, 'Origins of the Desmond Rebellion'; and *NHI, vol. III*, pp. 99–105.

88 *CSPI* (IMC), 1571–75, no. 1511, p. 918.

89 *Cal. Car. MSS*, 1575–88, no. 440, p. 285.

90 *Cal. Car. MSS*, 1575–88, no. 440, p. 285.

On campaign in 1580 an English captain called Fulke Greville wrote to Francis Walsingham recounting his travels on the Munster coast, including a stay at Berehaven, where he reported the mistreatment of the 'poor English fishermen' that visited the area yearly. According to Greville, 'many the poore mens great grief' were the charges imposed on them by the Irish. He asked how 'petty lords' could do such things in seas the English rightly owned.[91] The rates imposed on visiting fishers had been in place for centuries, and the arrangement seems to have been beneficial to all parties. But things had changed by the 1580s, and Greville, Pelham and others felt that England had a right to exploit these fisheries without the taxes imposed by the Irish. Moreover, the English now had the strength to act on threats to control the Munster coast.

Some of the most active fisheries in Munster were under the control of local towns, which remained loyal to the Crown during the rebellion and thus became prime targets for the rebels.[92] Some notable fishing ports, such as Waterford, emerged from the war relatively unscathed, but others fared far worse. In 1579 Desmond's forces attacked Youghal and burned many buildings.[93] The English report of this attack does not explicitly mention fisheries, but normal operations must have been disrupted. In the previous rebellion, the town relied on its herring fishery for food security, so the combination of agricultural destruction and disruption of the fisheries must have placed the population in danger of starvation. Only a year later, in 1580, the town was severely damaged again in another attack.[94]

One of the most significant consequences of the fighting was the depopulation of Munster, particularly in the west of the province. Irish warfare was traditionally based around cattle raiding, a tactic which was used heavily during the rebellion. Both sides lost tens of thousands of animals in the conflict, severely impacting food supplies.[95] Malby and other English commanders added to the depredation by adopting a 'scorched-earth' approach, indiscriminately burning fields and crops.[96]

The records of the English army can provide us with a perspective on the lack of supplies in the province. English commanders were forced to import large quantities of food to compensate for the lack of local production. In December 1579, Lord Justice William Pelham reported that 'beef was too dear' in Ireland due to a shortage and that food, including 'butter, cheese and fish', should be

[91] Fulk Greville to Secretary Walsyngham, 2 July 1580 (TNA, SP 63/74, f.4).

[92] Anthony M. McCormack, 'The Social and Economic Consequences of the Desmond Rebellion of 1579–83', *Irish Historical Studies*, 34 (2004–05), pp. 1–15, at p. 5.

[93] *CSPI*, 1574–85, no. 15, p. 196.

[94] McCormack, 'Consequences of the Desmond Rebellion', p. 5.

[95] McCormack, 'Consequences of the Desmond Rebellion', pp. 3–4.

[96] McCormack, 'Consequences of the Desmond Rebellion', p. 4.

sent from England.[97] The impact of war on the general population is clear from another of Pelham's reports, written in May 1580. He wrote that County Kerry had 'no inhabitants at all' and that the authorities should attempt to 'allure to the inhabitants, especially fishers and merchants, to return to their houses'.[98] Perhaps he felt that merchants and fishers had a better chance of operating without disruption and increasing the food supply in the province compared to farmers or overland traders.

The famine continued into 1583 when Edward Stanley, an English captain, reported that there was no food in Munster and that 'the poorest sort hath been driven to eat the dead men's bodies which was cast away in the shipwreck'.[99] This may have been an exaggeration for dramatic effect, but it highlights the desperate nature of the situation.

All the evidence suggests that the Munster fisheries were devastated by the war. Trade to England and the continent was disrupted, while foreign fishers, who had paid dues to the Irish lords, stopped coming to the coast. Anthony McCormack has estimated that one-third of the population in Munster, almost 50,000 people, died from the fighting and the resulting famines and outbreaks of plague.[100] This would represent a significant degree of depopulation and would have impacted massively on all economic activities and caused severe shortages of labour. Furthermore, the destruction of fishing infrastructure and vessels in towns and at sea must have led to a significant drop in production. The war in Munster had a profoundly negative effect on fisheries, which the victorious English would seek to rebuild for their own benefit.

On 11 November 1583 Gerald FitzGerald, 15th earl of Desmond, was murdered in a glen in County Kerry, ending four years of destructive war. There were plans to settle English colonists in Munster long before Desmond's defeat, but only in victory were the English able to enact formal plans for a Munster plantation in June 1586.[101] The confiscated lands of the earl would form the basis of the plantation and in order to establish what lands he had owned, the English commissioned several surveys in the years following the war. Collectively these have been known as the Desmond Survey, but in fact there were two distinct surveys over a decade apart. The first took place in 1584, in order to establish the extent of Desmond's landholding so that estates could be redistributed to English settlers, and is known as the Peyton Survey.[102] This has not survived intact, and

[97] *Cal. Car. MSS*, 1575–88, no. 231, p. 191.

[98] *Cal. Car. MSS*, 1575–88, no. 386, p. 254.

[99] *CSPI*, 1574–85, no. 49, p. 448.

[100] McCormack, 'Consequences of the Desmond Rebellion', p. 13.

[101] MacCarthy-Morrogh, *Munster Plantation*, p. 1.

[102] MacCarthy-Morrogh, *Munster Plantation*, p. 33.

what remains covers most of County Limerick.[103] A second survey was undertaken in 1598, covering the lands of Domhnall MacCarthy Mór, 1st earl of Clancarty, in southern Kerry and parts of Cork.[104] The MacCarthy lands were not confiscated directly after the rebellion because the family put up an effective legal argument against losing them. In 1592 Queen Elizabeth enquired if the 'fishing at Beare Haven, Bantry, Baltimore' belonged to the Crown.[105] The answer was that no record of her owning the land could be found, and it therefore remained in the hands of MacCarthy Mór.[106] However, Elizabeth did not have to wait long to gain ownership, as Domhnall MacCarthy Mór died in 1596, and after a new agreement his lands passed to the Crown.[107]

Map 4.1 shows fishing sites mentioned in both surveys.[108] The sites in north Kerry and County Limerick were all recorded in the Peyton Survey, while the sites in southern Kerry and Cork on the Atlantic coast were found in the Clancarty Survey. These sources tell us that Desmond controlled many river fishing sites, most of which were salmon and eel weirs. He also gained income from sea fishing along the Shannon estuary, but these sites were not as numerous. Part of the reason is that sea fishing had less physical infrastructure than riverine weirs, which made for difficulty in ascribing ownership. Evidently river fishing was a significant source of income for the earl, particularly around his capital of Askeaton, whose castle controlled prominent fishing weirs on the river Deel and had its own fishing ponds to furnish the household with fresh fish.[109] The survey also shows that religious houses controlled several fishing weirs in the earl's territory before the rebellion, which the English also confiscated.[110]

The Gaelic MacCarthys and their dependent families, such as the O'Sullivans, relied more on sea fishing. The Irish in south Kerry did not have access to fertile land or as many rivers as Desmond suitable for weirs. However, they did control some of the best harbours in Munster at Berehaven, Valentia and Sneem, all of which hosted sea fisheries. Moreover, the Gaelic families at Berehaven and elsewhere developed complex systems of charges to profit from their sea fishing grounds.

In 1587 an anonymous English observer noted that in O'Sullivan's country, there was 'no good farmland, but all valleys, cragged rocks and hills, can yield

[103] John A. Murphy (ed.), *The Desmond Survey* (CELT, 2013), p. 33.

[104] Clancarty Survey from Carew MS 625, in Murphy, *The Desmond Survey*, pp. 13–26.

[105] *CSPI*, 1588–92, no. 49, p. 515.

[106] *CSPI*, 1592–96, no. 38, p. 35, no. 8, p. 11.

[107] MacCarthy-Morrogh, *Munster Plantation*, p. 83.

[108] Murphy, *The Desmond Survey*; many of the sites on this map were found using B. J. Hodkinson, *Index of Placenames of the Desmond Survey* (Limerick, 2013).

[109] Murphy, *The Desmond Survey*, p. 156.

[110] See for example, Murphy, *The Desmond Survey*, p. 104.

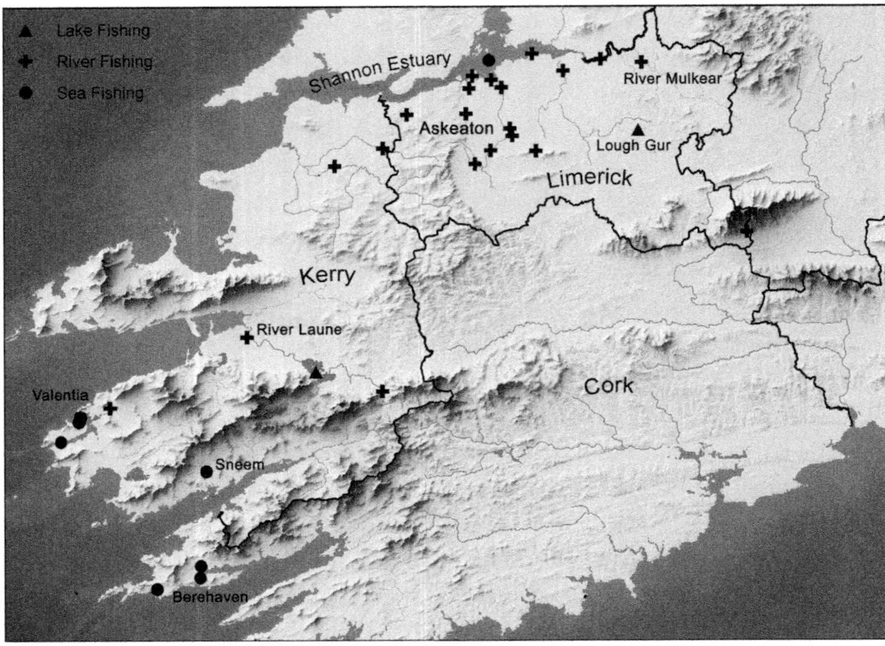

Map 4.1. Fishing sites in the Peyton and Clancarthy surveys, 1584–98.

no great commodity'.[111] Therefore, the clan 'liveth only by the sea, and the commodity thereof, as his fishing'. Again, the fragility of the fisheries is noted: 'fishing is a thing uncertain, for some years if fishing do fall upon the coast, then Dunboyne is worth much; if fishing fail, it cannot yield profit'.[112] This comment could be alluding to the fickle nature of herring shoals, which changed their migration patterns over time, or to the presence of foreign fishing fleets, which as the century progressed, visited the Irish coast less frequently.

These surveys provide a glimpse into the diverse range of fisheries operating in Munster. They also illustrate that geography determined economic structure: coastal Irish lords with inferior land looked to the sea, while inland lords focused on agriculture and also enjoyed access to prosperous river fisheries. However, the surveys were just the first step in the plantation process. The Munster Plantation would take decades to develop fully and never wholly met the aspirations of its original planners.[113]

Ultimately, conflict, and the plantations that followed, would transform fisheries in Munster. The population had been decimated, and many fishing

[111] *CSPI*, 1586–88, no. 9, p. 363.
[112] *CSPI*, 1586–88, no. 9, p. 363.
[113] *NHI, vol. III*, p. 115.

communities and towns had suffered irreparable harm. Trading and fishing relationships that had developed over centuries had been severed, and by the early seventeenth century, all the most prominent Gaelic Irish and many Old English fishing locations were now in English hands. This change in ownership had the most lasting impact, bringing a whole new system of fishing, one that no longer welcomed or relied on visiting fishers.

While local wars had a major impact, Ireland was not immune to the wider geopolitical conflicts of the sixteenth century. The English continuously feared a Spanish invasion of Ireland, and as tensions rose between England and Spain, the Spanish fishers who visited Ireland each year were increasingly perceived as a threat to English control. The launching of the Spanish Armada and the Anglo-Spanish war only amplified those fears and ultimately marked the end of Spanish fishing in Ireland. Both the Spanish and English used state-sponsored pirates, or privateers, to damage each other's maritime operations, leading to a dramatic rise in piratical activity on the Irish coast in the last decades of the sixteenth century, which severely disrupted Irish fisheries.

After months of speculation, the Armada was launched in August 1588. After losing some initial engagements with the English navy, a large portion of the Armada attempted to return to Spain by going around Ireland. During the voyage, violent Atlantic storms shattered many of the ships on the Irish coast.[114] The Armada was never destined for Ireland, and those Spanish who did land in Ireland were not invaders but castaways.[115] Nevertheless, the events surrounding the Armada did impact fisheries and affected the English authorities' attitudes towards fishing in Ireland. In the months after the first failed Armada, there were consistent reports of other Spanish fleets preparing to land in Ireland. In February 1589 it was reported that a Spanish fleet would invade Ireland and use 100 vessels from the Basque Newfoundland fishing fleet to land troops.[116] This report illustrates the extent of changes in fishing patterns. Boats that had formerly visited Ireland for catches had shifted to fishing in Newfoundland and were also implicated in a plot to invade. Even if the report was false, it shows how the English did not distinguish between civilian and military Spanish vessels: any Spanish ship, even a fishing boat, was a potential threat.

Ultimately, the Armada and related war between Spain and England made the position of Spanish fishing fleets in Irish waters untenable. English control of Ireland had increased significantly since the beginning of the sixteenth century, so that any war involving Spain and England now involved Ireland. Spanish

[114] K. S. Douglas et al., 'A Meteorological Study of July to October 1588: The Spanish Armada Storms', *Climatic Research Unit Research Publication*, 6 (1978), pp. 1–83.

[115] Laurence Flanagan, 'The Irish Legacy of the Spanish Armada', *Archaeology Ireland*, 2 (1998), pp. 145–49, at p. 145.

[116] *CSPI*, 1588–92, no. 6, p. 129.

ships were no longer safe in Irish waters, and many areas that had traditionally welcomed foreign fleets had come under English influence. But while the Spanish had been excluded, English fishing fleets continued to visit, particularly around Waterford. From the reports in the State Papers, it appears to have been these fishing fleets that the English were most concerned with protecting.

In preparation for a Spanish invasion, the English considered how they could avoid a repeat of the devastating famines that struck during the second Desmond rebellion. This included restricting the export of food from Ireland, including fish, in order to ensure the English army would have a ready supply if an invasion did take place. However, the restrictions had the unintended consequence of producing a massive oversupply of fish in Ireland. In January 1590 merchants from different parts of Ireland, including Waterford and Duncannon, begged to be allowed to export fish once again.[117] One petition stated that 'about 8 or 10 [thousand] barrels' of herring were going unsold, and they would 'not [be] fit to be eaten' by Lent.[118] The figure was repeated in another letter to the lord deputy around the same time, and it was also stated that the merchants would be 'utterly undone' unless they were allowed to export the fish.[119] The petition was successful, and the merchants were permitted to export fish 'into England or some parts of France'. However, they were forbidden to trade with Spain, and if the lord deputy required fish for the army, he could still purchase it on demand.[120] This episode shows that fish production was picking up in Ireland after the end of the Desmond rebellion. The figure of 10,000 barrels represents a significant catch and would equate to 1,578 TLW. However, the fishing industry still faced substantial barriers in the form of restrictions imposed by the authorities. These also partly explain why so little fish was recorded in English and Welsh ports in this period.

The expected Spanish invasion did not transpire during the sixteenth century. By 1590 English authority over the south and west of Ireland had been significantly extended, and only the northern province of Ulster remained mostly independent of English control. Previous plantation schemes for Ulster had all failed, but by the last decade of the sixteenth century there was a renewed drive among the English to bring this last remaining province under their control.[121]

Violence in Ulster increased from the late 1580s, and disputes over the control of fishing on the river Bann began to emerge again. In 1587 Turlough Luineach O'Neill requested patents to confirm his ownership of the fishing on the Bann and Lough Foyle 'as he now possesseth'.[122] While the O'Neills had maintained

[117] Letters of the Privy Council to the Lord deputy, 5 January 1589/90 (TNA, SP 65/12).

[118] Lord deputy to Burghley, 26 January 1589/90 (TNA, SP 63/150, f.40).

[119] *CSPI*, 1588–92, no. 23, p. 299.

[120] Privy Council to the Lord deputy, 8 March 1589/90 (TNA, SP 65/12, f.180).

[121] *NHI, vol. III*, p. 115.

[122] *CSPI*, 1586–88, no. 28, p. 375.

control over the Bann since it had been granted back to them in 1575, the English were not happy with this state of affairs. In 1588 Francis Stafford left Carrickfergus and crossed the Bann to kill Fordorough O'Cahan, whose alleged crimes included disrupting the fishing of Dublin merchants on the Bann. According to Stafford, O'Cahan had murdered some fishers and made others liable for 'outrageous exactions'.[123] Stafford's actions indicate that the English were not interested in more peaceful settlements in Ulster and were again willing to use violence to achieve dominance.

In this atmosphere, new plantation schemes began to be proposed again. In 1591, Captain Egerton, constable of Carrickfergus, suggested a plantation in lower Clandeboye on the shore of Lough Neagh, to foster the 'development of the fisheries there'.[124] Following previous bitter experiences, the English were more aware of the difficulty of traversing Ulster's terrain. In 1592 the archbishop of Cashel wrote that Ulster had 'three greater rivers where are shipping and fishing, viz, the Earne, Bann and Lough Foyle'.[125] He proposed building bridges over the Erne and Bann, which would significantly facilitate travel into and within the province.

Small-scale confrontations continued throughout the early 1590s as the English gradually encroached into Ulster. Around this time, Hugh O'Neill emerged as the figurehead of the Irish resistance. He was supported not only by his traditional clan allies but also by the O'Donnells of Tyrconnell, the second most powerful Irish lords in Ulster. In the summer of 1595 Sir Henry Bagenal led a force into the province to resupply an English garrison at Monaghan. O'Neill stunned the English by attacking and defeating this army at Clontibret, and by June of the same year, he had been declared a traitor by the Crown.[126]

The Anglo-Spanish war that began in 1585 also allowed the Ulster Irish to obtain Spanish aid against the English. This possibility caused immense anxiety to the English authorities, who had been anticipating a Spanish attack long before war broke out in Ulster. In 1596 Sir Geoffrey Fenton reported that the Irish lords were requesting help from the Spanish king. He guessed the Spanish would land in Munster, which would pose a threat 'to the English herring fishery off Waterford in November and December'.[127] His comment confirms that the English fishery in Ireland was still being carried on, and the government was concerned with protecting it.

The English now had more reason than ever to fear an attack from Spain. There are several accounts from 1594 onwards of Spanish ships harassing fishers

[123] *CSPI*, 1586–88, no. 36, p. 503.
[124] *CSPI*, 1588–92, no. 23, p. 423.
[125] *CSPI*, 1588–92, no. 47, p. 489.
[126] *NHI, vol. III*, p. 121.
[127] *CSPI*, 1596–97, no. 53, p. 153.

on the Irish coast. In May 1594 a merchant ship was carried out to sea by twenty Spaniards aboard a pinnace. The same ship threatened a fisher called Walter Devoraxe, but his life was spared because he was Irish: apparently, he would 'have been hanged if he had been English'.[128] The same Spanish pinnace captured another fishing boat near Dungarvan at the end of May. A second much larger Spanish vessel with a crew of 300 was reported in the area at the same time.[129]

English concern over the Spanish harassment of fishing boats may have been related to a preoccupation with ensuring food security in Ireland and supplying the Crown forces. A pamphlet of 1598 claimed to provide an Irish perspective on the war: the anonymous author wrote that the English could not prevail against 'Captain Hunger, Captain Sickness, [and] Captain Cold'.[130] The tract also described how the Irish drew English forces deep into Ulster and waited for hunger and exhaustion to set in before applying hit-and-run tactics. Scorched-earth tactics exacerbated the lack of supplies available for the English in Ulster, but the Irish also worked to deny them food.[131]

In 1595 the English army in Ireland consisted of 7,000 soldiers, a number that would rise to 21,000 by the war's end.[132] The records of army supplies and rations in Ireland provide glimpses of the wider availability of food during the conflict and the exceedingly high sums of money spent by the English. In this climate of dearth, alternative food sources came to the fore. The State Papers show a massive influx of fish imported from Newfoundland during the 1590s to feed the army and civilian population. In January 1597 Sir Henry Wallop wrote of a shortage of food, and of all prices being extremely high. He also reported a 'failure of the fishing at Waterford': unless herring was imported from Scotland 'the scarcity in Ireland will be great'.[133] In May of the same year the lord deputy, Thomas Lord Burgh, informed the English privy council that the Pale was in a state of 'poverty and waste' and that there was 'death and scarcity in all other places'.[134] In another letter Burgh wrote, 'our extremity is so great here by reason of the dearth and scarcity of victuals in all parts of the realm, but especially in the English Pale'.[135] Later he described how the country of the Pale was in waste, many villages depopulated, and crops unplanted.[136] His forces, and the people at large, were utterly dependent on food imports from England and suffering from

[128] *CSPI*, 1592–96, no. 5, p. 252.
[129] *CSPI*, 1592–96, no. 5, p. 252.
[130] *CSPI*, 1598–99, no. 75, p. 438.
[131] Colm Lennon, *Sixteenth Century Ireland: The Incomplete Conquest* (Dublin, 2005), p. 301.
[132] Canny Nicholas, *Making Ireland British, 1580–1650* (Oxford, 2001), p. 66.
[133] *CSPI*, 1596–97, no. 33, p. 213.
[134] *CSPI*, 1596–97, no. 66, p. 301.
[135] *CSPI*, 1596–97, no. 21, p. 285.
[136] *CSPI*, 1596–97, no. 58, p. 299.

famine.[137] The failure of the fisheries had evidently contributed to this shortage, underlining their strategic value.

O'Neill and the other Irish lords in Ulster do not seem to have experienced the same shortages for most of the war. In July 1597 an English captain, Sir Conyers Clifford, reported that O'Donnell was 'now stronger in men than any of his ancestors ever have been, and, by spoiling the country, better able to maintain them'.[138] Clifford was probably exaggerating, but it does seem that the Gaelic troops had greater food security during the war than their English counterparts. Part of that security came from the fisheries of Ulster. In May 1596, 'Tyrone and his young lady' were said to be at Castle Roe near Coleraine for the purpose of 'fishing salmon on the Bann'.[139] Control of the river also provided Tyrone with a valuable source of tax income and a ready supply of salmon for consumption and export.[140]

In 1594 O'Neill and O'Donnell had reportedly offered £8,000–10,000 to hire mercenaries from Scotland. Hiram Morgan posits that this money came from 'taxing salmon fishing and exporting grain and cattle to western Scotland'.[141] By 1598 it was said that the war cost O'Neill £500 a day; the level of expenditure speaks to his economic strength in that period.[142] Part of that strength came from fisheries. As we have seen, there was also an impressive export of 4,832 barrels of salmon from Ireland to Scotland in 1597/8.[143] A deposition from 1598 reports that a barrel of salmon was worth 40*s*. in Ireland that year, which means this fish had an estimated value of £9,664.[144] There is no way to know for certain this salmon came from Ulster, but it seems likely that it did and that O'Neill not only made money from taxing his fisheries, but also from exporting their products.

Outside of fish the Irish had other sources of food and income. An unknown English captain in 1599 suggested that garrisons be placed in Munster 'to hinder the rebels from sowing their corn in the plains'. The Irish usually planted from March until mid-May each year, and in addition had large cattle herds, which were kept in bogs and forests during the winter before being moved to graze on the plains during the summer. The cattle were at their weakest in spring, which was the same period the crops were planted.[145] This knowledge of the Irish food

137 *CSPI*, 1596–97, no. 26, p. 288.
138 *CSPI*, 1596–97, no. 143, p. 408.
139 *CSPI*, 1592–96, no. 27, p. 519.
140 Hiram Morgan, *Tyrone's Rebellion: The Outbreak of the Nine Years War in Tudor Ireland* (Dublin, 1993), p. 88.
141 Morgan, *Tyrone's Rebellion*, p. 180.
142 *NHI, vol. III*, pp. 125–26.
143 See Chapter 3, p. 000.
144 This value is in pre-decimal pounds with 240 pence. The price could have been different in Scotland where the salmon ended up. *Cal. HCA 1536–1641*, no. 431, p. 94.
145 *CSPI*, 1599–1600, no. 88, p. 329.

system was valuable to the English forces, enabling them to plan how they would disrupt Irish supplies to the greatest effect.

It also seems that trade between continental Europe and Ulster was disrupted in this period. In 1601 Captain Charles Plessington reported on his voyage to the west, that 'every year, till now' the coast had been 'much frequented with French-Men' who had carried wine and weapons to the ports of Killybegs, Donegal, Ballyshannon and Sligo. In return the ships brought 'hides, tallow and fish' back to France.[146] This suggests that continental trade and the exchange of goods (including fish) for weapons had been important for the O'Donnell war effort, but by 1601 had ceased.[147]

Meanwhile, the English army's supply issues in Ulster were continuing. The lords justices wrote in November 1597 that the herring fishery 'failed now here throughout the whole land'. The fish requested from Scotland had not arrived because the fishery had failed there as well. The failure of the fishing in Scotland suggests that the downturn in Ireland was not wholly linked to the war, and there were some wider natural fluctuations at play. The justices observed that herring was 'a most important provision and profitable for her majesty', and supplies were requested from England to meet the demand.[148]

During this period, English authors penned several descriptions of Ireland's land and seascapes, which give us a window into how the English perceived Ireland and justified fighting for it. One such account was written by John Bell, vicar of Christchurch, with the clever title, *How Irish rebels may be taught to be obedient to her Majesty*. He claimed Ireland was 'so full of fish, so full of cattle, and so full of fowl, that there is not a country upon the face of the earth more beneficial to the life of man'.[149] Edmund Spenser, in his *A View of the Present State of Ireland*, used similar language, writing that Ireland was 'replenished with all sortes of fishe most aboundantlie'. Interestingly, Spenser wrote that the western and southern shores were heavily trafficked with Spanish and French ships for fishing, which left a negative cultural influence on the Irish.[150]

In 1601 the lord deputy, Mountjoy, composed a 'Discourse concerning Ireland', in which he declared Ireland to be 'one of the goodliest provinces of the world', 'abounding with all the sustenances for life, as corn, cattle, fish'. Mountjoy even

[146] Plessington to the earl of Nottingham, and Secretary Cecil, 2 September 1601 (TNA, SP 63/209/1, f.148).

[147] Hiram Morgan has written on the economy of Ulster before the outbreak of war: Morgan, *Tyrone's Rebellion*, p. 113.

[148] Lords Justices Loftus and Gardener, and the Council of Ireland to the Privy Council, 29 November 1597 (TNA, SP 63/201, f.175).

[149] *CSPI*, 1598–99, no. 156, p. 498.

[150] Spenser's text was written between 1595 and 1596: Edmund Spenser, 'A View of the Present State of Ireland, 1596', ed. William M. Hennessy (*CELT*, 2008).

praised the Irish people, writing that they had an 'excellent temper'.[151] However, their leaders had led them astray and, in his opinion, Ireland would never reach its full potential until it was 'governed by virtue'. This rhetoric, and similar tropes from other English writers, should not be read as objective views of Ireland's natural resources, for these writers were trying to persuade the Crown to commit further to winning the war. In their eyes, the larger the prize appeared, the more motivated the Crown would be to invest. Ultimately, these tracts had the same conclusion: Ireland had bountiful fisheries and other natural resources, but these were not being exploited by the local inhabitants, and by 'civilising' the island, the English could unlock this potential.

In 1598 the war in Ulster spread to Munster, where there was yet another uprising. The earl of Ormond, who remained loyal, wrote to London in October asking for the maritime towns of the province to be protected, presumably to defend the fisheries and ensure supplies and soldiers could reach the region.[152] The threat of a Spanish invasion was also still present; in March 1598, another Spanish fleet was reported to be assembling, and in response, 'most of the shipping of all the ports in Ireland' and the 'trading there for wines and other commodities' had ceased. Additionally, there were many reports of Irish ships being seized in Spain, while pirates captured additional vessels.[153]

In 1599 the English were planning to use their superior naval strength to defeat the rebels on the west coast. They planned to take Inishbofin using mariners from the English West Country who were 'acquainted with the fishing at Ennisboffin [Inishbofin]'. The authors claimed that the fisheries provided 'great profit toward the relief of the garrison' in the west of Ireland.[154] In January 1600, a scheme was proposed to encourage fishers to travel to 'the Bann, Lough Foyle, and Lough Swilly', as 'salmon, herring, eels, cod' and many other sorts of fish were plentiful there. The fishers were to be provided with gear and given monetary incentives, although no provision was made to protect them from the Irish who controlled the area.[155]

Attention was focused on Lough Foyle where an assault force landed in the summer of 1600 and established a fort. The choice of Lough Foyle was in part motivated by abundant fishing resources. These fisheries would help feed the remote garrison and also deny a critical source of food and income to the Irish.[156] *A description of Lough Foyle and the country adjoining*, dated December 1600, reported that the region around the lough was 'full of salmon and in season all year', while

151 *CSPI*, 1600–01, no. 122, p. 251.
152 *CSPI*, 1600–01, no. 118, p. 293.
153 *CSPI*, 1600–01, no. 87, p. 84.
154 *CSPI*, 1599–1600, no. 53, p. 163.
155 *CSPI*, 1599–1600, no. 7, p. 377.
156 *CSPI*, 1600, no. 57, p. 194.

the river Foyle upstream from Derry was 'wonderfully replenished with trout's and flounders' that lasted all summer. In addition, Lough Swilly had 'excellent good herring fishing for two months in the latter end of summer' and plentiful oysters.[157]

After establishing their outpost on Lough Foyle, the English burned an Irish corn magazine, captured thousands of animals and killed 150 civilians.[158] The destruction of agricultural resources, which denied the enemy food, was a vital part of the English strategy. Concern with securing supplies is clear in a letter from Captain Humphrey Willis in January 1601, who wrote that holding the 'River Liffer', known today as the river Foyle, was of great strategic importance because 'whosoever shall farm it in any sort whatsoever may prepare himself presently against the season of the year'.[159] In other words, the river would supply whoever controlled it with food for the next campaigning season.

Similarly, in January 1601, Sir Arthur Chichester informed Robert Cecil that he should place garrisons at Coleraine on the river Bann and Lecale on the Strangford river. Chichester argued that this would take 'more assistance and relief' away from O'Neill 'than Spain affords him'. Moreover, by placing 'skilful fishermen' on the Bann, the garrison there would scarcely need to be resupplied, since they would be able to provide for themselves locally.[160] Of Lough Neagh and the Ulster coast, Sir Rafe Lane wrote in 1602 that they were 'so plentiful all the year round with cod and ling that a whole army might be victualled from the fishings there'.[161] As these examples illustrate, food supplies were of central strategic importance during the war, and thus fisheries became directly involved in the conflict and factored into decision-making for both sides.

Famine struck the island once again in mid-1600. The new lord deputy, Charles Blount, 8th Baron Mountjoy, informed the privy council in April of that year that the country was 'generally so wasted and eaten up' that the army had no supplies to draw on.[162] Subsequently, he wrote that a 'great famine' was growing, and storms hindered the approach of ships with food.[163] In October, news came from Sir George Carew in Mallow, County Cork, that the province of Munster had been 'so much wasted by this last summer's war' that there were no supplies at all left, and that unless supplies from England were sent to a convenient port, his soldiers would starve.[164] As with every other foodstuff, there was a great shortage

[157] *CSPI*, 1600–01, no. 85, pp. 92–95.
[158] *CSPI*, 1600–01, no. 85, p. 92.
[159] *CSPI*, 1600–01, no. 18, p. 158. The name Liffer does live on in the town of Lifford which lies on the river Foyle.
[160] *CSPI*, 1600–01, no. 5, p. 142.
[161] *CSPI*, 1601–03, no. 210, p. 317.
[162] *CSPI*, 1600, no. 74, p. 66.
[163] *CSPI*, 1600, no. 104, p. 91.
[164] *CSPI*, 1600, no. 113, p. 497.

of fish. The fact that fisheries were not functioning normally is confirmed by the complete lack of fish imports to Bristol at this time. Shipment of Newfoundland fish to Ireland seems to have also peaked in this period.[165] Evidently a lack of supplies at home made Newfoundland imports necessary.

By 1601 the war was reaching a climax. O'Neill and his Irish allies were under pressure in Ulster from continual English incursions, and after years of speculation and anticipation, the Spanish finally intervened, landing a force at Kinsale in September 1601.[166] The English reacted swiftly and trapped the Spanish army at Kinsale and soon after defeated the Irish army under O'Neill, who had marched to meet the Spanish. The battle of Kinsale did not immediately end the fighting, but the outcome decided the course of the war and what followed was only a long-drawn-out defeat for the Irish.[167]

The O'Sullivans of Berehaven had joined O'Neill earlier in the war, but in 1602 they were attacked by lord president Carew, and their castle at Dunboy destroyed.[168] According to Carew, O'Sullivan had maintained rebels on his land by supplying them with the proceeds of his fisheries. Carew estimated that the fishing had previously been worth £500 a year, but by 1602 foreign fishers no longer wanted to come because with the loss of the castle their protection was gone. Carew wrote that the fishers now had a 'place to defend him for the insolencies of the Irish', or where they could safely dry their fish on the shore.[169] Then in August came a report that the O'Sullivans had seized an English fishing boat, suggesting that some fishers did still wish to use the waters at Berehaven, but the English were no longer welcomed by the O'Sullivans.[170] At the same time, Spanish and French boats would have been open to attack by the English.

In a final push to bring the war to an end, the English again turned to scorched-earth tactics in Ulster, resulting in yet more famine in the province. In September 1602, Mountjoy reported that 'above three thousand starved in Tyrone' as a result of crop destruction.[171] Again we can assume that the English control of important fishing resources contributed to the dearth of food. The Irish confederacy, which had formerly rallied behind O'Neill, now began to fall apart, further affecting food supplies and fisheries. In 1602, the O'Cahans made peace with the Crown and agreed that the Bann fishery was 'wholly reserved' for the maintenance of the English garrison at Coleraine.[172]

165 *CSPI*, 1600, no. 134, p. 270.
166 *NHI, vol. III*, p. 134.
167 *NHI, vol. III*, p. 135.
168 Lennon, *Sixteenth Century Ireland*, p. 301.
169 *Cal. Car. MSS*, 1601–03, no. 254, p. 254.
170 *CSPI*, 1601–03, no. 468, p. 474.
171 *NHI, vol. III*, p. 135.
172 *CSPI*, 1601–03, no. 211, p. 302.

Seeing that he could not continue the war, Hugh O'Neill himself signed a peace agreement with lord deputy Mountjoy on 30 March 1603. Mountjoy knew that Queen Elizabeth had died six days before, but O'Neill did not.[173] Her successor, James I, endorsed the treaty, and O'Neill was pardoned. The Irish defeat marked the end of the Tudor re-conquest. The new English hegemony would change all facets of Irish life, including fisheries. O'Neill eventually fled Ireland in 1607, and his confiscated lands were subsumed into the new Ulster Plantation. The English had spent an estimated £2,000,000 on the war over its duration,[174] the equivalent of nearly half a billion in today's money.[175]

It is impossible to completely quantify the impact of war on Irish fisheries over the sixteenth century. However, we do know from a wealth of qualitative accounts and trade data that fisheries suffered a massive decline in the 1590s, and there can be no doubt that war and violence contributed heavily to this decline. Terrestrial warfare disrupted healthy economic life, destroying towns and economic infrastructure, and the violence and related famines devastated the population. One of the most visible consequences can be seen in Irish fishing towns and settlements. The town of Carlingford had been one of the most prosperous herring fishery towns in Ireland. Although hit hard by the Kildare rebellion, its herring fishery recovered in the mid-century. By 1575, however, the town was described as 'extremely impoverished'.[176] The Nine Years War brought some reprieve, as the English used Carlingford as a hub for importation. After the war, the army left the town, and without its traditional herring fishery, it was largely abandoned, and by 1606 was described as one of the 'poor towns' of Ulster.[177]

A similar decline was experienced by many other Irish fishing towns. Ardglass was essentially 'abandoned at the end of the sixteenth century with the collapse of the Irish Sea fisheries'.[178] The stone fish traps of Strangford Lough were similarly abandoned by local fishers.[179] Further evidence of decline can be found in a survey of customs duties from 1611 written by an English official, Robert Cogan.[180]

[173] Lennon, *Sixteenth Century Ireland*, p. 303.

[174] Hiram Morgan, 'Hugh O'Neill and the Nine Years War in Tudor Ireland', *Historical Journal*, 36 (1993), pp. 21–37, at p. 22.

[175] 1600 was used as the reference year and based on purchasing power the exact figure is an estimated £474,500,000 in relation to 2021: The Measuring Worth Foundation, 'Measuring Worth' (https://www.measuringworth.com/calculators/ukcompare/relativevalue.php) (accessed 7 August 2021).

[176] Simms et al., 'Carlingford', p. 3.

[177] *CSPI*, 1603–06, no. 719, p. 464.

[178] Aidan O'Sullivan and Colin Breen, *Maritime Ireland: An Archaeology of Coastal Communities* (Stroud, 2007), p. 188.

[179] Thomas McErlean et al. (eds), *Strangford Lough: An Archaeological Survey of the Maritime Culture Landscape* (Belfast, 2002), p. 148.

[180] *Cal. Car.* MSS, 1603–25, no. 101, pp. 174–77.

Cogan provided short descriptions of many fishing towns, describing Kinsale as 'a poor town ruined by the last rebellion' that was still reliant on its diminished fisheries for income;[181] Dungarvan was 'a very poor fisher town', Dingle a 'poor ruined place', and New Ross 'a poor ruined town'.[182] Not every town described by Cogan was poor; some important fishing ports like Waterford were still actively trading. But the export trade of the late sixteenth and early seventeenth centuries focused on wood, hides and beef, with fish featuring little. Fisheries no longer held the significance they had in the sixteenth century.

Perhaps most impacted of all were Gaelic Irish settlements and fisheries. Not only were the Irish directly affected by conflict, many Gaelic Irish fisheries also changed ownership in the years following defeat in the Desmond rebellion and Nine Years War. The victorious English were now able to implement far more ambitious and broad plantation schemes. The largest of these was in Ulster, but change also occurred across Ireland. Obviously, the migration and plantation of settlers did not put an end to fisheries in Ireland, and new and prosperous fisheries did emerge later, but these settler fisheries were distinct from those that came before. In 1606, Baltimore, previously one of the prime Gaelic Irish fishing settlements, was leased by the lord, Sir Fineen O'Driscoll, to Thomas Crooke, an English settler.[183] What had been one of the centres of Gaelic fishing became an English town. In the seventeenth century Baltimore would come to be synonymous with piracy. It would also host fisheries again, but these were pilchard fisheries controlled by the new earl of Cork, Richard Boyle.[184] Baltimore is just one example: similar stories could be told for many fishing locations formerly controlled by the Gaelic Irish.

A century of conflict seriously reduced the sustainability of fisheries in Ireland. War damaged infrastructure, ruined towns, and depressed trade. The resulting famines also reduced the population and destroyed normal economic life. Subsequently, after 1603, many of Ireland's most important fishing locations changed ownership, which fundamentally shifted the character and operation of the fisheries. The Nine Years War also ended the operations of continental fisheries in Ireland. As the English took over, Ireland's waters became hostile to Spanish and French fleets, resulting in a sharp rise in piracy.

181 *Cal. Car. MSS*, 1603–25, no. 101, p. 176.
182 *Cal. Car. MSS*, 1603–25, no. 101, pp. 174–77.
183 MacCarthy-Morrogh, *Munster Plantation*, p. 151.
184 Nicholas Canny, 'Boyle, Richard' (*DIB*, 2009) (https://doi.org/10.3318/dib.000853.v1) (accessed 29 July 2022).

The Impact of Piracy

Today the United Nations Convention on the Law of the Sea defines piracy as 'any illegal acts of violence or detention, or any act of depredation, committed for private ends by the crew or the passengers of a private ship or a private aircraft'.[185] Piracy in the early modern period is more difficult to define and existed in shades of legal greyness. Many nations had their own maritime or admiralty courts to deal with matters of piracy, and pirates would often move between ill-defined international jurisdictions. To complicate matters, in the sixteenth century a distinction was made between acts of piracy, reprisal and prize-taking. 'Letters of reprisal' issued by the government authorised vessels to capture enemy ships and take them back to a friendly port. They were issued in response to acts of piracy and were meant to give victims redress, ideally by capturing those who had attacked them.[186] In this way, an act of prize-taking could be viewed as a legitimate source of justice.[187] Additionally, letters of marque could be issued, allowing a commissioned ship to sail into foreign waters and capture the ships of other nations.[188]

In wartime, letters of reprisal were issued with the intent of targeting the trade of enemy states. The English, Dutch and French all sent ships to target the Spanish treasure fleets without cause for reprisal.[189] In this sense, reprisal was similar to state-sponsored privateering, but the word privateering itself did not enter common parlance until the seventeenth century.[190] As Connie Kelleher has written, any act of 'piracy' in the early modern period has to be 'defined within the context of what is taking place economically, politically and socially at any given time'.[191]

Defining piracy in the sixteenth century is difficult; the legality of each act of piracy must be established on a case-by-case basis. In the words of J. T. Anderson, 'private purpose (animus furandi) has generally been held to be a necessary element of piracy, to distinguish the crime from political activity'.[192] Even when pirates claimed to have the backing of a state, they often acted out of self-interest. The reprisal system thrived on ambivalence, and 'The path from that action

185 UN General Assembly, 'United Nations Convention on the Law of the Sea' (1982), p. 61.

186 Connie Kelleher, *The Alliance of Pirates: Ireland and Atlantic Piracy in the Early Seventeenth Century* (Cork, 2020), p. 44.

187 David J. Starkey, 'Voluntaries and Sea Robbers: A Review of the Academic Literature on Privateering, Corsairing, Buccaneering and Piracy', *The Mariner's Mirror*, 97 (2011), pp. 127–47.

188 Kelleher, *Alliance of Pirates*, p. 44.

189 Kelleher, *Alliance of Pirates*, p. 45.

190 John C. Appleby, *Under the Bloody Flag: Pirates of the Tudor Age* (Stroud, 2009), p. 13.

191 Kelleher, *Alliance of Pirates*, p. 44.

192 J. T. Anderson, 'Piracy and World History: An Economic Perspective on Maritime Predation', *Journal of World History*, 6 (1995), pp. 175–99, at p. 177.

[reprisal] to random piracy is short and clear.'[193] Examples from the Admiralty court records make it abundantly clear that, in many cases, the law was used as a cover by pirates to rob indiscriminately.

Before looking in detail at how piracy impacted fisheries, this section will attempt a broad overview of piracy events related to Ireland in the sixteenth century, based on Admiralty court records.[194] These are the best sources to explore broader trends across the sixteenth century. The data presented is intended to provide an indication of the chronological incidence of piracy related to Ireland and the spatial distribution of that piracy. The data is available online and methods used to create it are outlined in Appendix 1.[195]

Figure 4.1 shows the frequency of piracy events around Ireland per year based on the evidence in John C. Appleby (ed.), *A Calendar of Material Relating to Ireland from The High Court of the Admiralty Examinations 1536–1641.* Eighty-nine piracy events in total were recorded, the first event happening in 1546 and the last in 1603. Immediately we can see that instances of piracy remained relatively consistent until the 1580s when a substantial spike can be observed; of the eighty-nine events recorded, fifty-eight occurred between 1580 and 1603. Unfortunately, the survival of records is somewhat haphazard: in some years, no records have survived and thus no events were reported. However, enough evidence has survived to indicate general trends and peaks of violence.

In presenting the data an attempt has been made to distinguish between lawful acts of prize-taking and illegal acts of piracy. Pirates could operate in legal grey areas, so these determinations were based on how the English court would have viewed the legality of a piracy event. A French or Spanish court might have had a different interpretation. Making decisions around legality is made more difficult because many piracy cases were disputed and the final judgments kept in another record series.[196] With this caveat, we may conclude that the vast majority of these events were defined by the English court as illegal acts of piracy against innocent merchants and fishers (seventy-two), with only nine appearing to be legal and a further eight impossible to determine.[197]

Reviewing the witness testimony, in fifty-seven of the events the pirate ships' crews were predominantly English. In a further nine events, the crews were a

[193] Anderson, 'Piracy and World History', p. 177.

[194] *Cal. HCA 1536–1641*; HCA 13, The High Court of Admiralty: Instance and Prize Courts: Examinations and Answers, 1536–1826 (TNA).

[195] Patrick Hayes, 'Piracy Events, 1564 to 1603' (2022) (10.6084/m9.figshare.20367972) (25 July 2022).

[196] Judgments were kept in a separate record series, which are not included in the Examinations series. It is very difficult to cross-reference cases among the different record series.

[197] Legal as defined in the jurisdiction of the deponents; several were reported to be legal prizes in France for example.

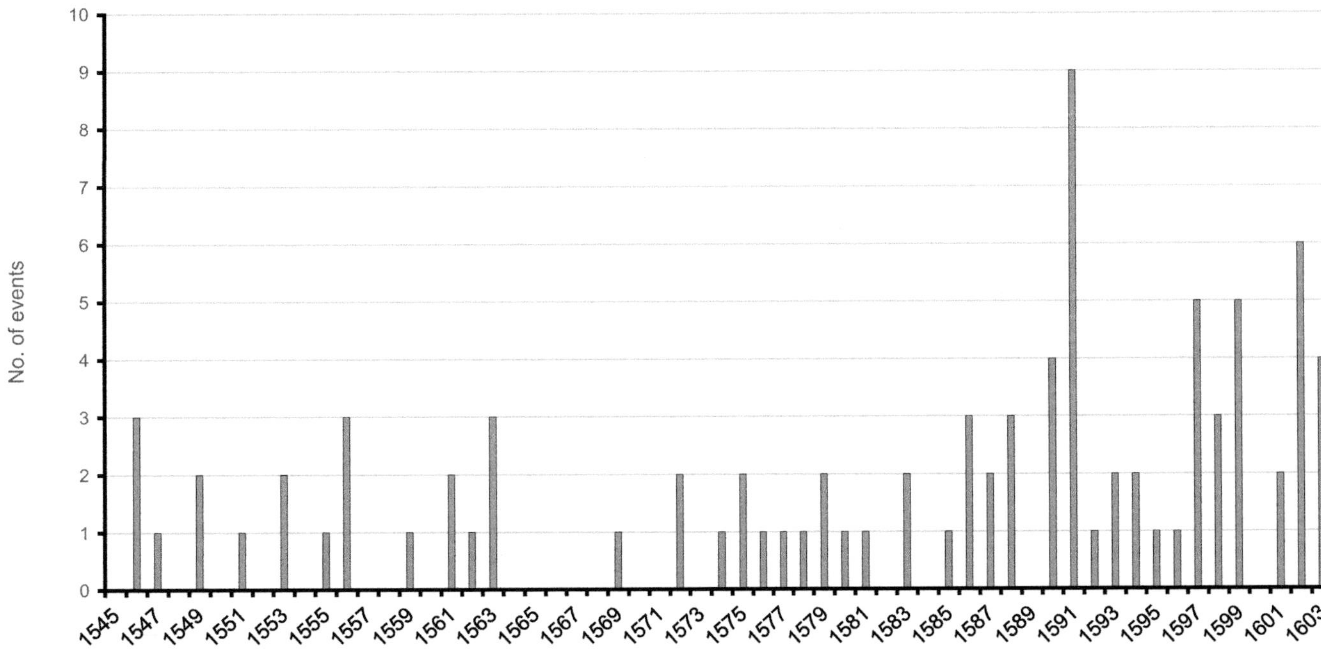

Figure 4.1. Piracy events around Ireland, 1546–1603.

blend of Irish and English, and yet another nine ships had French crews. The remainder encompassed a mixture of Scottish, Irish, Welsh, and Spanish. The victims' crews were most commonly of French nationality, with nineteen French ships being assailed in total. A further seventeen ships with Irish crews and fifteen with English crews were attacked by pirates. Nine German-crewed ships were also assaulted, along with six from Spain.[198] English pirates attacked Irish and French crews eleven times each. English pirates also attacked English victims in five instances. Overall, it would appear that pirates were opportunistic, attacking ships from every nationality, friend or foe. In many instances, pirates with letters of marque against one particular nation attacked ships indiscriminately, including their fellow countrymen.

Targets varied: in seventy-two cases, those attacked were merchant traders; in twelve cases fishers; and in three instances, other pirates.[199] Not all these examples directly impacted Irish fisheries, but piracy was a constant threat throughout the sixteenth century to all merchants and fishers. Additionally, by mapping out these instances of piracy, we can build a picture of which areas of Ireland and adjacent waters were most affected. Map 4.2 shows the eighty-seven instances of piracy where the location of the attack was mentioned in the witness statements.

These events were all related to Ireland in some way but did not necessarily take place in Irish waters. Attacks outside Ireland were included because the capture of Irish ships in England or Spain would still have a direct economic consequence in Ireland. We have also to account for the fact that the Irish maritime industry could be influenced by events far from Irish coasts; national borders did not constrain pirates, and during this period piracy was a problem across Europe. Nevertheless, the data does still show a high concentration of piracy events around the Irish coast, particularly in Munster, with another cluster appearing on the coast of the English West Country. Events were recorded further away on the coasts of France, Portugal, and Spain.

The coasts of Munster and the West Country shared some common features. Both were rugged and peppered with numerous small bays and ports, were close to busy shipping lanes and provided markets for pirates to sell goods. Furthermore, in the sixteenth century, neither benefited from the control of a central power able to police the seas.[200] In his study of piracy in nineteenth-century Morocco, C. R. Pennell concluded that piracy was most likely to flourish where 'political and economic structures on the one hand and physical geography of the region on the

[198] A few ships from the Netherlands (4), Scotland (4), and several other ships had mixed crews.

[199] One further victim of piracy was a town, and another was unidentifiable.

[200] Appleby, *Under the Bloody Flag*, p. 59; Kelleher, *Alliance of Pirates*, p. 31.

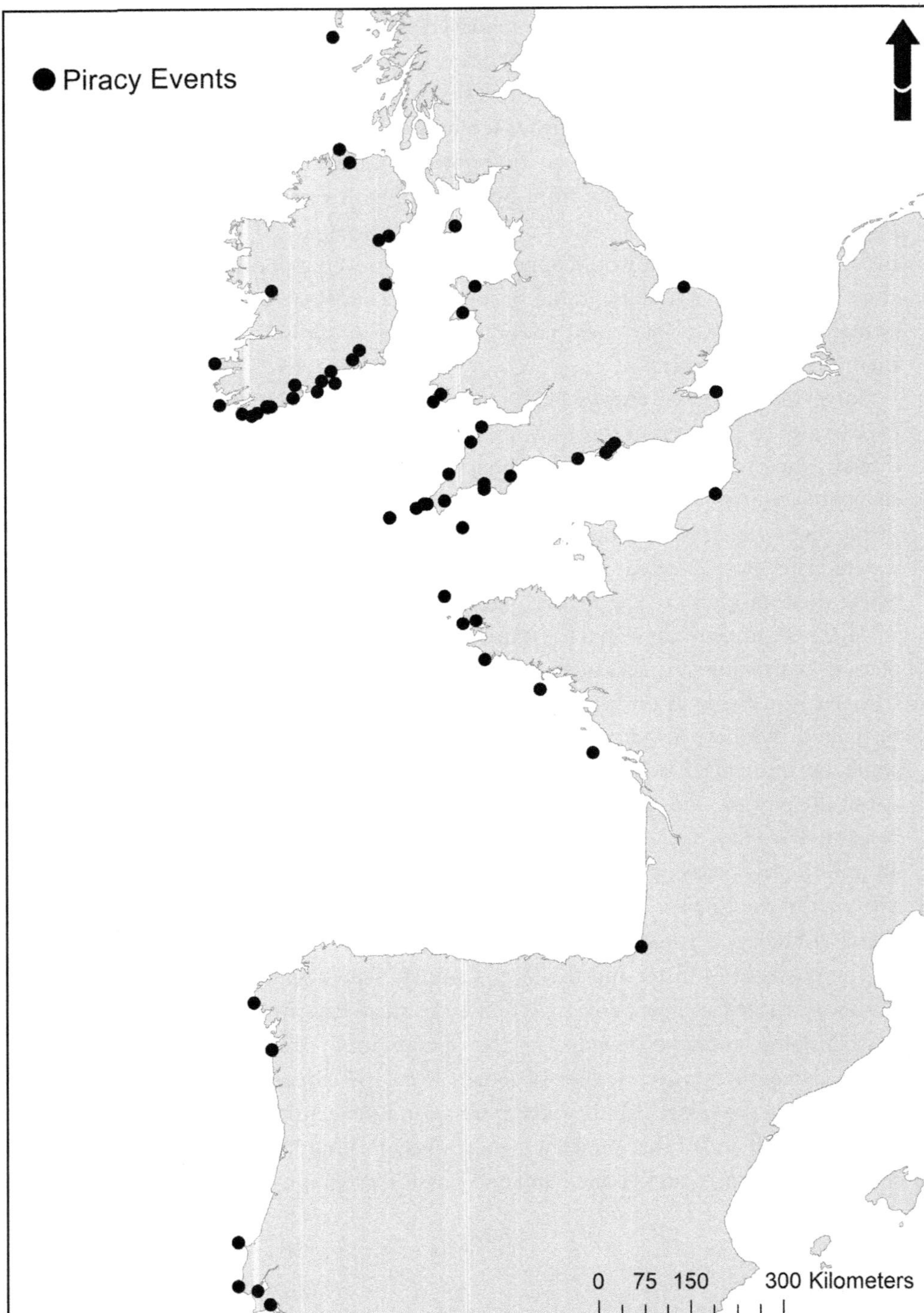

Map 4.2. All piracy events relating to Ireland, 1546–1603.

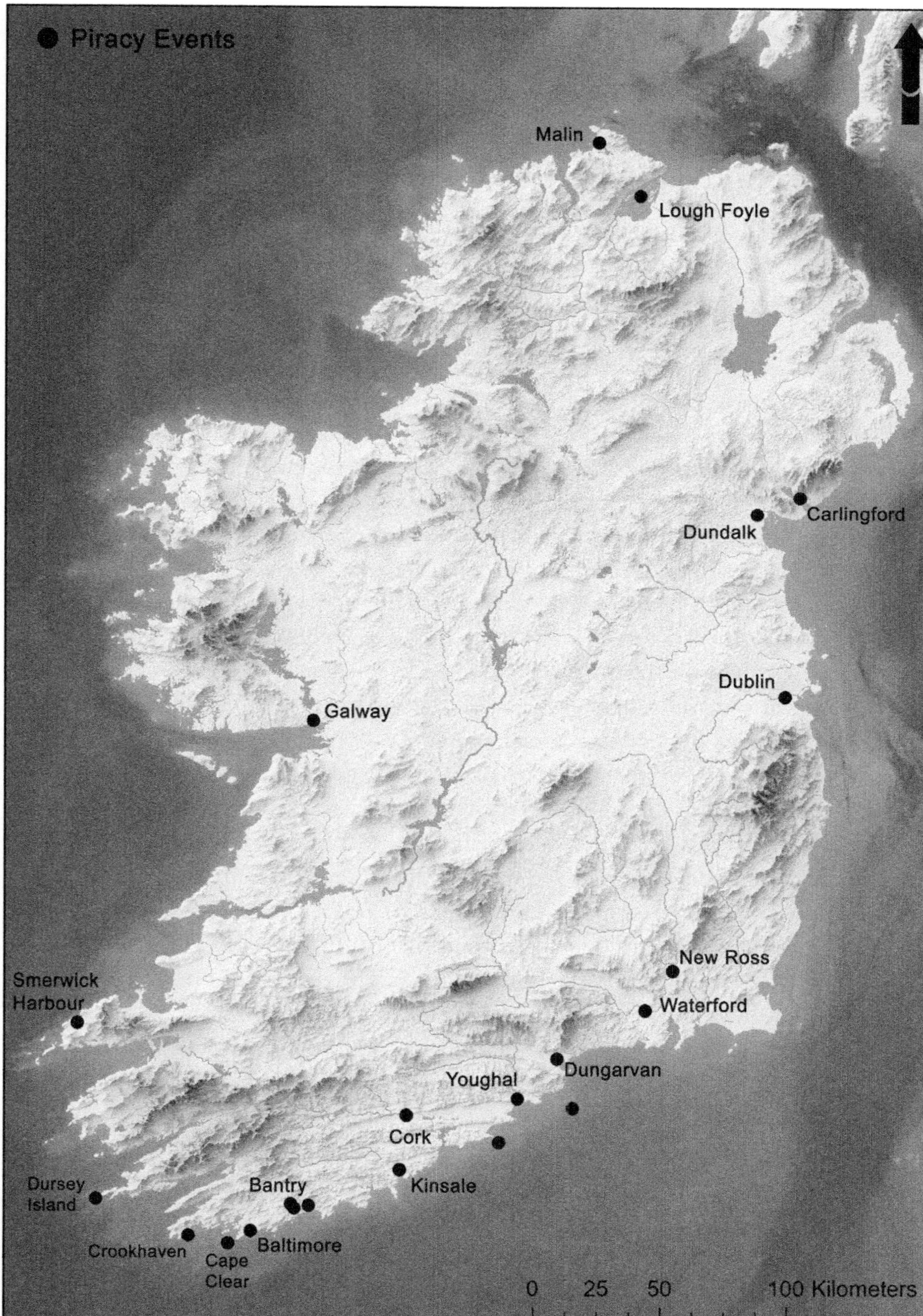

Map 4.3. Piracy events on the Irish coast, 1546–1603.

other gave pirates the most opportunity'.[201] Munster and the West Country had the ideal mixture of weak government control and suitable physical geography.[202]

Pirates also often waited near capes and other choke points to make capturing ships easier.[203] Land's End in Cornwall, Cape Clear in Ireland and Cape Finisterre in northern Spain all show up as significant hot spots. Other choke points, such as St George's Channel and the Straits of Gibraltar, also saw a high frequency of attacks. Map 4.3 shows instances of piracy that occurred in or around the coast of Ireland. Again, Munster saw the highest number across the century, with Cork being the most frequent location, followed by Kinsale, Cape Clear and Waterford. We also see far more instances of piracy around ports; it was far easier for pirates to wait for ships to come to them rather than chase vessels on the open sea.

If pirates were targeting a particular town, all trade would often cease there. Merchants would avoid the area if it were known that pirates were operating, opting to trade in safer waters. Piracy produces the immediate economic effect of reducing the availability of goods reaching markets, but endemic piracy also reduces the incentives for merchants or fishers to operate or for others to invest in those industries.[204] For a small market like Ireland, the losses from piracy could have a devastating impact on individual merchants and whole communities.

Map 4.4 further shows the routes taken by vessels before or after pirates assailed them. The routes on this map are indicated by straight lines between points of call for each ship. This map gives us a better understanding of which shipping routes were most vulnerable to pirate assault. As we might expect, there is a high level of traffic between the Munster ports and Bristol. We also see many voyages going to the smaller ports of the West Country. Specific ports outside this region, such as London and La Rochelle, also emerge as foci for voyages. Ships travelling south to Spain and Portugal were more frequently involved in piracy incidents than ships travelling east to the Low Countries or Germany. This pattern may reflect the extent of traffic, but it is also true that conflict between England, Spain and France made this southern channel particularly dangerous at different times throughout the century. There were few instances of piracy on ships crossing the Atlantic, in part due to the rarity of such voyages for Irish vessels. As it turns out, every transatlantic ship attacked by pirates was going to or returning from Newfoundland. Almost all were French fishing vessels attacked on the Irish coast as they returned home.[205]

[201] C. R. Pennell, 'The Geography of Piracy: Northern Morocco in the Mid-Nineteenth Century', in C. R. Pennell (ed.), *Bandits at Sea: A Pirate Reader* (New York, 2001), pp. 55–68, at p. 66.

[202] Kelleher, *Alliance of Pirates*, p. 260.

[203] C. M. Senior, *A Nation of Pirates: English Piracy in Its Heyday* (Newton Abbot, 1976), p. 21.

[204] Anderson, 'Piracy and World History', pp. 179–80.

[205] See, for example, *Cal. HCA 1536–1641*, no. 109, p. 21, no. 240, pp. 46–47, no. 308, p. 64.

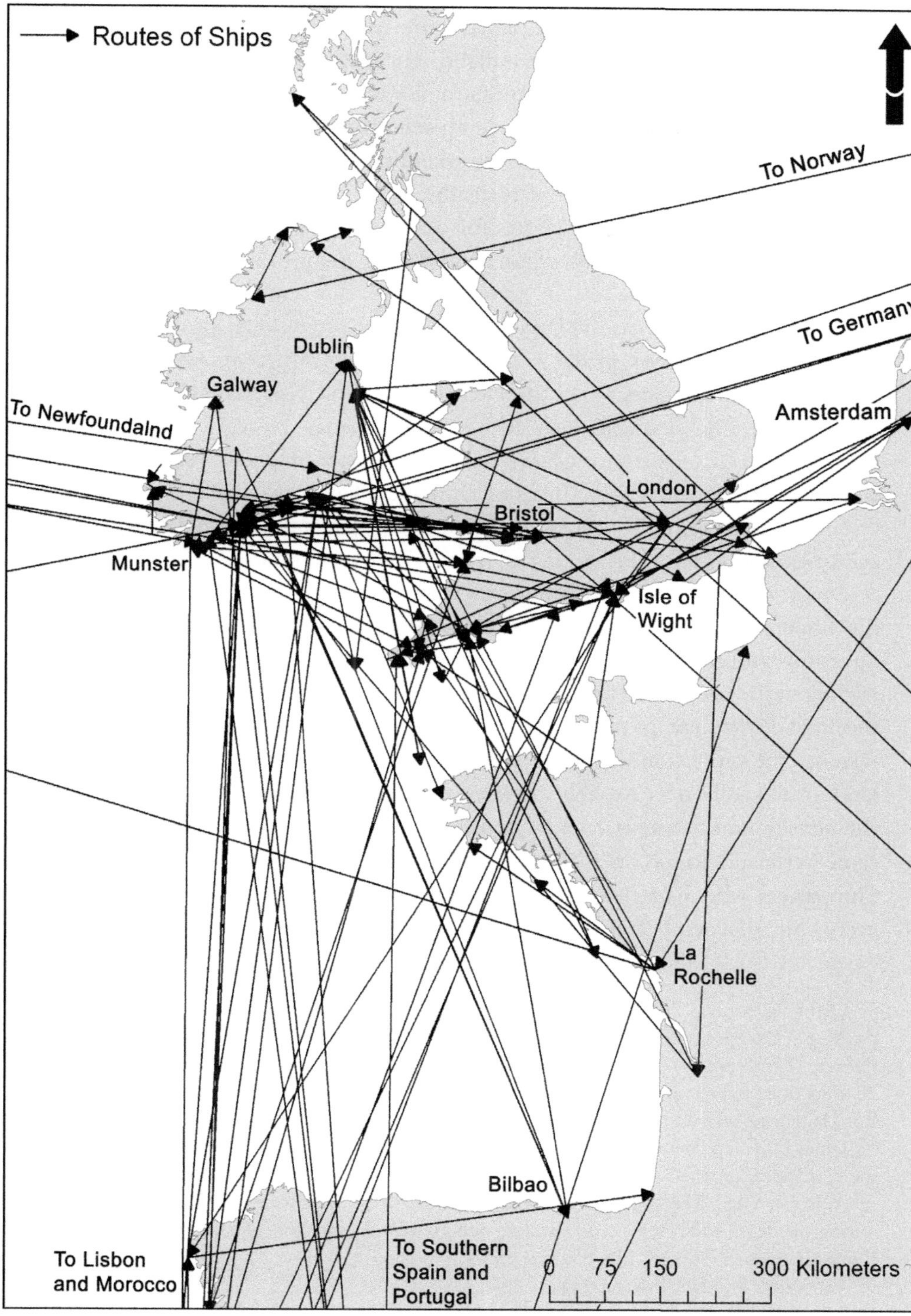

Map 4.4. Routes of ships assailed by pirates, 1546–1603.

Piracy in Irish waters may have played a role in the decision of foreign fishers to shift their focus toward Newfoundland, but the waters there were far from tranquil. According to Olaf Uwe Janzen, piracy was 'endemic' in Newfoundland during the closing decades of the sixteenth century, with Portuguese and Basque fishers significantly affected.[206] Nor were other nations immune to piracy. In 1586 the French port of Le Havre sent thirty-three ships to Newfoundland, three of which were captured by English pirates and one by rebel pirates from La Rochelle.[207] Still, the profitability of the Newfoundland fishery seems to have outweighed the potential risk.

Other fisheries, such as the Dutch North Sea herring fishery, also experienced heavy losses from piracy in the sixteenth and seventeenth centuries.[208] In 1583 Spain captured harbours along the Flemish coast and began sending pirates to target Dutch, English and French shipping and fisheries. From 1583 to 1607 the Spanish sent out an average of fifteen to twenty ships; from 1621 to 1647, over 100 ships annually.[209] The fishing fleets of Dunkirk, Blankenberge, Nieuwpoort and Ostend all experienced a substantial decline in the sixteenth and early seventeenth centuries as a result of the actions of Spanish-sponsored pirates.[210]

Piracy was a problem over the whole period covered in this book; even in the fifteenth century, increasing numbers of Breton and Spanish pirates were operating around Ireland.[211] This increase was partly driven by an upturn in trading traffic between Ireland, Brittany and Spain.[212] In response to the situation, the Irish Parliament passed a statute in 1460 encouraging Irish shipowners to take up 'the protection of the sea and the merchants and fishermen of the said land, from spoil hurt or robbery of their merchandises and fish'. Parliament felt the measure was necessary because Breton, Spanish and French pirates frequently came to Ireland 'to rob, spoil and take and slay, as well merchants as fishermen'.[213] Shipowners who patrolled for pirates were promised a payment of 5*s.* from every ship that went 'faske fishing' and 6*s.* 8*d.* from every small fishing boat.

[206] Olaf Uwe Janzen, 'The Problem of Piracy in the Newfoundland Fishery in the Aftermath of the War of the Spanish Succession', in Poul Holm and Olaf Janzen (eds), *Northern Seas Yearbook* (Esbjerg, 1998), pp. 57–75, at p. 58.

[207] Data obtained through correspondence with Dr Bernard Allaire, based on notarial records from La Havre, and yet to be published.

[208] James D. Tracy, 'Herring Wars: The Habsburg Netherlands and the Struggle for Control of the North Sea, ca. 1520–1560', *Sixteenth Century Journal*, 24 (1993), pp. 249–72.

[209] A. P. van Vliet, 'The Influence of Dunkirk Privateering on the North Sea (Herring) Fishery during the Years 1580–1650', in J. Roding and Heerma L. van Voss (eds), *The North Sea and Culture* (Leiden, 1996), pp. 150–65, at p. 155.

[210] Van Vliet, 'Dunkirk Privateering', p. 164.

[211] Timothy O'Neill, *Merchants and Mariners in Medieval Ireland* (Dublin, 1987), pp. 119–22.

[212] O'Neill, *Merchants and Mariners*, p. 122.

[213] *Statute Rolls, Ireland*, vol. II, 38. Henry VI, A.D. 1460, p. 671.

Additionally, it was ordered that 'sea wardens' should be put out to watch from Easter to Michaelmas, instructed not to allow 'merchants or fishermen to be hurt thereon'.[214] The extent to which shipowners took up this offer is unclear, but the necessity for such measures illustrates the problem piracy posed.

We cannot quantify piracy in the fifteenth century in the same way we can for later parts of the sixteenth; data is only available from 1546, as there are no extensive Admiralty court records available. But other evidence, including trade data, suggests that there was no major disruption from piracy in the late fifteenth or early sixteenth centuries. Piracy seems to have increased in Ireland from the 1540s, the upsurge mainly driven by the war between France and England, which broke out in 1542. Lord Justice William Brabazon reported in 1544 that 'pirates and Bretons are now keeping on these coasts'.[215] As the war progressed, both sides issued letters of marque empowering pirates to capture enemy ships.

A French example can be found in the Irish State Papers. Written in 1546, it licensed Captain Jehan le Gras and the crew of the *Petit Regnard*, a 35-ton vessel, 'to make war upon the King's enemies'.[216] The English would have issued similar letters to their own ships. However, pirates often ignored the constraints of such letters. In July 1548 the mayor of Youghal wrote to the lord deputy, recounting that an English pirate named Smith had 'attacked our returning fishing fleet'.[217] Ostensibly Smith and the Youghal fishers were on the same side, but that did not prevent such attacks from taking place. Similarly, in July 1548, the people of Kinsale informed the lord deputy that a plague had caused 'great mortality' in the town. The port 'has been our lifeline, but now it is blocked by Eagle's pirates who prevent food supplies from reaching us'.[218] Again, it seems that the pirate captain called 'Eagle' was English; certainly, there is no mention of him or his crew being French. In 1557 a mariner called John Wallwyn gave a deposition in the English Admiralty court, that his former captain, Thomas Borrowe from Dublin, prepared his ship, the *Anne*, 'to spoil and rob friends and enemies' following the outbreak of war between France and England in 1556. Borrowe attacked an English ship called the *Moon* near Land's End in May 1557 but was captured and taken to Plymouth to be tried as a pirate.[219] Another witness claimed that Borrowe and his crew had 'acted as rovers' and had attacked several friendly Irish ships around Waterford.[220] Borrowe's crew had a different version of events, which also provides

214 *Statute Rolls, Ireland*, vol. II, 38. Henry VI, A.D. 1460, p. 672.
215 *CSPI* (IMC), 1509–47, no. 691, p. 387.
216 *CSPI* (IMC), 1509–47, no. 717.7, p. 400.
217 *CSPI* (IMC), 1547–53, no. 27, pp. 9–10.
218 *CSPI* (IMC), 1547–53, no. 29, p. 10.
219 *Cal. HCA 1536–1641*, no. 87, pp. 16–18.
220 *Cal. HCA 1536–1641*, no. 97, p. 20. 'Rover' was another word for pirate, derived from the Dutch word *zeeroover*; Kelleher, *Alliance of Pirates*, p. 31.

us with some noteworthy details on how these pirate ships operated. One of them, William Dennam, deposed that the *Anne* was formerly a Spanish ship that had sunk while coming into Drogheda in 1557. Borrowe and others had bought, floated, and repaired the ship before sailing it to Waterford, where they heard that war was declared between England and France, so they fitted the ship out with ordnance and recruited a crew of '60 soldiers and mariners'. They were then commissioned by the lord deputy to 'spoil and rob' all French and Scottish ships coming near the Irish coast.[221] In Dennam's account, the *Anne* did not attack any friendly ships and was accosted by the *Moon* without provocation,[222] but what makes this implausible is that war with France was declared in June 1556, while the *Moon* had captured Borrowe's ship in May.[223] Thus the *Anne* was acting illegally. This example illustrates the lack of clarity in the circumstances under which pirates operated and the widespread violence and confusion prevailing in time of war.

Of course, French pirates also attacked Irish vessels during the mid-sixteenth century, and the general situation put many fishing towns in difficulty. In October 1548 the mayor of Waterford wrote, 'no shipping dares to come to this coast because of the menace of piracy is to the detriment of this city, the lifeblood of which is trade'.[224] Then in August the mayor observed that 'piracy has cost us twice as much as the losses we sustained in the recent wars with France'.[225] The problem continued into the next year when he wrote another letter to the lord deputy, stating that 'during these past three years we have been greatly troubled by pirates' and that foreign trade, on which the town relied, had 'been greatly hindered also'.[226] While none of these accounts specifically mention fisheries, it is safe to assume the piracy impacted them too, as well as the trade of fish from the town. As we have seen, the export of fish declined markedly in the 1540s; the high level of piracy at this time may help explain why.

Other towns on the south and east coasts were also suffering from piracy around the same time. In January 1549 Walter Crowley reported to the lord deputy that many ships had been taken in Cork harbour and that 'piracy has greatly disrupted trade'. He then requested that 'royal shipping be dispatched to counter the pirates and apprehend them'.[227] John Goldsmith wrote in August 1548 that Captain Logan, a Scot, was in the vicinity of Lambay and Howth Head near Dublin. A ship had already been captured, along with 'two or three other

[221] *Cal. HCA 1536–1641*, no. 91, pp. 18–19.
[222] *Cal. HCA 1536–1641*, no. 91, pp. 19.
[223] *Cal. HCA 1536–1641*, no. 87, pp. 16–18.
[224] *CSPI* (IMC), 1547–53, no. 112, p. 31.
[225] *CSPI* (IMC), 1547–53, no. 59, p. 17.
[226] *CSPI* (IMC), 1547–53, no. 211, p. 64.
[227] *CSPI* (IMC), 1547–53, no. 189, p. 52.

small boats that were bound for the fisheries at Carlingford'. 'Fishermen or travellers to England dare not venture out of port.'[228]

Piracy not only affected Irish fishers but those who visited Irish waters. A deposition from Dominic de Gryber in 1551 told how in December 1550 two pirate ships – one Irish and one English – came into Youghal and brought with them a ship they had captured at sea. The pirates claimed that they found the ship 'with no creature aboard her' and therefore could claim it as a legal prize.[229] However, de Gryber said he knew it was a Spanish ship from Biscay because he had seen it fishing at Baltimore the previous year.

Instances of piracy decreased during the 1560s and 1570s, primarily due to the state of peace prevailing between England and the major European powers. However, there were still scattered incidents impacting fisheries. In 1577 an English ship captained by William Smithe of Bristol went on a fishing voyage to Waterford, only to be captured in December by English pirates and taken to the Isle of Wight.[230] At this time plans for plantations in Munster were being regularly proposed, and many such schemes listed among their benefits the prevention of piracy. They included the proposal from Humphrey Gilbert and others, who argued that establishing English towns on the coast of Munster would discourage 'European pirates'.[231] In 1573 another plan for settling Munster was put forward, and again the authors claimed that 'numbers of ravening pirates (which be there for the most part succoured) shall be utterly scattered and discouraged'.[232] Several more plantation schemes proposed similar benefits. It would not be until the following century that these plans were actually enacted, but, far from discouraging piracy, many of the towns established by the English settlers in Munster became centres for piracy.[233] As mentioned, Baltimore became a haven for piracy in the early seventeenth century, which may in part have been responsible for its sacking by Algerian pirates in 1631.[234]

Although in general we know comparatively little about piracy beyond the English parts of Ireland, it is clear that piracy was a key part of the maritime operations of some Gaelic clans. Connie Kelleher concluded that 'predation on water

228 *CSPI* (IMC), 1547–53, no. 78, p. 22.

229 *Cal. HCA 1536–1641*, no. 38, p. 8.

230 *Cal. HCA 1536–1641*, no. 218, p. 43.

231 *CSPI* (IMC), 1568–71, no. 273, p. 112, and no. 327, p. 131.

232 *CSPI* (IMC), 1571–75, no. 507, p. 302.

233 John C. Appleby, 'A Nursery of Pirates: The English Community in Ireland in the Early Seventeenth Century', *International Journal of Maritime History*, 2 (1990), pp. 1–27, at p. 2.

234 For a popular account of events, see Theresa Denise Murray, 'Baltimore to Barbary: The 1631 Sack of Baltimore', *History Ireland*, 14, no. 4 (2006), pp. 14–18; see also K. Pluymers, '"Pirates" and the Problems of Plantation in Seventeenth-Century Ireland', in Peter C. Mancall and Carole Shammas (eds), *Governing the Sea in the Early Modern Era* (San Marino, CA, 2015), pp. 79–107.

was a feature of Gaelic-Irish rule throughout the millennia'.[235] By the sixteenth century, the main Gaelic Irish pirates were operating on the more remote west coast. The O'Brian clan controlled the Aran Islands and used their strategic position to extract a tribute of wine from the town of Galway in exchange for keeping Galway Bay free of pirates. A report from 1400 also tells us that the clan engaged in piracy themselves, using galleys to engage in 'spoiling of the king's legies'.[236] In 1568 George Thornton, an English captain, informed William Cecil that, if circumstances allowed, he would fit out his ship for the 'apprehending the remaining pirates in the west of Ireland'.[237]

The O'Malleys of Mayo were also known for their piratical activity.[238] In 1601 Sir George Carew complained of Irish pirates around Limerick and on the river Shannon: they had 'blocked up the river from trade' and prevented merchants from trading.[239] In response, it was suggested that English galleys be sent to the west of Ireland to 'keep the sea coasts from the O'Malies and other rebels'. It was asserted that the clan made a living 'by robbing poor fishermen and others that pass in small vessels, to the great hindrance of the common wealth'.[240] The O'Malleys also produced the most famous of all Irish pirates, Grace O'Malley, who in modern times has been characterised as the 'Pirate Queen'.[241] That title may be an exaggeration, but it is clear that Grace was an exceptional leader who managed to rise to a position of power in an otherwise male-dominated political world. She signed a surrender and regrant agreement with the Crown in 1576 but continued raiding at sea and maintained control over local fisheries for years afterwards.[242] For the O'Malleys and other Gaelic clans, piracy was a means of exerting control over their coastal territory and helping them survive in the face of English encroachment.[243] Although it is impossible to quantify the piratical activities of the Gaelic Irish on the west coast, it seems likely that piracy in the Celtic and Irish seas had a greater economic impact overall, since they were far more heavily trafficked by ships and pirates operating there simply had more opportunities to make valuable captures.

It must also be said that many shipowners and captains who took part in piracy did not always operate as pirates. Pirate crews and vessels could at other

[235] Kelleher, *Alliance of Pirates*, p. 36.

[236] O'Neill, *Merchants and Mariners*, p. 127.

[237] *CSPI* (IMC), 1568–71, no. 149, p. 67.

[238] O'Neill, *Merchants and Mariners*, p. 127.

[239] *CSPI*, 1600, no. 113, p. 245.

[240] *CSPI*, 1601–03, no. 265, p. 259.

[241] Mary O'Dowd, 'O'Malley, Gráinne [Grace] (fl. 1577–1597)' (*Oxford Dictionary of National Biography*, 2008) (https://doi.org/10.1093/ref:odnb/20753) (accessed 13 June 2020).

[242] Kelleher, *Alliance of Pirates*, p. 73.

[243] Kelleher, *Alliance of Pirates*, p. 73.

times be fishers or merchants. The captain or owners of a boat often decided how to use their vessel year on year depending on what seemed most opportune and profitable. The case of John Callice, a Welsh pirate, illustrates this. Callice was implicated in numerous instances of piracy during the 1570s, including the capture on the Irish north coast in 1577 of the *Red Lion* from the Scottish port of Irvine and its cargo of salmon, herring, and salt hides.[244] However, Callice was also a fishing captain. A year earlier, he had been 'bound to Newfoundland on a fishing voyage'.[245] This multi-purpose approach may have suppressed fisheries further in times of war and unrest at sea. When piracy became more profitable and safer than fishing, some captains switched over, compounding the danger for those who remained in peaceful enterprises.

During the early 1580s instances of piracy between England and Spain began to increase rapidly, and this had an impact on Ireland. In 1581 a Spanish ship boarded a fishing boat near Kinsale and captured 'half of the skilful fullest mariners to use for pilots'.[246] Evidently fishers had value to pirates beyond the monetary value of their cargo or ship. Fishers knew their home coasts very well and could be pressed into service. In fact, after being captured by the authorities, many suspected pirates claimed they had been forced into the service of pirate captains unwillingly and had no choice but to participate. In 1610 a mariner by the name of Roger Salter, apprehended for piracy, claimed that while he was at Baltimore a pirate crew forcibly carried him on to their ship and got him drunk. By the time he had woken up, the ship had sailed, and he had no choice but to join the pirate crew.[247] Similar stories were told by other arrested pirates.[248]

During the second Desmond rebellion and Nine Years War, famine struck different parts of Ireland several times. The problem of piracy made it much more difficult to bring in supplies to relieve hunger. Lord deputy Arthur Grey reported to London in 1582 that famine prevailed in Munster and that he feared ships sent to relieve the shortage had 'perished in the tempests or by pirate'.[249] In 1586, it was said that North Wales was 'a place lately of great resort of pirates' which continued to hinder supplies to Ireland.[250] Piracy again severely impacted fishing towns on the Munster coast during this period. In 1586 an English captain wrote that 'many harms have been done of late to the town of Kinsale by pirates and men of war'.[251]

[244] *Cal. HCA 1536–1641*, no. 206, p. 40.
[245] *Cal. HCA 1536–1641*, no. 201, p. 40.
[246] Christopher Walter, Mayor, to the Lord deputy, 18 March 1581 (TNA, SP 63/81, f.75).
[247] *Cal. HCA 1536–1641*, no. 1223, p. 296.
[248] *Cal. HCA 1536–1641*, see no. 1205 and its related entries for more.
[249] *CSPI*, 1574–85, no. 22, p. 353.
[250] *CSPI*, 1586–88, no. 1, p. 51.
[251] *CSPI*, 1586–88, no. 71, p. 190.

From the mid-1580s onwards, English merchant ships began to be seized in Spanish harbours. In response, the Admiralty in England issued letters of reprisal to reclaim the losses inflicted on English traders. Acts of piracy by both Spanish and English intensified, eventually escalating into all-out war and the launch of the Armada in 1588.[252] In Ireland Old English and English leaders responded by improving the defences of fishing and trading towns. Waterford initiated a tax on fishers in 1583 for building fortifications to protect shipping coming and going from the town.[253] The tax included 600 fish from every herring boat and the 'principal' fish from every other vessel.[254] Only in December 1587 was the fort's construction completed amidst fear of Spanish attack. Its declared purpose was to defend 'the harbour of Waterford, and the fleet resorting thither for trade of fishing'.[255] In the same month, there was a report that 'the Spaniards intend to come to this realm this winter': if they 'had sent but six or eight ships to the haven of Waterford at any time since Michaelmas they would have taken above 200 sails of barks that came thither a fishing'.[256]

The new fort does not seem to have been sufficient and, in September 1589, the privy council asked that ships be sent to Waterford for the 'defence of the ships and vessels fishing about the haven'.[257] Then in October, the lord deputy, William Fitzwilliam, feared that the new fort 'could not resist a strong pirate' and that the herring fishing at Waterford 'doth not endure two months' because of piracy. He too asked for ships to be stationed there to 'keep away pirates, of whom there are too many'.[258] A further demand for ships was made in October, on the grounds that Waterford was only 'four days sail' from Spain with a fair wind.[259]

The new defences built at Waterford are visible on a map from 1591 (Plate 4.1), which shows the approach to the city with a sizeable new fortification built at Duncannon, close to the fishing village of Ballyhack. It also picks out the new tower at Hook Head.[260] The example of Waterford illustrates the threats fishing towns faced from piracy and the lengths the authorities would go to protect them. However, in this case, the fishers themselves were taxed to pay for the defences. We also know that English fleets still frequented the area to fish, which may have

[252] Kenneth R. Andrews, *Elizabethan Privateering: English Privateering during the Spanish War, 1585–1603* (London, 1964), p. 3.

[253] *Cal. Car. MSS*, vi, p. 467.

[254] The original used the measure mease; *CSPI*, 1601–03, no. 583, p. 584.

[255] *CSPI*, 1586–88, no. 27, p. 441.

[256] *CSPI*, 1586–88, no. 35, p. 447.

[257] *CSPI*, 1588–92, no. 67(1), p. 243.

[258] *CSPI*, 1588–92, no. 40, p. 254.

[259] *CSPI*, 1588–92, no. 34, p. 252.

[260] These features are described in the map's description, 'The New Description of the Cittie of Waterford with the New Fortifications There...' (Hardiman atlas) (The Board of TCD, IE TCD MS 1209/64, 1591).

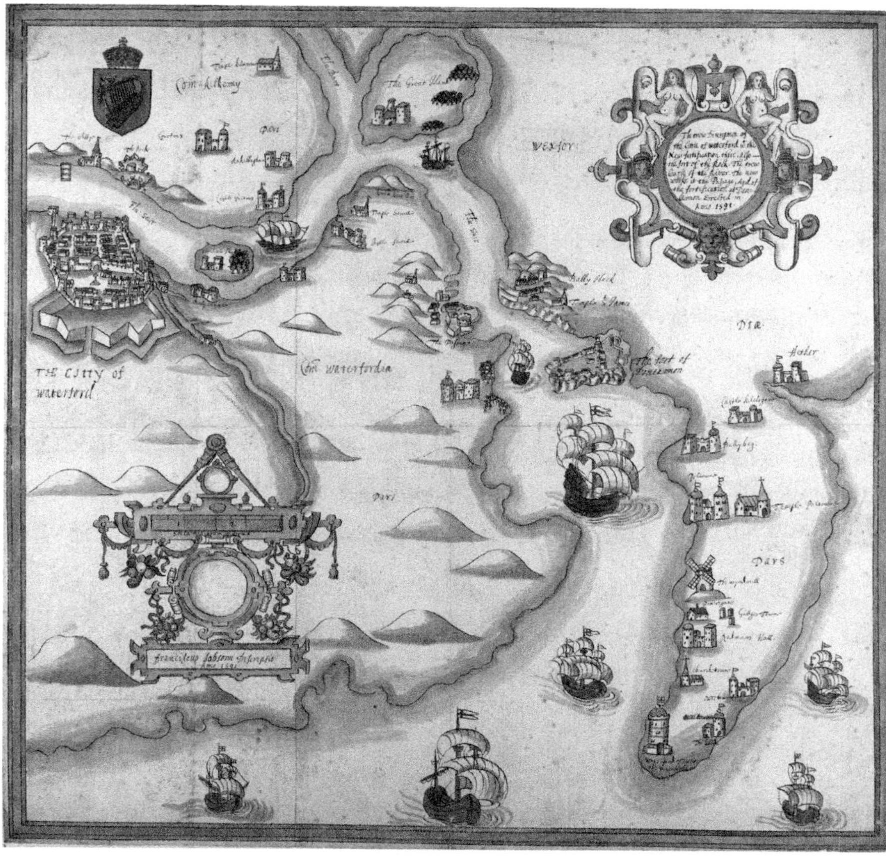

Plate 4.1. 'Cittie of Waterford with the New Fortifications There', 1591.

motivated the English to prioritise the protection of Waterford over other fishing towns.[261] Additionally, food security was a major concern, and Waterford was one of the largest producers of fish in Ireland.

The conflict between Spain and England continued into the 1590s, as did the instances of piracy on the Irish coast. In May 1594 a Spanish pirate ship was loose on the Munster coast, capturing and robbing several fishing boats near Dungarvan and Waterford.[262] During the following year, two barques were taken by Spanish pirates near Dublin.[263] In January 1595, another Spanish ship captured and interrogated fishers around Dungarvan. This last attack seems to have been a scouting mission more than an act of piracy, but we can imagine how fear of Spanish attack

[261] *CSPI*, 1586–88, no. 35, p. 447.
[262] *CSPI*, 1592–96, no. 5, p. 251, and no. 5, 252.
[263] *CSPI*, 1586–88, no. 45, p. 331.

deterred fishers from operating.[264] In January 1597, the herring fisheries across Ireland failed to produce any catches, a failure which may have been caused by natural fluctuations in climate or the herring stocks, but also perhaps by continual attacks by pirates which prevented or dissuaded fishing fleets from going to sea.[265]

As well as affecting whole towns and fishing fleets, piracy could have a devastating impact on the fortunes of individual merchants and fishers. In January 1597, Nicholas Weston loaded his ship, the *Lion*, in Dublin with hides, linen cloth and 4,000 rotten fish. He sailed for Bilbao, a potentially dangerous choice given the war between England and Spain. The ship was intercepted off the Scilly Isles by an English pirate, Captain Norris, who stole some goods and let the ship carry on to Bilbao, where it unloaded and took in 55 tons of iron for Dublin. Before the journey could be completed, however, the ship was captured again between Cork and Youghal by another English pirate, Alexander Vaughan. Vaughan let the crew go free near Kinsale and took the ship off to Wales.[266] More of Weston's ships suffered piratical attacks in the following years, and the losses eventually forced him out of the fish trade altogether.[267] His case shows the high level of risk in fishing and merchant ventures: if a merchant suffered several such losses, his whole business could be put in jeopardy.

The misfortunes of one merchant could also discourage others. In a later case, from 1612, a Wexford merchant was captured in Newfoundland by another English pirate. After this incident, the lord deputy, Sir Arthur Chichester, urged the privy council to compensate the merchant because this event was likely 'to deter all others, of this Country, from such like adventures thither'.[268] Weston's is one of the better-recorded cases, but other fishers and merchants would have experienced similar losses from piracy throughout the sixteenth and early seventeenth centuries.

This chapter has shown that it is possible to formulate a fair understanding of the patterns and frequency of piracy events by compiling records from qualitative sources. It is clear that piracy in Irish waters was a constant, if sporadic, threat through the fifteenth and sixteenth centuries, and that piracy became far worse during periods of European warfare. Piracy was a weapon utilised by all countries in this period. Unfortunately for ordinary merchants and fishers, in times of war they were as likely, if not more likely, to be attacked by their own countrymen. The

264 *CSPI*, 1586–88, no. 24, p. 293.

265 Sir Henry Wallop to Burghley, 18 January 1596/7 (TNA, SP 63 197, f.8); The Lords Justices Loftus and Gardener and Council of Ireland to the Privy Council, 29 November 1597 (TNA, SP 63/201, f.175).

266 *Cal. HCA 1536–1641*, no. 427, pp. 92–93.

267 John C. Appleby, 'The Fishing Ventures of Nicholas Weston of Dublin', *Dublin Historical Record*, 39 (1986), pp. 150–55, at p. 153.

268 Appleby, 'Fishing Ventures of Nicholas Weston', p. 154.

problem of piracy became particularly acute in the late sixteenth century during the Anglo-Spanish war. Individual acts of piracy could ruin the fortunes of any merchant or fisher, and on a broader scale, the presence and actions of pirates could destroy the trade of whole towns and discourage fishers and merchants from operating. Ultimately, the lack of a safe maritime environment in which to operate hampered the economic sustainability and development of Irish fisheries in the sixteenth century and contributed to the decline of the industry. The situation would deteriorate even further during the first decades of the seventeenth century.[269]

Considered together, the dual pressures of terrestrial warfare and piracy at sea had a negative impact on fishing in Ireland and constrained the industry at a lower level of development and production. No part of Ireland escaped the disruption and tragedy caused by warfare and piracy. Some areas, like Dublin, were more protected than others, but even in the capital, trade was disrupted, ships were lost to pirates and, perhaps most important, food supplies were severely hampered. It was probably the hunger and disease caused by war and piracy that inflicted the greatest amount of human suffering in sixteenth-century Ireland. Fish played a vital role in the food security of many populations in Ireland and the disruption of normal fishing and trading operations contributed significantly to food shortages. The next chapter will explore in detail the role fish played in the diet of both ordinary people and soldiers in early modern Ireland.

[269] For more on this, see Senior, *Nation of Pirates*; Appleby, 'Nursery of Pirates'; Pluymers, '"Pirates" and the Problems of Plantation'; and Kelleher, *Alliance of Pirates*.

5

Fish Consumption and Provisioning

Iscach mara muiridi
A críchaib Baí is Béire
Medbhán Faíde firghlaine
Duileasc a cuanuibh Cléire

[Deep sea fishing
From the regions of Dursey and Bere
Laver of clean and fresh Whiddy
Dulse from the harbours of Clear Island]

From the twelfth-century text *Acallam na Senórach*

Hitherto, this book has focused on the mechanics, economics, and ecology of fisheries in fifteenth- and sixteenth-century Ireland, so it can be easy to forget the prime reason why fishing was happening: to provide food. Of course, there are non-food uses for marine products; farmers would have scattered waste fish bones and meat over the fields to add nutrients to the soil. Seaweed too has been used on fields as fertiliser for centuries.[1] But there is no indication that these practices supported large industries in the late medieval or early modern periods. In the twentieth century, the production of fishmeal for animal feed became a major industry; indeed, in some European countries, the consumption of fishmeal even outmatched the demand for fish as food.[2] However, the fishmeal industry did not exist in the fifteenth or sixteenth centuries, and Ireland has never been a major producer, even in modern times.[3]

Food production was almost the sole focus of fisheries in fifteenth- and sixteenth-century Ireland; and like any foodstuff, elements of culture, taste,

[1] Heike K. Lotze et al., 'Ecosystem-Based Management of Seaweed Harvesting', *Botanica Marina*, 62 (2019), pp. 395–409; Leonel Pereira and João Cotas, 'Historical Use of Seaweed as an Agricultural Fertilizer in the European Atlantic Area', in Leonel Pereira et al. (eds), *Seaweeds as Plant Fertilizer, Agricultural Biostimulants and Animal Fodder* (Boca Raton, FL, 2019), pp. 1–22.

[2] This data is available from FAO, 'Food Balances (2010-)' (2022) (https://www.fao.org/faostat/en/#data/FBS) (accessed 20 January 2022); FAO, 'Food Balances (-2013, Old Methodology and Population)' (2021) (https://www.fao.org/faostat/en/#data/FBSH) (accessed 21 May 2021).

[3] Kåre Fasting, *En industri av verdensformat. Utviklingen av sildolje og sildemelindustrien i Norge* (Bergen, 1954); Poul Holm et al., *Proteiner til verden. Fiskemelsindustrien i Esbjerg 1948–1998* (Esbjerg, 1998); Charles L. Cutting, *Fish Saving* (London, 1955), p. 325.

tradition and ritual factor into how people chose and consumed fish. Cured fish was also a vital supply for armed forces in this period and was commonly eaten by soldiers and mariners alike. If we want to fully understand fluctuations in the production and demand of fish, we have to try to understand these elements of food culture and consumption too.

Until recently, Irish historians have shown little interest in studying food or consumption in the early modern period.[4] The task is a challenging one; it is difficult to determine what foods people from different social classes consumed daily, and even more difficult to know how consumption patterns changed over time. This situation is not helped by a dearth of documentary sources detailing daily diet.[5] In 1992, Louis Cullen wrote that Irish consumption history had been too focused on the role of the potato and the Great Famine of the 1840s.[6] He also contended that diet has been used by Irish historians to emphasise poverty and class distinctions, to the detriment of more analytical or detailed studies. Much of what Cullen wrote remains valid today, but the field has also developed since the 1990s.

L. A. Clarkson and Margaret Crawford added greatly to our knowledge of the subject in their monograph covering Irish food consumption from 1500 to 1920.[7] They argued that fish was a 'useful adjunct to diets', but that the bulk of the Irish diet was composed of food from cattle, and also asserted that fisheries in Ireland were not developed substantially until English settlers arrived later in the sixteenth and seventeenth centuries, when catches were mainly exported and had little effect on local consumption.[8] By contrast, this book has shown that fisheries in many parts of Ireland were highly developed before the arrival of English settlers, and the fisheries developed by those settlers were not necessarily more productive.

Máirtín Mac Con Iomaire has briefly traced the history of fish consumption from the Bronze Age onwards.[9] While he presents valuable information on nineteenth-century diets that provides a helpful perspective on consumption in earlier centuries, there is a lack of detail on the early modern period; on the sixteenth century, he notes that Irish fishers had 'up-to-date ships' in 1540 and were quick to exploit the Newfoundland fisheries. No evidence or sources

4 L. M. Cullen, 'Comparative Aspects of Irish Diet, 1550–1850', in Hans J. Teuteberg (ed.), *European Food History, a Research Review* (Leicester, 1992), pp. 45–55, at p. 52.

5 Susan Flavin, 'Food, Drink and Society in Sixteenth-Century Ireland: Cultures of Consumption', in Eve Campbell et al. (eds), *Becoming and Belonging in Ireland c.1200–1600* (Cork, 2018), pp. 212–43, at p. 214.

6 Cullen, 'Comparative Aspects of Irish Diet', p. 52.

7 L. A. Clarkson and Margaret E. Crawford, *Feast and Famine – History of Food and Nutrition in Ireland, 1500–1920* (Oxford, 2001).

8 Clarkson and Crawford, *Feast and Famine*, p. 24.

9 Máirtín Mac Con Iomaire, 'The History of Seafood in Irish Cuisine and Culture', *History Studies*, 5 (2004), pp. 61–76.

are provided to substantiate this claim, and the present author has found little evidence to suggest that Irish fishers were heavily involved in Newfoundland in the 1540s. He goes on to say that Hugh O'Neill 'saw the importance' of commercial fishing and considered forming an 'Irish navy' in the sixteenth century.[10] Again, there is no evidence for this statement.

The absence of research into fish consumption is not just an Irish phenomenon. There is no systematic assessment available of the contribution of seafood to the European diet through the last few centuries. A recent review of consumption history, for example, does not discuss fish except for a fleeting reference to the cod trade, and another voluminous standard world history of food only gives a very cursory reference to seafood.[11]

Fortunately, the situation is beginning to change, with new research projects filling in the gaps in our knowledge. The 4-Oceans project, which will run from 2021 to 2026, has compiled a new dataset of seafood consumption estimates in fourteen European countries from 1500 to 2019, due to appear in 2023.[12] The Irish figures in this dataset only begin after 1700, as there was not enough evidence to make estimates before this point. Nevertheless, this research is a major step forward in our understanding of how seafood consumption changed over time, and gives a far better appreciation of just how much biomass was extracted from the oceans to feed people in Europe. The FoodCult project at TCD also promises to broaden our knowledge of food consumption; it is focused on Ireland from 1550 to 1650, and the project has already produced valuable new insights.[13] FoodCult is taking a very interdisciplinary approach, supporting both archaeological and documentary research streams.

While the present book is focused on documentary evidence, archaeology can also be a valuable asset in understanding past fish consumption. Sheila Hamilton-Dyer has recently discussed the evidence available from fishbone assemblages uncovered in Ireland.[14] This shows a wide range of sea fish present at both coastal and inland sites. The dominant species uncovered in most medieval and late medieval sites were cod and hake. In most instances, whole fish were found

10 Mac Con Iomaire, 'The History of Seafood', p. 63.

11 F. Trentmann (ed.), *The Oxford Handbook of the History of Consumption* (Oxford, 2012); Sheldon Aaronson, 'Algae', in K. F. Kiple and K. C. Ornelas (eds), *The Cambridge World History of Food* (New York, 2000), pp. 231–48.

12 '4-Oceans Project' (TCD, CHAM Centro de Humanidades, Norwegian University of Science and Technology, 2023) (https://www.tcd.ie/tceh/4-oceans/); Poul Holm et al., 'Historical Marine Footprint for Atlantic Europe, 1500–2019', paper in progress (2023), pp. 1–26.

13 Susan Flavin, 'FoodCult: Food, Culture and Identity in Ireland c. 1550–1650' (TCD, 2020) (https://foodcult.eu/); 'Food and Social Politics in Early Modern Ireland: Representing the Peasant in The Parliament of Clan Tomas', *Food and History*, 20 (2022), pp. 9–40, at p. 30.

14 Sheila Hamilton-Dyer, 'Exploring the Contrast: Fish-Bone Assemblages from Medieval Ireland', in *C&H*, pp. 231–38.

at these locations, presumably consumed locally rather than processed for export. A lack of sieving means that, in many cases, smaller bones of fish like those of herring were not collected, although even on sites where sieving was conducted few herring bones were found. This led Hamilton-Dyer to conclude that herring was unlikely to have been an item of everyday consumption in Ireland, and instead most herring was caught by foreign vessels and exported.[15] However, the lack of sieving and the low number of early modern excavations makes it difficult to accept a hypothesis based on archaeological data alone, and current research argues the reverse: that while foreign vessels did catch herring for export, it was also a major element in domestic fisheries and consumption.

Other archaeological remains, such as those of shell middens, indicate that coastal people in Ireland had consumed oysters since the Mesolithic period,[16] collected by hand along with all kinds of other shellfish and crustaceans from the intertidal zones.[17] The consumption of shellfish continued into the sixteenth century; we know that Dublin had an active oyster fishery which supplied the local fish market at that time.[18] Furthermore, the remains of a shell midden in Dublin have been dated to the seventeenth century. Similar post-medieval middens have also been found in Kerry, Galway, Carlingford, and Sligo.[19] There are no records to suggest shellfish was exported in large numbers in the sixteenth century, so we have to assume that much of what was collected was consumed locally, even if, as we will see in the next chapter, stocks of shellfish may have been in decline.

Newer forms of archaeological analysis have the potential to further our understanding of past diets and consumption patterns. Stable isotope analysis measures the levels of carbon and nitrogen isotopes in human remains.[20] In turn, these can tell us the balance of an individual's diet and, specifically concerning marine food, can reveal 'if, when and where marine fish became a dietary staple and by whom such fish were preferentially consumed'.[21] Researchers have conducted stable isotope studies in many parts of Europe, such as the UK, Sweden and Denmark, and have even used this method to trace the spatial dimensions of the medieval

[15] Hamilton-Dyer, 'Exploring the Contrast', pp. 236–37. Although Hamilton-Dyer did state that the interpretation of the archaeological data might well change as new information is generated, and wrote of the 'great potential for further research'.

[16] Aidan O'Sullivan and Colin Breen, *Maritime Ireland: An Archaeology of Coastal Communities* (Stroud, 2007), p. 29.

[17] O'Sullivan and Breen, *Maritime Ireland*, p. 41.

[18] A. E. J. Went, 'Oyster Fisheries', *Dublin Historical Record*, 18 (1963), pp. 56–63.

[19] O'Sullivan and Breen, *Maritime Ireland*, p. 223.

[20] R. H. Tykot, 'Stable Isotopes and Diet: You Are What You Eat', in M. Martini et al. (eds), *Proceedings of the International School of Physics "Enrico Fermi"*, Course CLIV (Amsterdam, 2004), pp. 433–44.

[21] Gundula Müldner, 'Marine Fish Consumption in Medieval Britain: The Isotope Perspective from Human Skeletal Remains', in *C&H*, pp. 239–49, at p. 239.

European cod trade.[22] However, similar research has yet to be done in Ireland. Other methods, such as the DNA analysis of fish and other organic remains, can show us the origin of products traded long distances.[23] In the Irish case, this could be used to trace the importation of foreign fish and perhaps identify fish in other countries that originated in Ireland. Again, little of this research has touched on Ireland, but the FoodCult project intends to apply isotopic analysis and other methods, such as organic residue analysis, to the Irish context.[24]

While future archaeological studies promise new insights, we can still learn a great deal from the limited documentary sources available. The most extensive records concerning diet in our period are the ration and supply records of the English army in Ireland. These are far from perfect; for the most part, they do not reflect the diet of 'ordinary' people; and when it came to rations, what was on paper was not always what the soldiers received in the field. But despite their obvious drawbacks, these records can be valuable. For one thing, they are numerous and cover almost the whole sixteenth century. Reports on supplies to the army can give an idea of which foodstuffs were abundant or in short supply. They also provide an insight into what contemporaries perceived as a proper diet for a soldier, soldiers' tastes and attitudes toward food, and finally, when certain food products arrived in Ireland.

The consumption of Irish fish by the English army dated back to at least the thirteenth century when fish was sent from south-east Ireland to supply English forces on campaigns in Scotland and Wales.[25] Additionally, Irish Chancery records from the mid-fourteenth century demonstrate that fish was a regular part of the household consumption of the Lord Justiciar at that time.[26] In 1393 the

22 Müldner, 'Marine Fish Consumption'; D. C. Orton et al., 'Stable Isotope Evidence for Late Medieval (14th–15th C) Origins of the Eastern Baltic Cod (Gadus morhua) Fishery', *PLoS One*, 6 (2011), pp. 1–11; Gundula Müldner and Michael P. Richards, 'Fast or Feast: Reconstructing Diet in Later Medieval England by Stable Isotope Analysis', *JAS*, 32 (2005), pp. 39–48; Anna Kjellström et al., 'Dietary Patterns and Social Structures in Medieval Sigtuna, Sweden, as Reflected in Stable Isotope Values in Human Skeletal Remains', *JAS*, 36 (2009), pp. 2689–99; Cassady Yoder, 'Let Them Eat Cake? Status-Based Differences in Diet in Medieval Denmark', *JAS*, 39 (2012), pp. 1183–93; James H. Barrett et al., 'Detecting the Medieval Cod Trade: A New Method and First Results', *JAS*, 35 (2008), pp. 850–61; James H. Barrett et al., 'Interpreting the Expansion of Sea Fishing in Medieval Europe Using Stable Isotope Analysis of Archaeological Cod Bones', *JAS*, 38 (2011), pp. 1516–24.

23 W. F. Hutchinson et al., 'The Globalization of Naval Provisioning: Ancient DNA and Stable Isotope Analyses of Stored Cod from the Wreck of the Mary Rose, AD 1545', *Royal Society Open Science*, 2 (2015), pp. 1–13; Bastiaan Star et al., 'Ancient DNA Reveals the Chronology of Walrus Ivory Trade from Norse Greenland', *Proceedings of the Royal Society B: Biological Sciences*, 285 (2018), pp. 1–9.

24 Flavin, 'FoodCult'.

25 *CIRCLE*, Henry III, no. 17, no. 18, no. 20.

26 *CIRCLE*, Edward III, Patent Roll 32, no. 76.

earl of Ormond spent Christmas with his household at New Ross, and a vast amount of food was ordered to keep the household fed, including not only 2,000 gallons of ale, 40 bullocks and 100 pairs of rabbits, but also five meases of herring, 100 of cod and ling, 100 of salted salmon and 100 of whiting.[27]

Before the build-up of English forces in Ireland later in the sixteenth century, fish was consumed by the soldiers of the earl of Kildare. In 1534 Kildare agreed to supply his soldiers with a fixed ration that included both 'flesh days' when meat would be eaten and 'fish days' when either fish or butter was consumed.[28] The practice of instituting meat and fish days was common among armed forces, and the population at large.[29] The soldiers would also get staples of bread and ale every day of the week; their protein source being alternated daily.[30] Other records pertaining to the consumption of fish by armies before the 1560s are rare. This absence of documentation is not necessarily related to the frequency with which fish was eaten but because fewer records were kept relating to supply. As the English army became more professionalised, and as its presence in Ireland increased, we find a growing number of records relating to its supply. However, there are some scattered mentions of fish as victuals from earlier decades. In 1548, an enquiry was made about the supply of fish to English forts in Ireland; and in 1551, there was a brief mention of fish in the supplies of an English expedition to Baltimore.[31]

It was during the conflict in Ulster in the 1560s that the number of records relating to supplies increased dramatically. A document from November 1566 set out that garrisons in Ulster were to have 'biscuit, meal and drink every day', then 'bacon, butter cheese or fish' three days a week.[32] In practice, these set fish days did not always work out as planned. In November 1566 an English captain, Geoffrey Vaughan, was camped in Derry and critically low on supplies. He was expecting a delivery of live cattle from the O'Donnells in Donegal but had to also send a ship off to Scotland to collect fish.[33] The fact that Vaughan could not get fish locally suggests a lack of supply, or a lack of enthusiasm on behalf of the Irish to part with their own.

27 One mease was 600 fish, so this was 3,000 herring total. *CIRCLE*, Richard II, Close Roll 17, no. 71.

28 The agreement was necessary to prevent the forced purveyance of goods which, as mentioned earlier, was a practice hated by the population.

29 Patrick W. Hayes et al., 'European Naval Diets in the Sixteenth Century: A Quantitative Method for Comparative and Nutritional Analysis', *Historical Methods*, 52 (2019), pp. 195–212.

30 *Cal. Car. MSS*, 1515–74, no. 24, p. 28.

31 *CSPI* (IMC), 1547–53, no. 151, p. 41, and no. 286, p. 119.

32 *CSPI* (IMC), 1566–67, no. 307, p. 138.

33 *CSPI* (IMC), 1566–67, no. 339, p. 152.

The primary concern with any foodstuff used in military rations was preservation. In the sixteenth century the technologies available to preserve food were limited. Food could be dried out, preserved in alcohol, salted or brought alive, like the cattle Vaughan expected from Donegal. The staples of the army diet – biscuit, beer, cheese, and salted meat and fish – were all preservable but required expertise or external resources like salt to make them last. Ireland had no domestic supply of salt at this time, which limited the possibility of the English acquiring supplies of preserved fish in certain parts of the island. Of course, soldiers could forage, scavenge, and steal food from the general population to supplement their diet.[34] So we cannot be entirely sure of what they ate, but despite this, ration records are still useful for what they tell us about the preservable foods available and how this may have changed over time.

When we consider food consumption in the early modern period, we also have to think about the pseudo-scientific theories through which contemporaries viewed what they ate. The most important was the Galenic principle of bodily 'humours'. In very simple terms, this assigned hot, cold, moist, and dry properties to different foods.[35] The combination of a food's properties determined how healthy it was but was also intertwined with its moral and social status. Meat was a hot food that 'enflamed the cardinal sin of lust and led to the loss of sexual innocence', while fish was considered cold and moist.[36] Social class also determined what foods were best for an individual: labourers and common soldiers were said to have better digestion and stronger stomachs; they also produced more vital heat, which gave them a greater capacity to handle foods like salted fish.[37] So while preserved fish was a staple of the army for practical and economic reasons, contemporaries also believed that soldiers could handle such foods. Members of the aristocracy, on the other hand, were believed to be suited to more delicate food and advised against coarse preserved fish. Surprisingly, this theory was pervasive across Europe; it permeated all communities within Ireland, from English settlers to Gaelic Irish.[38]

The diet of ordinary people in general remains far more obscure than that of the soldiery, owing to the limitations of the sources. A few examples do exist, however. In the summer of 1552, a group of English adventurers set up a mine

34 Clarkson and Crawford, *Feast and Famine*, p. 180.

35 Eleanor Barnett, 'Reforming Food and Eating in Protestant England, c. 1560–c. 1640', *Historical Journal*, 63 (2020), pp. 507–27, at p. 510; for a full explanation of the theory and its history, see Ken Albala, *Eating Right in the Renaissance* (Berkeley, CA, 2002).

36 Barnett, 'Reforming Food and Eating', p. 514; Craig Muldrew, *Food, Energy and the Creation of Industriousness: Work and Material Culture in Agrarian England, 1550–1780* (Cambridge, 2011), p. 33.

37 Muldrew, *Creation of Industriousness*, p. 38; Flavin, 'Food and Social Politics', p. 14; for more on this in relation to beer, see Susan Flavin et al., 'Understanding Early Modern Beer: An Interdisciplinary Case-Study', *Historical Journal* (2023), pp. 1–34, at p. 3.

38 Flavin, 'Food and Social Politics', p. 15.

near Waterford; their food supplies included 'corn, herring and hake'.[39] Due to their geographic proximity to Waterford, we can assume that they could obtain some fresh fish, but they probably ate preserved products too. Preservation was still an important part of the general food system too, especially for fish, which would rot very quickly without processing. If anyone living inland wanted to eat sea fish, it needed to be salted or dried. Thus, it is not difficult to see why preserved fish became associated with the lower classes; it was mass-produced, relatively cheap and associated with labour.

Meanwhile, the higher classes had greater access to fresh fish; inland landowners controlled rights to river fisheries, giving them freshwater fish like salmon. While coastal lords, such as the O'Driscolls, could demand the rights to prized sea fish like halibut that were caught in their waters.[40] This practice suggests the clan valued the fish for its rarity, and it probably ended up on the lord's table. A core part of Gaelic lordship was the right to levy *cuid oidhche*, which was the privilege of hospitality and food tributes from subjects.[41] The claiming of valuable fish could be viewed as part of this tradition, but one that was adapted by Gaelic families with a maritime focus.

The extant household accounts of William Fitzwilliam, lord deputy of Ireland from 1572 to 1575 and from 1588 to 1594, also provide a rare and detailed view into the consumption of an elite household in the Pale. A recent detailed analysis of these accounts suggested that Fitzwilliam's household at Dublin Castle was not only the largest elite household in Ireland, but possibly the second largest in the English dominion after the royal court itself. The household consumed vast quantities of all foodstuffs, including an impressive variety of fresh and preserved fish.[42]

One of the most detailed accounts comes from 1574, and it provides detailed information on fish consumption. The most consumed fish, in both fresh and preserved states, was herring. In 1574, 8,162 salted herring, 1,280 red herring and 4,274 fresh herring were eaten at the castle. Other preserved fish eaten included eels, cod, and ling. The most popular fresh fish after herring were flounder, whiting and plaice.[43] In total, twenty-four distinct species of fish were consumed, but this does not include generic listings like 'ray' and 'eel', which could have comprised multiple different species.

39 *CSPI* (IMC), 1547–53, no. 346, p. 159.

40 John O'Donovan (ed.), *The Genealogy of Corca Laidhe* (*CELT*, 2011), pp. 100–07.

41 Luke McInerney, 'A Sixteenth Century Bardic Poem Composed for Seán Mac Conmara, Lord of West "Clann Chuiléin"', *Seanchas Ardmhacha*, 23 (2010), pp. 33–56, at p. 46; Katharine Simms, 'Guesting and Feasting in Gaelic Ireland', *JRSAI*, 108 (1978), pp. 67–100, p. 68.

42 Charlie Taverner and Susan Flavin, 'Food and Power in Sixteenth-Century Ireland: Studying Household Accounts from Dublin Castle', *Historical Journal*, 66 (2023), pp. 1–26, at p. 18. For a detailed study of the beer consumed at the castle, see Flavin et al., 'Early Modern Beer'.

43 See the full breakdown at Taverner and Flavin, 'Food and Power', p. 14.

This pattern of consumption generally follows what has been outlined throughout this book; commercially exploited species like herring were by far the most popular. However, other items in the household accounts, like the 3,354 flounder and 1,849 plaice that were eaten, are a surprising addition, as we know virtually nothing about the fishery that caught these flatfish. That said, Dublin was well positioned to draw on marine and freshwater fish supplies, as it was flanked by several smaller fishing towns, which could have supplied marine species, while fisheries on the Liffey could have provided salmon, eels, and trout. Overall, the number of species available and the numbers of preserved fish speak to both the sophistication of the commercial operations that produced large volumes of preserved herring, as well as the bountiful nature of the surrounding sea that provided this elite household with a cornucopia of other marine species.

This variety of consumption was only available to a household such as this, and there was likely a class divide in terms of who ate what. At least 108 people were employed by Fitzwilliam in 1574, and when his family is included, the total size of the household is even greater. It is probable that the servants and guards consumed more of the preserved fish, while the lower volume of fresh fish was reserved for the lord's family, chief offices and guests.[44] When Fitzwilliam was in residence, spending on fish increased by 30 per cent, obviously a result of more mouths to feed, but also likely a reflection of the need to buy more fresh fish.[45]

Records of expenditure alone show that purchases of fish were far below those of meat (and many other foods), with beef being by far the most popular protein source.[46] However, cost alone does not reflect the nutritional value of food, nor does it speak to its cultural or religious significance. The accounts illustrate that the household adhered to regular fast days throughout the year, with over a third of all days in 1574 being reserved for abstinence. This included almost every Friday and Saturday and the duration of Lent. Fish consumption increased markedly during these days, but poultry and some meats, such as mutton, were also consumed.[47]

In contrast to this elite household, we can find details about the diets of skilled workers in Dublin from the 1564/5 accounts of Peter Lewys, proctor of Christ Church cathedral in Dublin. The documents describe the food purchased to feed a group of masons working on the cathedral over that period.[48] The staples of the masons' diets were not dissimilar to those of soldiers or elites; they consumed bread and ale daily and regularly had fresh and salted beef. Also included was

44 Taverner and Flavin, 'Food and Power', p. 13.

45 Taverner and Flavin, 'Food and Power', p. 14.

46 Taverner and Flavin, 'Food and Power', p. 19.

47 Taverner and Flavin, 'Food and Power', pp. 21–22.

48 Raymond Gillespie (ed.), *The Proctor's Accounts of Peter Lewis 1564–1565* (Dublin, 1996); Cullen, 'Comparative Aspects of Irish Diet', p. 50.

a variety of fresh food, including milk, vegetables, and eggs.[49] They ate a range of fish, most commonly herring and cod. Both fresh and salted herring were purchased from merchants near Dublin.[50] On 5 April 1565, the clerk purchased one barrel of herring from George Forland in Dalkey and had it carried to Dublin by horse.[51] On other occasions the clerk purchased thornback rays and cod at Clontarf. A variety of 'fresh fish' was also purchased at different points to feed the workers.[52] The variety of fish eaten by the masons was not nearly as diverse as that of the elite household in Dublin Castle, but it was still varied, albeit dominated by the mass-produced herring.

The masons most commonly ate fish on Fridays, Saturdays, and Wednesdays. The proctor's accounts show that during Lent, weekly expenditure on fish more than doubled, and the masons had more fish during this period than at any other time in the year.[53] This pattern is of course similar to that observed in the army and within Fitzwilliam's household. A similar pattern can be observed in sixteenth-century institutional accounts from England. Fish was consumed all the year round, but for certain institutions it replaced meat entirely during Lent.[54] Together these different accounts show that fast days were still common practice among the English inhabitants of Ireland in the mid-sixteenth century, although the justification and context of that fast were changing. Abstinence from meat and the consumption of fish on certain days was a Jewish tradition adopted by the early Christian church. By the sixteenth century, the practice remained a Catholic tradition but was being contested by Protestant reformers across Europe.[55] Protestant thinkers argued against abstinence from meat, citing scripture to show that both St Paul and Jesus taught that no foods were taboo. Writers like Luther and Calvin went further by claiming that abstinence was a 'reprehensible remnant of Judaism' from which Jesus had liberated his followers.[56] The prevailing view among Protestants is summed up by the English writer Edward Jeninges, who wrote, 'eating of fleshe or forbearing to eate fleshe, is not anie matter or thing concerning saluation of man'.[57] Thus Lutheran and other Reformed churches abandoned compulsory fast days across most of continental

[49] Clarkson and Crawford, *Feast and Famine*, p. 19.

[50] Gillespie (ed.), *Proctor's Accounts*, pp. 25, 61, 64.

[51] Gillespie (ed.), *Proctor's Accounts*, p. 64.

[52] This fish was not named by species; Gillespie (ed.), *Proctor's Accounts*, p. 54.

[53] Gillespie (ed.), *Proctor's Accounts*, p. 15.

[54] Alison Mary Locker, 'The Role of Stored Fish in England 900–1750AD; the Evidence from Historical and Archaeological Data' (PhD thesis, University of Southampton, 2000), p. 90.

[55] Christopher Kissane et al., *Food, Religion and Communities in Early Modern Europe* (London, 2018), p. 54.

[56] Barnett, 'Reforming Food and Eating', p. 515.

[57] Edward Jeninges, *A briefe discouery of the damages that happen to this realme by disordered and vnlawfull diet* (London, 1590).

Europe; however, this was not the case in Protestant England and, by extension, the English parts of Ireland.[58]

Under the reign of Edward VI, many Catholic traditions were discarded, but in 1548 the English Parliament passed an Act that maintained fast days two days a week and during Lent. The Act upheld the belief that meat-eating led to vice, saying 'that due and godly abstinence is a means to virtue and to subdue men's bodies to their soul and spirit'. However, it also offered practical reasons for keeping fish days: that they would preserve fishers in employment and increase the supplies of meat and other foods in the kingdom.[59] The fish days were reinforced by legislation passed in Queen Elizabeth's reign, but were reframed as a secular observance.[60] In 1563 the English Parliament passed a bill on the initiative of William Cecil, who argued that fish days, including a new one on Wednesdays, were necessary to sustain the English fishing industry, produce skilled mariners and so maintain a strong English navy.[61] The fast day on Wednesday proved unpopular, and while it seems to have been observed by the Dublin masons in 1564/5, it was not followed in Dublin Castle in the 1570s.[62] However, secular fish on Friday and Saturday, and during 'political Lent', as it came to be known, continued in England until the Long Parliament of 1642.[63]

English law regarding fast days applied to the English-controlled parts of Ireland but could not have been enforced across the whole island. The English army generally included three fish days per week, as did the navy in the same period.[64] An account of rations from 1578 also mentions that soldiers would receive eight herring a day four times a week during the 'Lent season', suggesting that Lent factored into army food too.[65] These practices were driven by tradition and by the new secular fish days, but it may also have been a practical measure to ensure variety, as fish was one of the major preserved protein products available.

There is additional evidence for civilian observance of fish days in Dublin. In a letter written in 1580, Thomas Might, the 'purveyor to Her Majesty's garrisons' in Dublin, promised Thomas Cecil that he would source what herring he could for the army from the 'Lent store' in Dublin.[66] Evidently the city needed to stockpile

58 Stephen Hampton, '"Welcome Dear Feast of Lent": River Understandings of the Forty-Day Fast in Early Stuart England', *Journal of Theological Studies*, 63 (2012), pp. 608–48, at p. 646.

59 Hampton, 'Welcome Dear Feast of Lent', p. 624.

60 Barnett, 'Reforming Food and Eating', p. 516; Hampton, 'Welcome Dear Feast of Lent', p. 626.

61 R. C. L. Sgroi, 'Piscatorial Politics Revisited: The Language of Economic Debate and the Evolution of Fishing Policy in Elizabethan England', *Albion*, 35 (2003), pp. 1–24, at pp. 5–6.

62 Taverner and Flavin, 'Food and Power', p. 22.

63 Barnett, 'Reforming Food and Eating', p. 515.

64 Hayes et al., 'European Naval Diets'.

65 *Cal. Car.* MSS, 1575–88, no. 80, p. 122.

66 Might's title is mentioned in an earlier letter found in *CSPI* (IMC), 1566–67, no. 368, pp.

fish to meet the far higher demand during Lent. Ultimately we do not know if the masons and other citizens of Dublin viewed eating fish as a religious or secular practice. Probably some upheld Catholic traditions and perceptions while others conformed to the new English laws.

At other times, fish consumption peaked for non-religious reasons. In 1533 an English priest in Dublin reported that the citizens were eating fish five days a week outside of Lent, a trend 'he attributed to a scarcity of meat and not to an excess of piety or a fondness for fish'.[67] This was clearly not a normal level of consumption but was driven by necessity. Many communities in sixteenth-century Ireland had to deal with hunger and shortage. In these instances, people based their food choices on availability and price rather than tradition. The examples discussed here also illustrate that Dublin's fisheries had the capacity to meet the city's demand for fish, both during Lent and at times when other foodstuffs were in short supply. The consumption-side evidence also reinforces a point made before, that Dublin's fisheries were more focused on supplying local demand than the more export-driven fisheries of Munster.

Understanding fish consumption outside the English parts of Ireland is more difficult, especially for the Gaelic Irish community. The perennial issue is a lack of evidence, and one potentially valuable source, Gaelic Irish poetry, contains very few mentions of sea fish.[68] One potential reason is that Gaelic poets generally focused their attention on the Irish nobility, who were their patrons, and would have been averse to making lowly sea fishers or sea fish the subject of their work.[69] Gaelic poetry does, however, contain many mythical and metaphorical allusions to freshwater fish, particularly salmon, Cormac mac Airt choking on a salmon bone, for example. In *Mairg do ghrios Giolla Padraig* the poet Aithirne insults the river Mourne, which then rises up against him, and when the waters subside, many salmon are left stranded.[70]

The few poems that do deal with seafood can provide some useful insights. The poem *Mo-chean do theacht, a sgadáin* (My affection for your coming, o herring), composed around 1630 by an unknown Irish author, deals with the arrival of the herring shoals on the coast and their consumption during Lent. The poem is intended as a moral

161–62; the Lent store is discussed in Thomas Might to Burghley, 28 September 1580 (TNA, SP 63/76, f.146); while a similar statement was made by Might in October, see Thomas Might to Burghley, 18 October 1580 (TNA, SP 63/77, f.104).

67 Clarkson and Crawford, *Feast and Famine*, p. 24.

68 This assertion is based on an extensive search of the Bardic Poetry Database and personal correspondence with Dr Katharine Simms; Katharine Simms and Mícheál Hoyne (eds), 'Bardic Poetry Database' (https://bardic.celt.dias.ie/) (accessed 25 August 2020).

69 I owe this suggestion to Dr Katharine Simms.

70 Thanks are again due to Dr Simms for providing these references. See also Katharine Simms, 'The O Hanlons, the O Neills and the Anglo-Normans in Thirteenth-Century Armagh', *Seanchas Ardmhacha*, 9 (1978), pp. 70–94, at p. 79.

and religious lesson, but also has some practical elements that match up with what we already know about sixteenth-century herring fisheries and ecology.

Mo-chean do theacht, a sgadáin![71]

Mo-chean do theacht, a sgadáin!
druid liom, a dhaltáin uasail;
do chéad beatha 's do shláinte!
do thuillis fáilte uamsa.

Dar anam h'athar, a sgadáin,
gidh maith bradáin na Bóinne,
is duit labhras an duain-se,
ó's tú is uaisle 's is óige.

A fhir is comhghlan colann,
nách déanann comann bréige,
cara mar thú ní bhfuaras;
ná bíom suarach fá chéile.

Dá bhféachdaois uaisle Banbha
cia is mó tarbha don triúr-sa,
iasg is uaisle ná an sgadán,
idir bradán is liúsa.

Is é ar bhféachain gach cósta
go crích bhóchna na Gréige,
iasg is uaisle ná an sgadán
ní bhfuair Conán Chinn-tsléibhe.

A sgadáin shéimh shúgaigh.
a chinn chúmhdaigh an Chárghais,
a mhic ghrádaigh mo charad,
leam is fada go dtángais!

Gidh mór do thuit a-nuraidh
dod ghaol bunaidh fán méis-se,
ná cuimhnigh fíoch ná fala,
ó's tú cara na cléire.

A sgadáin shailltigh shuilbhir
nach bíonn go duilbhir dúinte,
leamsa do theacht ní hanait,
súil ar charaid an tsúil-se.

I dtús an Charghais chéasta,
a fhir le ndéantar comhól,
ortsa, go teacht na Cásga,
is mór mo ghrása 's is romhór.

71 *Mo-chean do theacht, a sgadáin*, poem 1, in Thomas F. O'Rahilly, *Measgra Dánta: Miscellaneous Irish Poems, Part I* (Cork, 1970). My thanks to Dr Katharine Simms for providing a copy of this poem and aiding with the translation, and my sincerest thanks to Professor Sylvester ó Muirí for doing the main part of the English translation.

[*My affection for your coming, o herring*
Sit near me you noble student
For your good health and life
You have earned a welcome from me.

On your father's life, o herring
Although salmon of Boyne are good
For you I compose this verse
Because you are the youngest and most noble.

O man of the completely clean body
Who does not form false friendships
I never had a friend like you
Let us not take each other for granted.

If the nobles of Ireland set out to discover
Which of these three is most useful
A fish more noble than the herring
Let it be a salmon or a pike.

It is after searching each coast
As far away as the seas of Greece
A fish more noble than the herring
Was not found by Conán Cinn tSléibhle.

O herring that is meek and happy
That protects us at the beginning of Lent
O my loving companion and friend
For too long I await your coming!

Although there was a fall last year
In the count of your cousins on this platter
Do not be angry or hateful
Because you are a friend of the clergy.

O herring which is salty and generous
That does not be sad or sorrowful
I never doubt your coming
These eyes wait for a friend.

At the beginning of Lent of the Crucifixion
O man that one takes a drink with
For you, till the coming of Easter
Great is my love and very great.

Herring were targeted by fishers when they came close to shore in large shoals during spawning periods (from September to February on different parts of the coast).[72] Clearly, the title and the first line of the poem, 'My affection for your coming', refer to the yearly arrival of the shoals, a point supported by the line 'For too long I await your coming!'. The spawning and recruitment patterns of

[72] David O'Sullivan et al., 'An Inventory of Irish Herring Spawning Grounds', *Marine Institute, Irish Fisheries Bulletin*, 42 (2013), pp. 1–31, at p. 26.

the herring are unreliable and can change over time; thus, the reference to a 'fall last year/ In the count of your cousins on this platter' suggests a natural drop in the abundance of herring in the years before 1630 when the poem was probably written. This observation is also supported by the trade data discussed earlier. Figure 3.22, for example, shows herring exports in decline in 1626 and being eclipsed by exports of pilchards, a trend that continued into the 1630s.[73]

The author also writes of 'the beginning of Lent of the Crucifixion', suggesting that, as in Dublin, herring was a staple food for the Lenten fast. However, in the Gaelic Irish case, this was still a religious tradition associated with the crucifixion. Susan Flavin has observed of a seventeenth-century Gaelic poem, *The Parliament of Clan Tomas*, that an association with fish symbolised 'religious observation, spirituality, [and] social compliance' for contemporaries.[74] In *Mo-chean do theacht, a sgadáin*, the arrival and consumption of herring is directly related to the church, the author writing the herring is 'a friend of the clergy', perhaps partly because it fed them during the many fast days of the religious calendar, and also because it marked and emphasised their piety.

In addition, the consumption of herring around Lent served a practical purpose. Lent came at the end of winter, a time when other food supplies were running low. The arrival of the herring in the depths of winter would have provided many communities with a much-needed boost to food supplies. This seems to be confirmed by the comment that the herring 'protects us at the beginning of Lent'; a line that may have the literal meaning of guarding against starvation at a time of scarcity. Those herring that were not consumed fresh could be preserved to last over the whole of Lent and later into the year. The availability of preserved herring is clear from the reference to 'salty' herring in the poem.

Interestingly, the herring is presented as a noble fish, despite it being a mass-produced and relatively cheap food for the time, compared to other fish such as halibut or salmon, which had higher status and cost: 'salmon of Boyne are good', but 'For you [the herring] I compose this verse'. The author also writes that 'For your good health and life' the herring have 'earned a welcome'. Representing herring in this way reflects the political context at the time when the poem was written. It captures the changes that were taking place in early seventeenth-century Ireland, a society transitioning from Gaelic rule to English control and plantation. All Gaelic poets in this period were understandably concerned with

[73] Donald Woodward, 'Irish Trade and Customs Statistics, 1614–1641', *Irish Economic and Social History*, 26 (1999), pp. 54–80. For more on why this occurred, see later in this chapter, and A. J. Southward et al., 'Fluctuations in the Herring and Pilchard Fisheries of Devon and Cornwall Linked to Change in Climate since the 16th Century', *Journal of the Marine Biological Association of the United Kingdom*, 68 (1988), pp. 423–45.

[74] Flavin, 'Food and Social Politics', p. 30.

their position in this new world.[75] The author of *Mo-chean do theacht, a sgadáin* asks the 'nobles of Ireland' to discover which fish is most useful of salmon, pike, and herring. Salmon and pike are river fish, a commodity controlled and licensed by landowners. Herring, on the other hand, is a sea fish caught by a class of people of lower status. For this reason, herring itself was a humble fish, consumed by the people and strongly associated with Catholic tradition. Furthermore, herring were in decline in Irish waters in the 1620s and 1630s, so the author may also be reflecting ecological change in his evocation of the herring and in his lament that there was a 'fall last year' in the number of herring on his plate.[76] Praise for herring consumption was thus a reaction to change, and a practical and symbolic way to maintain tradition and keep the past alive.

In another poem from the early seventeenth century, composed by Eochaidh Ó hEoghusa (d. 1613), the narrator sardonically instructs a former cattle-raider, now a butler, on how to restore his digestion. It is the rich food of his new circumstances that is causing ill health, and he should return to the diet he enjoyed as a young raider, mainly sour milk cheese, beans, undercooked meat, and whiting.[77] This poem is also a reaction against change, the alteration in the butler's circumstances symbolising wider shifts in Irish society, and once again, the cure is a return to older and simpler ways.[78] However, we must be careful about assuming that this text, or any other Gaelic poetry, reflects the reality of daily consumption. As Flavin has written, there was a tendency to 'romanticize rustic life as a way of critiquing the state of contemporary society and apportioning blame for social decline'.[79] Following Galenic principles, it was a common trope in much of European literature to describe peasant and lower-class diets as including beans, dairy, and old meat. The seemingly ubiquitous nature of these dietary descriptions calls into question their accuracy and may place them more in the realm of a literary device rather than a real descriptor of dietary habits.[80] The presence of whiting in Ó hEoghusa's poem, however, is not seen in other descriptions of peasant diet, which might suggest it was an authentic detail. Unfortunately, we do

75 Katharine Simms, 'References to Landscape and Economy in Irish Bardic Poetry', in H. B. Clark et al. (eds), *Surveying Ireland's Past: Multidisciplinary Essays in Honour of Anngret Simms* (Dublin, 2004), pp. 145–68, at pp. 159–60.

76 Southward et al., 'Fluctuations in the Herring and Pilchard Fisheries'; Woodward, 'Irish Trade and Customs Statistics'.

77 *A dhuine na heaslainte*, discussed in Simms, 'Landscape and Economy', pp. 158–59. The word used in the original poem is 'faoitel', which Dr Simms believes refers to whiting. The poem has also been published in an untranslated form in Damian McManus and Eoghan Ó Raghallaigh (eds), *A Bardic Miscellany: Five Hundred Bardic Poems from Manuscripts in Irish and British Libraries* (Dublin, 2010), no. 10.

78 Simms, 'Landscape and Economy', p. 159.

79 Flavin, 'Food and Social Politics', p. 23.

80 Flavin, 'Food and Social Politics', p. 20.

not have enough evidence to be certain. In contrast, *Mo-chean do theacht, a sgadáin* does not contain the same obvious tropes concerning peasant diet. Fish fulfils a symbolic role in the text, but the poem also contains many real and accurate details related to early modern herring fishing, which suggest that it may reflect a real dietary provision, albeit one that was fading by the 1630s.

A similar reflection of real economic trends can be found in the long satirical poem *Muintir Fhiodhnach na mionn*, composed around 1600 by Aonghus Ruadh Ó Dálaigh. The poet describes visiting the noble and ecclesiastic families of Gaelic Ireland and complains that they were 'so stingy or poverty-stricken' that they could not feed him or their own families properly. He also makes a curious objection to being served meat by the coarb of Fenagh, who was a religious leader. Katherine Simms suggests that this may have occurred during Lent, and Ó Dálaigh was offended by the failure to observe the fast.[81] Alternatively, Ó Dálaigh's observation may have been simply an allegory for general moral and economic decline rather than a reflection of a real event, but given what we know about the general trajectory of fisheries at this time, we should expect fish to be rare and expensive.[82]

All the poems discussed here reflect the current period of change, not just broader societal change but specific changes in fisheries and food culture. The fact that fish eating is associated in these texts with a return to the past fits with what we know was happening with fisheries in general – not just the shift toward pilchards in the 1620s, but the general decline of Gaelic fisheries over the late sixteenth and early seventeenth centuries. The poems support the assertion that when Gaelic power was intact and fisheries were prosperous, fish like herring constituted a common item of consumption for some Gaelic Irish communities, particularly around spawning time and during Lent. We know that the export trade of fish was important, but before their failure, fisheries in Gaelic Irish areas would have had the capacity to feed local populations, a fact seemingly reflected in the limited poetic representations that survive.

One major deficiency of the poetic sources is that they do not help us to estimate the amount of fish consumed or how its consumption compared to other foodstuffs. As far as the Gaelic Irish are concerned, this cannot be done, but we can say something about the composition of soldiers' diets. In 1573, the earl of Essex was preparing to establish his plantation in Ulster, and the records of his expedition list the rations that soldiers would receive. Every day they could expect a bottle of beer and either a 24-ounce (oz) loaf of bread or 2 pounds (lbs) of biscuit. On fish days, the soldiers would be given half a pound of butter and

81 Simms, 'Landscape and Economy', pp. 157–58.

82 Simms mentions a medieval text that described meat eating during Lent as 'habitual among the twelfth-century Irish' that may support Ó Dálaigh's text, but even assuming this were true, we cannot presume it was the same in Ireland four centuries later; Simms, 'Landscape and Economy', pp. 157–58.

a pound of cheese, or '8 white herrings'.[83] Not surprisingly, this is remarkably similar to the diet provided to English sailors at this time, because it was built around what was locally available and what would survive for a reasonable period without rotting.[84]

Despite what the ration lists said, many references in the State Papers make it clear that supplying these basic needs was a constant challenge for the English army in Ireland.[85] These supply problems also tell us something about the state of fisheries during these periods of conflict. In the later decades of the sixteenth century, the English expanded military operations into more remote parts of Ireland, creating more complex supply lines and requiring the transportation of large quantities of food. In 1574, the English garrison at Athenry in County Galway had to be supplied with substantial amounts of food from Dublin, including the ingredients to brew ale, 250 beeves and 30 barrels of herring.[86] In the same year, a group of merchants sent 364 barrels of herring from Dublin to the earl of Essex in Ulster.[87] Herring from Carlingford was also sent to forces throughout Ulster in 1574.[88] The 1570s is a period where trade data is lacking, so these examples tell us that the herring fisheries around Dublin and Carlingford were still operating effectively. The capacity of those fisheries is further demonstrated in a memorandum from 1575 that stated 'corn and fish will be had in Ireland, butter and cheese must come out of England and Wales'.[89] Even with a shortage of some supplies, fish remained available. However, the army buying up fish would have reduced the availability for the local population. The prices paid by the army were also far below market value, which would have damaged the business of fishers and merchants.[90]

In the late 1570s, the focus of conflict in Ireland shifted from Ulster to Munster, and fish remained part of the army rations in the Munster campaign, too. A report by William Pelham in 1579 recounts that fish was not available locally that year and had to be shipped from England to meet the army's needs in Munster. While the fact the army imported fish from England suggests a shortage in Ireland, other issues may have influenced the decision. Pelham writes that the

83 *CSPI* (IMC), 1571–75, no. 798, p. 485.

84 N. A. M. Rodger, *The Safeguard of the Sea: A Naval History of Great Britain, Volume 1, 660–1649* (London, 1997), p. 235.

85 *NHI, vol. III*, p. 94.

86 'Beeves' refers to beef. *CSPI*, 1571–75, no. 1093, p. 675.

87 *CSPI*, 1571–75, no. 888.1, p. 545.

88 *CSPI*, 1571–75, no. 884.1, p. 542.

89 *CSPI*, 1571–75, no. 1253, p. 768.

90 The example discussed earlier showed herring prices were 33 per cent lower for the army compared to the typical market rate.

merchants supplying the army in Waterford and Cork were 'skilful enough', not in supplying food, but in 'making their own profit'.[91]

In January 1580, seven lasts (or 84 barrels) of stockfish were imported to Ireland to supply 2,000 soldiers for a whole month.[92] This stockfish was imported from England, but had not been produced there; more likely, it originated from Iceland, Norway or the Shetland Islands, the major producers of stockfish at this time.[93] Later that year, the first recorded shipment of Newfoundland fish arrived in Ireland to supply the army.[94] Stockfish and Newfoundland fish were both preserved cod products; stockfish were air-dried cod, while most of the cod from Newfoundland was heavily salted and dried. Europeans had been fishing in Newfoundland since its discovery in 1497, and while pirates had captured a few French ships returning from Newfoundland and brought them to Irish ports, these were not purposeful imports.[95] The army shipments in 1580 are the first purposeful imports of Newfoundland fish to Ireland to have been discovered. The arrival of Newfoundland fish marks one of the biggest shifts in food consumption patterns in the sixteenth century.

In the same letter in which Thomas Might discussed a Lent store of herring for Dublin, he requested that 'butter and very good stockfish and Newfoundland fish' be sent from Liverpool to Dublin to 'serve the fish days' of the army.[96] The importation of stockfish and Newfoundland cod in the 1580s seems connected to a lack of locally available fish. Although Edward Baeshe wrote in September 1580 that stockfish was not needed because there were 'plenty of herrings and other fish' in Munster, the weight of qualitative evidence and quantitative trade data suggests a lack of local fish supplies in Ireland, and that these shortages motivated the importation of alternative fish products for the army.[97]

91 *Cal. Car. MSS*, 1575–88, no. 231, p. 191.

92 One last was equal to 12 barrels. Proportion of victual for 2,000, 19 January 1580 (TNA, SP 63/71, f.46).

93 Orri Vesteinsson, 'Commercial Fishing and the Political Economy of Medieval Iceland', in *C&H*, pp. 71–79; Wendy R. Childs, 'England's Icelandic Trade in the Fifteenth Century: The Role of the Port of Hull', in Poul Holm et al. (eds), *Northern Seas Yearbook* (Esbjerg, 1995), pp. 11–31; J. Jónsson, 'Fisheries off Iceland, 1600–1900', *ICES Marine Science Symposia*, 198 (1992), pp. 3–16; Arnved Nedkvitne, 'The Development of the Norwegian Long-Distance Stockfish Trade', in *C&H*, pp. 50–59; Helge Sorheim, 'The Birth of Commercial Fisheries and the Trade of Stockfish in the Borgundfjord, Norway', in *C&H*, pp. 60–70; Hance D. Smith, *Shetland Life and Trade 1550–1914* (Glasgow, 1984).

94 Thomas Might to Burghley, 28 September 1580 (TNA, SP 63/76, f.146).

95 *Cal. HCA 1536–1641*, no. 109, p. 21, and no. 240, pp. 46–47.

96 Thomas Might to Burghley, 28 September 1580 (TNA, SP 63/76, f.146). The term 'Newfoundland fish' almost always refers to cod.

97 *CSPI*, 1574–85, no. 16, p. 205; Edward Baeshe to Burghley, 6 September 1580 (TNA, SP 63/76 f.17).

Later, in November 1580, a large purchase of 20,000 Newfoundland cod at 20s. per one hundred fish was made at Chester for the army in Ireland. The author of this order stated that the purchase was motivated by the high price of victuals in Ireland. Clearly, Might's efforts in September to source herring in Dublin had not been successful.[98] When this cod reached Ireland, the 'Commissioners for the victuals at Dublin' wrote that it was rotten.[99] When Newfoundland cod was dried and salted correctly, it could last for years, but if the process was rushed or carried out incorrectly, it was liable to rot far more quickly.[100] Much later, in 1750, when the Victualling Board of the English Royal Navy conducted a study on the reliability of different supplies, they found that dried fish was the least reliable foodstuff purchased regularly, rotting on average 7.9 per cent of the time, compared to less than 1 per cent for most other foods.[101] The preservation techniques used in the eighteenth century remained unchanged, so these figures are a useful indication of how reliable supplies of fish in the sixteenth century were as well.

In 1581, army suppliers were attempting to purchase herring for the soldiers in Munster, efforts that fell through due to a lack of money.[102] Around the same time, army officials reported that 'the soldiers in Munster are dying of famine'.[103] Another report from January 1582 noted a complete lack of victuals in Munster.[104] Supplies in Dublin were healthier, and multiple shipments of herring were sent to towns in Munster as the year progressed.[105] Supplies of fish were also brought from Wales.[106] Shipments continued into the next year, but the English problems with supply continued as many ships were destroyed by 'tempests or by pirates' on their way to Munster.[107]

A detailed report dating from January 1583 showed large quantities of herring being shipped from Dublin to Munster, with 20 tons going to Limerick, 13 tons

[98] Chester. W. Glaseour to Burghley, 13 November 1580 (TNA, SP 63/78, f.81).

[99] *CSPI*, 1574–85, no. 69, p. 272.

[100] Peter Pope, *Fish into Wine: The Newfoundland Plantation in the Seventeenth Century* (Chapel Hill, NC, 2004), p. 11. The seller in Chester later asked permission to sell the returned fish on to another buyer: see Chester. W. Glaseour to Burghley, 19 January 1581 (TNA, SP 63/80, f.42).

[101] N. A. M. Rodger, *The Wooden World: An Anatomy of the Georgian Navy* (Glasgow, 1986), pp. 84–85.

[102] Wallop and Waterhouse to Burghley, 14 March 1581 (TNA, SP 63/81, f.65); *CSPI*, 1574–85, no. 20, p. 332.

[103] *CSPI*, 1574–85, no. 20, p. 332.

[104] Wallop and Waterhouse to Burghley, 13 January 1582 (TNA, SP 63/88, f.36).

[105] A rate for herrings, 1582 (TNA, SP 63/98, f.205).

[106] *CSPI*, 1574–85, no. 20, p. 349.

[107] In relation to shipwrecks, see *CSPI*, 1574–85, no. 22, p. 353; Wallop to Burghley, 27 July 1583 (TNA, SP 63/103, f.122); *CSPI*, 1574–85, no. 49, p. 448. On shipments of fish to Munster, see no. 76, p. 480.

to Cork and 32 tons to Waterford.[108] Further shipments were carried over by other ships as well.[109] Since towns like Waterford and Cork usually hosted productive fisheries of their own, these shipments imply a continuing lack of productivity. Another report from February 1583 shows some fish being purchased locally in Wexford and Waterford, but small amounts compared to what was still being shipped from Dublin.[110] A general famine prevailed across Munster at this time, which provides further evidence that fisheries were not operating normally.[111]

A drop in the number of army supply records followed the conclusion of the Desmond rebellion in 1583. However, a report from 1585 outlines changes to the rations that would be supplied to soldiers in subsequent conflicts. The rations remained almost identical to those given to troops in 1573, except for the new inclusion of Newfoundland fish. On fish days in 1583, the soldiers could be fed with eight herring as they were before, or half a 'Newland fish' if it was available.[112]

In the late 1580s the export of herring from Ireland was restricted by the English in order to increase food security and provide more supplies to the army. In 1590 a group of Irish merchants began to petition the authorities to be allowed to export again. In one letter, they asked to export up to 10,000 barrels of herring that would 'not [be] fit to be eaten' by Lent if remaining in Ireland.[113] Lent was the prime season for herring consumption and fisheries were more productive in the relatively peaceful years of the early 1590s than they had been in the 1580s.

Unfortunately, the peace of the early 1590s would not last, as the Nine Years War began in 1593 and continued until 1603. In November 1595, the lord deputy, William Russell, wrote that the harvest had failed due to extremely wet weather. He was running out of money to purchase the 'staples' of 'herrings and beef'.[114] Fisheries were still operating to some extent, for despite this lack of money, an order of herring from Waterford and Wexford was placed in late November 1595. At the same time, beef had to be sent from England as there was none available locally.[115]

War, and the continuing abnormal weather conditions, produced another poor harvest in 1596, when for once hunger reached Dublin too.[116] In October of that year, Sir Henry Wallop requested '600 or 700' barrels of herring to be sent to

108 Note of beef and herrings sent from Dublin, 13 January 1583 (TNA, SP 63/99, f.42).

109 Wallop to Burghley, 13 January 1583 (TNA, SP 63/99, f.40).

110 Note of the charges of beef and herrings sent and to be sent into Munster, 1583 (TNA, SP 63/99, f.112).

111 *CSPI*, 1574–85, no. 49, p. 448.

112 *CSPI*, 1601–03, no. 586, p. 595.

113 Lord deputy to Burghley, 26 January 1589/90 (TNA, SP 63/150, f.40).

114 *CSPI*, 1592–96, no. 10, p. 429.

115 Wallop to Burghley, 30 November 1595 (TNA, SP 63/184, f.131).

116 *CSPI*, 1596–97, no. 50, p. 62, and no. 31, p. 116.

Dublin from Munster to save the troops from starvation.[117] Around the same time the archbishop of Dublin wrote that herring was 'now in season' and requested funds to buy the fish to help feed the population.[118] However, these herring did not materialise. In January 1597 Wallop expected a 'good likelihood of fishing at Waterford' to help feed the country, but the fishery failed.[119] Then, by the end of the year, the herring fishery collapsed throughout Ireland and Scotland, severely limiting supplies.[120]

In response to shortages and the failure of traditional fisheries, some enterprising Irish merchants sought out new sources of food. In January 1596, Nicholas Weston, a Dublin alderman and one of the city's most prominent merchants, made a voyage in his own ship to Danzig and brought back rye, giving 'great relief both to the country here and to the army'.[121] He then sent two more ships to Newfoundland to catch cod.[122] Newfoundland cod had been imported to Ireland since the 1580s, and some Irish ships had carried it to England as early as 1566,[123] but Weston was the first Irish entrepreneur directly involved in the Newfoundland fishery for whom detailed records have survived.[124]

As well as sending his own ships across the Atlantic, Weston also purchased Newfoundland cod from English merchants. In November 1596 he sent four ships to Poole to bring back dried Newfoundland fish to supply Dublin.[125] Weston's efforts helped keep the populace fed in a time of need and contributed to his victory in the Dublin mayoral election of 1597.[126] The following year he put forward a new plan to supply the whole of Dublin and its garrison 'with good Newfoundland fish' for the duration of Lent. He also proposed to supply the garrisons in Ulster with the same product.[127] This plan signalled a significant shift

117 Sir Henry Wallop to Burghley, 18 October 1596 (TNA, SP 63/194, f.142).

118 *CSPI*, 1596–97, no. 27, p. 171.

119 Sir Henry Wallop to Burghley, 18 January 1596/7 (TNA, SP 63/197, f.87).

120 Lords Justices and Council of Ireland to the Privy Council, 29 November 1597 (TNA, SP 63/201, f.175).

121 *CSPI*, 1592–96, no. 13, p. 450; Ruth A. Canning, 'Profits and Patriotism: Nicholas Weston, Old English Merchants, and Ireland's Nine Years' War, 1594–1603', *Irish Economic and Social History*, 43 (2016), pp. 85–112; John C. Appleby, 'The Fishing Ventures of Nicholas Weston of Dublin', *Dublin Historical Record*, 39 (1986), pp. 150–55; Terry Clavin, 'Weston, Nicholas' (*DIB*, 2009) (https://doi.org/10.3318/dib.008977.v1) (accessed 29 July 2022); Colm Lennon, *The Lords of Dublin in the Age of Reformation* (Dublin, 1989).

122 *CSPI*, 1592–96, no. 13, p. 450.

123 K. P. Wilson (ed.), *Chester Customs Accounts, 1301–1566* (Record Society of Lancashire and Cheshire, vol. CXI, Liverpool, 1969), pp. 88–89.

124 Appleby, 'Fishing Ventures of Nicholas Weston', p. 151.

125 Sir John Norreys and Sir Geffrey Fenton to Burghley, 1 November 1596 (TNA, SP 63/195, f.1).

126 Canning, 'Profits and Patriotism', p. 88.

127 *CSPI*, 1598–99, no. 123, p. 297.

in the consumption of fish in Ireland. It constitutes the first recorded instance of a population being sustained with a new product from North America and shows that Ireland was not an exception to the broader trends in the consumption of fish across Europe.[128]

A deposition taken by the Admiralty court in 1602 provides a wealth of details on how Weston's voyages to Newfoundland operated and how he was able to bring this new product to the Irish market. The deponent was Nicholas Parris, a crew member of the *Lion*, one of three ships owned by Weston that sailed from Poole to Newfoundland in April 1601. The ship was 60 tons and described as a 'bad sailor' with poorly fitted sails.[129] Parris came from Poole, but other crew members hailed from Weymouth, Cornwall, London, Ireland, France and the Netherlands.[130] Apparently Weston had an agreement with the queen's purveyors in Ireland to send three ships to Newfoundland that year and bring back fish 'for the provision of the army'.[131] He did not make the voyage himself but appointed a master to command each ship. In this sense, it is debatable if this was an 'Irish' voyage to Newfoundland. Weston financed it and used his own ships, but the English and European crew provided the necessary skills and expertise for the voyage.

Once at Newfoundland, the crew of the *Lion* caught 'eighty and odd thousand' codfish, along with five hogsheads of train oil, which was oil pressed from fish.[132] Unfortunately for Weston and the crew, on the return voyage, the ship was captured near Kinsale by the Spanish fleet. They had spent roughly six months travelling to and fishing in Newfoundland. The *Lion* was towed toward Lisbon and, after breaking free in a storm, was captured by yet another pirate, this time an Englishman, Henry Nott. Ignoring the fact that the *Lion* was a friendly ship, Nott took it back to England and sold the cargo of fish for profit. We know of this because of Weston's attempts to claim compensation in the Admiralty court. Because of all this misfortune, Weston was unable to fulfil his contract to supply the army in Ireland.

In total, the cargo carried by the *Lion* was worth between £515 and £520, based on the fish prices provided by the crew.[133] According to another crew member, Michael Browne, the crew were going to be paid £7, 'a hogshead of train oil, a cloth cloak and two shares' for taking part in the voyage.[134] Parris estimated that

[128] For more on the broader European story, see Poul Holm et al., 'The North Atlantic Fish Revolution (ca. AD 1500)', *Quaternary Research*, 108 (2019), pp. 1–15; and 'Historical Marine Footprint'.

[129] *Cal. HCA 1536–1641*, no. 455 and 456, p. 100.

[130] *Cal. HCA 1536–1641*, no. 450, 454, 455, 456, 457, pp. 98–101.

[131] *Cal. HCA 1536–1641*, no. 450, p. 98.

[132] Most of the fish they caught was dried, with only 4,000 carried back 'wet'.

[133] *Cal. HCA 1536–1641*, no. 450, p. 99, and no. 457, p. 101.

[134] *Cal. HCA 1536–1641*, no. 451, p. 99.

Weston sustained damages in excess of £150, not including the lost value of the ship (between £250 and £300) and the value of the fish.[135] If everything is factored in, then Weston lost almost £1,000, a massive sum for the period.[136] Parris also recounted that, after dropping the fish off in Ireland, the *Lion* was due to pick up salt in La Rochelle and engage in another voyage to Newfoundland that year. This loss of future potential revenue only compounded Weston's misfortunes. He had been the victim of another piracy incident in 1597, and in 1600 we know that he imported more rye and Newfoundland fish but also lost more ships.[137] The combined impact of these losses seems to have forced Weston out of the Newfoundland trade, and by 1603 he was only importing Newfoundland cod from Chester.[138]

Despite Weston's own personal troubles, Newfoundland cod and other fish remained vital supplies for the English army during the closing years of the Nine Years War. An updated ration list from 1599 had one fish day per week on which each soldier would get eight herring or one and a half Newfoundland fish. Alternatively, four soldiers could share either one 'Holland ling' or one large Newfoundland fish.[139] While eight herring remained a consistent ration across the century, the amount of Newfoundland cod provided to each soldier seems to have increased: from half a fish to one and a half, to four men sharing one large fish. These adjustments suggest that there was a more significant variation in the size of cod from Newfoundland compared to herring, meaning that it was harder to establish a consistent portion. This account also suggests that fish from the Netherlands was being imported to Ireland. Another ration list from November 1600 prescribed fish two days a week. Likewise, 'dry fish', likely from Newfoundland, was eaten two days a week by the garrison of Dundalk in October 1600.[140]

Large shipments of herring and Newfoundland fish continued to be imported into Ireland to meet the demands of these rations. Herring mainly came from Scotland to supply troops in Ulster.[141] In 1601 a shipment of 5,000 Newfoundland fish was sent from England to Donegal to supply the army.[142] The English army came to rely on fish in this period as an essential item to keep their troops fed.[143]

[135] Unfortunately, we do not know the size of the crew so we cannot work out the profit Weston expected to make.

[136] This equates to approximately £237,200 today.

[137] Gerald Young, Mayor of Dublin, to Sir Robert Cecil, 10 September 1600 (TNA, SP 63/207/5, f.35).

[138] Appleby, 'Fishing Ventures of Nicholas Weston', p. 153.

[139] *CSPI*, 1599–1600, no. 26, p. 12.

[140] *CSPI*, 1600, no. 109, p. 490.

[141] *CSPI*, 1598–99, no. 22, p. 31, and no. 75, p. 80; *CSPI*, 1600–01, no. 9, p. 10.

[142] Sir Henry Docwra to the English Privy Council, 10 August 1601 (TNA, SP 63/209/1, f.50).

[143] Appleby, 'Fishing Ventures of Nicholas Weston', p. 152.

However, just because the soldiers were kept alive with fish does not mean they enjoyed eating it. Another ration list from 1600 dictated that the men should get one ling or one 'Poor John' on one day in the week.[144] Poor John was English vernacular for dried Newfoundland cod; the name suggests little appreciation for this fish.[145] Newfoundland cod seems to have been despised by many of the soldiers who were provided with it.

In 1599 the earl of Essex requested supplies for his soldiers but specified they should contain 'little or no fish, for it neither keepeth well, nor pleaseth the soldier, who by such victual hath so much to provoke his thirst, and no provision to quench it'.[146] Newfoundland fish was heavily salted for preservation purposes, which would produce thirst, and it did spoil more often than other supplies.[147] The method used to cook the Newfoundland fish is expanded on in a letter written by Thomas Sackville, Lord Buckhurst: every captain should be provided with a kettle in which to seethe (boil) the fish.[148] But before the fish was boiled, it would have to be soaked in water for an extended period to soften it up.[149] Sackville suggests seething the fish with 'herbs and some oatmeal to make porridge'. While this does not sound appetising to the modern palate, he suggests it will provide 'great sustenance and relief to the soldier'.[150] Another contemporary pointed out the problems posed by cooking fish in this manner: the common soldier 'cannot possibly carry fish with him' and had 'neither time to water it, nor vessels to dress it in'.[151] Lord Mountjoy complained that his men were selling away their ration of fish for a fraction of the original cost due to a lack of equipment and fuel necessary to cook it.[152] The English commander in Carrickfergus made similar complaints, requesting that his troops be 'no more victualled with fish' because they had no means to cook it in the field.[153] According to Sir George Carew, biscuit and cheese were far easier to carry because they did not require the 'carriage of pots and pans', and for this reason, fish was 'very unpleasing to the soldier'.[154] However, the dislike extended beyond the practical issues with cooking

[144] *Cal. Car.* MSS, 1601–03, no. 23, p. 500; John Traves, Commissary for Ulster, to Lord Buckhurst, 30 June 1600 (TNA, SP 63/207/3, f.332).

[145] D. B. Quinn, *Explorers and Colonies: America, 1500–1625* (London, 1990), p. 309.

[146] *CSPI*, 1599–1600, no. 54, p. 31.

[147] Rodger, *The Wooden World*, pp. 84–85.

[148] *CSPI*, 1599–1600, no. 7, p. 380.

[149] Terje Inderhaug, 'Stockfish Production, Cultural and Culinary Values', *Food Ethics*, 5 (2020), pp. 1–40, at p. 20.

[150] *CSPI*, 1599–1600, no. 7, p. 380.

[151] *CSPI*, 1600–01, no. 43, p. 42.

[152] *CSPI*, 1600, no. 109, p. 491.

[153] *CSPI*, 1600–01, no. 59, p. 419.

[154] *CSPI*, 1601–03, no. 55, p. 52. See also *CSPI*, 1600–01, no. 68, p. 334, and no. 71, pp. 209–10; *CSPI*, 1601–03, no. 262, p. 265.

the fish. In July 1601, Mountjoy and others complained that their only remaining supplies were dried fish, of which the soldiers were 'exceeding weary', refusing to eat it more than once a week. In the summer heat, the fish sickened the soldiers because of its poor condition and high level of salt.[155] Eating heavily salted foods in warm weather would lead to greater thirst, but it is also important to consider Galenic principles.[156] Mountjoy seems to be suggesting that eating 'hot' salted fish in hot weather was unbalancing his soldiers' health.[157]

A memorandum from August 1601 shows the scale of fish imports needed to feed the troops in Ireland. It lists 5,000 troops at three different locations who needed to be provided with fish two days a week for months at a time.[158] Other accounts paint a similar picture. In May 1601, 90,000 fish were sent to Lough Foyle to keep the troops supplied, and a ship had to stay in the local bay to store the fish. After a time, however, the troops 'utterly refuse to eat of it at all'.[159] Another report from that year alleged that soldiers in Ireland had lived on nothing but 'biscuit and Poor John' for several months.[160] In January 1603 Sir George Carew insisted that, unless other provisions were sent, his soldiers had to 'live this Lent upon poor John' alone.[161] The oversupply of Newfoundland fish must have also contributed to its unpopularity.

When they had variety and the fish was of decent quality, the soldiers could sometimes prefer it. In August 1600 the Treasurer in Dublin asked for the troops in Leinster and Connacht to be supplied with nothing but fish, although it should be of 'excellent good' quality: 'being good fish, the soldier will like it better than always butter and cheese'.[162] For soldiers, variety and quality were the most important characteristics of their rations. However, there was little of either element in the food they were provided at many points during the war.

Corruption in the supply system could partially explain why fish was provided so often. Detailing transgressions among the suppliers of army rations in Ireland in 1602, the lord chancellor and others claimed that the victuallers provided more fish than their contracts specified. They were trying to 'make delivery in fish only' because it was 'a provision most profitable unto them'.[163] Newfoundland fish obviously had high profit margins, but the soldiers despised it. Their dislike

[155] *CSPI*, 1600–01, no. 83, p. 439.
[156] Muldrew, *Creation of Industriousness*, p. 33.
[157] Muldrew, *Creation of Industriousness*, p. 38.
[158] Memorandum on the Provision of Victuals for Ireland, 22 August 1601 (TNA, SP 63/209/1, f.108).
[159] *CSPI*, 1600–01, no. 71, p. 338.
[160] *CSPI*, 1601–03, no. 265, p. 254.
[161] *CSPI*, 1601–03, no. 549, p. 563.
[162] *CSPI*, 1600, no. 63, p. 342.
[163] *CSPI*, 1601–03, no. 210, pp. 304–15.

was compounded by the fact that much of the fish was low-quality or rotting. Merchants in Ireland blamed this on their English suppliers sending bad provisions. Meanwhile, the English suppliers claimed Irish import agents were hoarding good supplies until prices increased, sometimes even allowing food to rot. These instances of corruption may help explain why Newfoundland fish was so common as an English supply during the Nine Years War. While it is difficult to unravel the truth, it seems likely there were high margins to be made on the sale of Newfoundland fish to the army, even if it was inferior in quality.

There were other reasons for the prevalence of fish. Newfoundland cod was cheap and in ready supply; several nations, including the English, had large fleets engaged in its harvesting. We also know that Nicholas Weston and possibly other Irish merchants were involved in the fishing and transportation of the fish to Ireland. The failure of Irish domestic fisheries must have further motivated the importation of supplies, and Newfoundland had the most productive fishery at the time. What is not clear is the extent to which the general population consumed Newfoundland fish in the sixteenth century. Nicholas Weston planned to feed the population of Dublin with Newfoundland cod in 1598 and people there probably could have purchased the fish in the market. But no other records detailing everyday consumption have been found. On the whole, the consumption of Newfoundland fish was very unlikely outside the Pale, especially among the Gaelic Irish, with limited access to markets where the fish was imported.

A lack of documentary evidence makes it difficult to fully understand the role of fish in late medieval and early modern Ireland. But by using the records that are available, we can appreciate a great deal about the role of fish in diet and how this interacted with food culture, religion, societal change, and economics. It is becoming increasingly clear that broader European food perceptions and changes affected all communities in Ireland. Food in Ireland, including fish, was imbued with complex meaning and intertwined with theories about diet and health. Members of both the Gaelic Irish and English communities in Ireland applied Galenic principles in their writings about fish. However, there were also divergent views between these communities when it came to the religious significance of fish in the sixteenth century. Gaelic Irish writers used fish as a practical and symbolic representation of change. On the one hand, they were reacting to the decline of their caste and the collapse of Gaelic power; on the other, their references to fish relate to the overall decline of fisheries and ecological change. Different species of fish were also associated with social class: preserved fish, like salted herring and cod, were associated with the lower classes, while river fish and rare sea fish like halibut were reserved for the well-to-do.

Ultimately, it is very difficult to say just how important fish was to the Gaelic diet in the sixteenth century or before. Ample evidence shows that Gaelic fisheries were significant enough to supply the population, but neither archaeological nor documentary evidence provides a firm basis to say how much fish was eaten in comparison to other foods. However, it does seem clear that fish retained religious significance through the sixteenth century and into the seventeenth. The poem, *Mo-chean do theacht, a sgadáin*, outlines how the arrival of the herring shoals was a significant event that both protected the population from hunger and supplied fish for Lent. For the Gaelic Irish, the fast and the consumption of fish remained part of their identity and religious practice. For others in sixteenth-century Ireland, fasting was being reframed as a secular exercise unrelated to salvation. These practices came from the Westminster Parliament, which, while approving of Protestant objections to fasting, wanted to maintain the tradition for political and military reasons. We know that William Fitzwilliam's household observed fast days and that Dublin city maintained a store of herring to cover demand at Lent, but we do not know how many in the city perceived fish eating as a secular or religious practice.

Army ration and supply records are the most copious sources for fish consumption in sixteenth-century Ireland and also the most problematic. In simple terms, they illustrate that fish formed a critical element in the rations given to soldiers in Ireland over the whole sixteenth century. They do not reflect the fish-eating habits of ordinary people in general, but can highlight periods when supplies of fish were abundant or low, and when new products reached Irish markets. For much of the sixteenth century, the army provided its soldiers with local Irish fish, the most common being herring, with eight herring being a standard portion in rations throughout the period. When fisheries were not doing well, fish had to be imported, especially to meet the army's needs. In the early decades of the century, much of this imported fish came from Scotland and England. However, in the 1590s, when Irish fisheries were suffering their most pronounced collapse, far more was imported from Newfoundland, with stockfish from northern Europe and ling from the Netherlands. The arrival of Newfoundland fish in Europe was in fact one of the most significant shifts in food consumption across Europe in the sixteenth century. Army supply records confirm that Ireland did not stand apart from these broader trends. Some Irish merchants even took part in the trade, with the likes of Nicholas Weston sending his own ships to catch cod and bring it back to Ireland to keep the English army, and at times the general population, supplied with cheap protein. However, Newfoundland cod was not popular among soldiers in the sixteenth century, largely because of the difficulty of cooking cod in the field, but also because of contemporary perceptions of humours and health, with preserved fish unwelcome in hot conditions and associated with lower classes and strong stomachs. Owing to cheapness, and also corruption, soldiers were given little choice but to consume low-quality fish regularly. This overconsumption was also born from necessity; at many points during the Nine Years War, cod kept the English army supplied when all other foodstuffs were lacking. It is no exaggeration to suggest it saved the English war effort, or at least brought about an English victory sooner than would have happened otherwise.

Much more remains to be uncovered about fish consumption in the sixteenth century. We have no quantitative estimates of per capita seafood consumption in Ireland before 1700, and the figures after that point show Ireland was well behind most other European nations, with its consumption closer to mainly landlocked Germany than coastal Portugal, Iceland, or Scotland.[164] It was not until the Celtic Tiger era of the 1990s that Ireland's seafood consumption caught up with the rest of Europe.[165] To explain this would require a much larger study covering a longer period. But it is safe to say that Ireland's fish consumption habits and practices would have looked very different without the turbulent events of the sixteenth century. The decline of fisheries, the exodus of foreign fishers and the transfer of ownership from Gaelic Irish to English settlers altered not only the economics of the industry but consumption too. There was a greater disconnection between the Gaelic Irish and the sea from the seventeenth century onwards. Sea fish consumption continued into the seventeenth century, but the Gaelic Irish had lost control of its production, and the few poets who did discuss sea fish associated it with a lost past. New products too, like Newfoundland cod, did not catch on in Ireland or become a common dietary item after the sixteenth century. Today there are no Irish recipes for salted cod, and it is not commonly found on Irish menus.[166] Meanwhile, in other European nations, such as Spain and Portugal, salted cod became a delicacy and is still commonly eaten today.[167] Had Ireland's fisheries maintained a greater connection to Spain, recipes and tastes for salted cod may have persisted in Ireland too. English soldiers took a severe dislike to the Newfoundland cod at the end of the sixteenth century, and it is possible these attitudes had a lasting impact on its popularity in Ireland and England. Equally, it may simply have been more profitable for English fishers to take their cod to Spain rather than bring it back to England or Ireland.[168]

164 Holm et al., 'Historical Marine Footprint'.

165 FAO, 'Food Balances (-2013, Old Methodology and Population)'.

166 This has not been studied in detail, but Newfoundland fish does not receive a single mention in Clarkson and Crawford, *Feast and Famine*.

167 Inderhaug, 'Stockfish Production', p. 11; Regina Grafe, *Distant Tyranny: Markets, Power, and Backwardness in Spain, 1650–1800* (Princeton, NJ, 2012), p. 52.

168 Grafe, *Distant Tyranny*; Regina Grafe, 'The Globalisation of Codfish and Wool: Spanish-English-North American Triangular Trade in the Early Modern Period', *Economic History Working Papers*, 71 (2003), pp. 1–37; Pope, *Fish into Wine*.

6

Oceanographic Variation and Environmental Change

> There was a fish in the sea found, formed like a man, and was kept with raw flesh and fish six months upon the land, and because it would not speak they threw it into the sea again, whereof it did much rejoice and make a great noise.
>
> Christopher St Lawrence, *The Book of Howth*, compiled between 1567 and 1571[1]

Oceanographic Change and Fisheries

Marine fish are intrinsically linked to the environment and ocean conditions they inhabit. Changes in sea temperature, currents, and chemical balance, among many other factors, can profoundly affect the abundance and behaviour of marine species and, in turn, impact the fisheries that target these species. Today, human-induced climate change is rapidly altering the state of our oceans and threatening the sustainability and viability of marine ecosystems and fisheries across the globe.[2] Human activity was not the main driver of climate or oceanographic change in the past, but natural fluctuations in ocean conditions occurred in the early modern period and almost certainly had an important influence on fisheries in fifteenth- and sixteenth-century Ireland.

To understand the full impact of ocean conditions on historical fisheries is not at all easy. This stems in part from the problem of isolating the impact of different anthropogenic and natural variables. As we have seen, fisheries in fifteenth- and sixteenth-century Ireland were disrupted by many human-related events such as war, piracy, and changing market conditions. Understanding how these events interacted with and combined with natural fluctuations, like changes in ocean temperature, is problematic. It is nevertheless a worthwhile exercise.

The first step is to establish as best we can the trajectory of fisheries over the fifteenth and sixteenth centuries. In other contexts, researchers have been able to generate enough historical data to trace catches over centuries of historical

1 *Cal. Car. MSS*, vi, p. 238.

2 *The Ocean and Cryosphere in a Changing Climate: Special Report of the Intergovernmental Panel on Climate Change* (Cambridge, 2022).

fisheries.[3] In the Irish case, we simply do not have enough data to generate accurate catch-estimates, but by combining quantitative and qualitative data, it is possible to estimate trends in the health of fisheries. This is what is shown in Figure 6.1.

This estimate works on a five-point scale, with a score of one indicating that fisheries were doing poorly in a particular year and a score of five signalling they were doing well. A full explanation can be found in Appendix 3. In brief, the scores are based on the trade data presented in Chapter 3, with higher scores indicating better overall exports of fish. In years without trade data, scores were estimated on the basis of qualitative information and expert knowledge of the fisheries. The scores have been averaged over five-year periods from 1400 to 1604.

We have no quantitative, and very little qualitative, evidence for the period 1405–59, meaning that the trajectory over these years is less accurate. In other years we can be more certain; the trade data and other qualitative accounts strongly suggest that from 1480 to 1540, fisheries were booming. We also know with some certainty that fisheries suffered a sharp decline toward the end of the sixteenth century. Of course, figures for trade do not represent the full extent of fishing in Ireland, especially in Gaelic Ireland, but they still represent the best quantitative information we have, and the pattern in the trade data is also reflected in the qualitative record. This may not be a perfect methodology, but it provides a solid foundation for a comparison with past ocean and climate conditions.

Three paleoceanographic reconstructions of sea surface temperature (SST) were chosen for comparison, selected for their robustness and their temporal and geographic specificity. The Atlantic Multidecadal Variability (AMV) is the most general, covering summer temperature variability for the whole North Atlantic from 800 to 2010. The AMV is based on forty-six series of proxy data from around the North Atlantic, including temperature-sensitive annual tree-ring reconstructions.[4] The other two series come from the same study, which includes a generalised reconstruction of SST for the north-east Atlantic from 1000 to 1974, based on a composite of ten other SST series from around the region.[5] Of

[3] For example, a long-term series of Dutch herring catches was used in Cordula Scherer et al., 'A Historical Plankton Index: Zooplankton Abundance in the North Sea since 800 CE', paper in review (2023), pp. 1–36; that Dutch series and other historical catch data can be found in Poul Holm et al., 'Accelerated Extractions of North Atlantic Cod and Herring, 1520–1790', *Fish and Fisheries*, 23 (2022), pp. 54–72.

[4] Jianglin Wang et al., 'Internal and External Forcing of Multidecadal Atlantic Climate Variability over the Past 1,200 Years', *Nature Geoscience*, 10 (2017), pp. 512–18.

[5] Laura K. Cunningham et al., 'Reconstructions of Surface Ocean Conditions from the Northeast Atlantic and Nordic Seas during the Last Millennium', *The Holocene*, 23 (2013), pp. 921–35; Laura K. Cunningham et al., 'NOAA/WDS Paleoclimatology – 1000 Year Composite Sea Surface Temperature Record from the North Atlantic Ocean, NENA Comp & S1', (2013) (https://doi.org/10.25921/p44h-7t71) (accessed 27 July 2022).

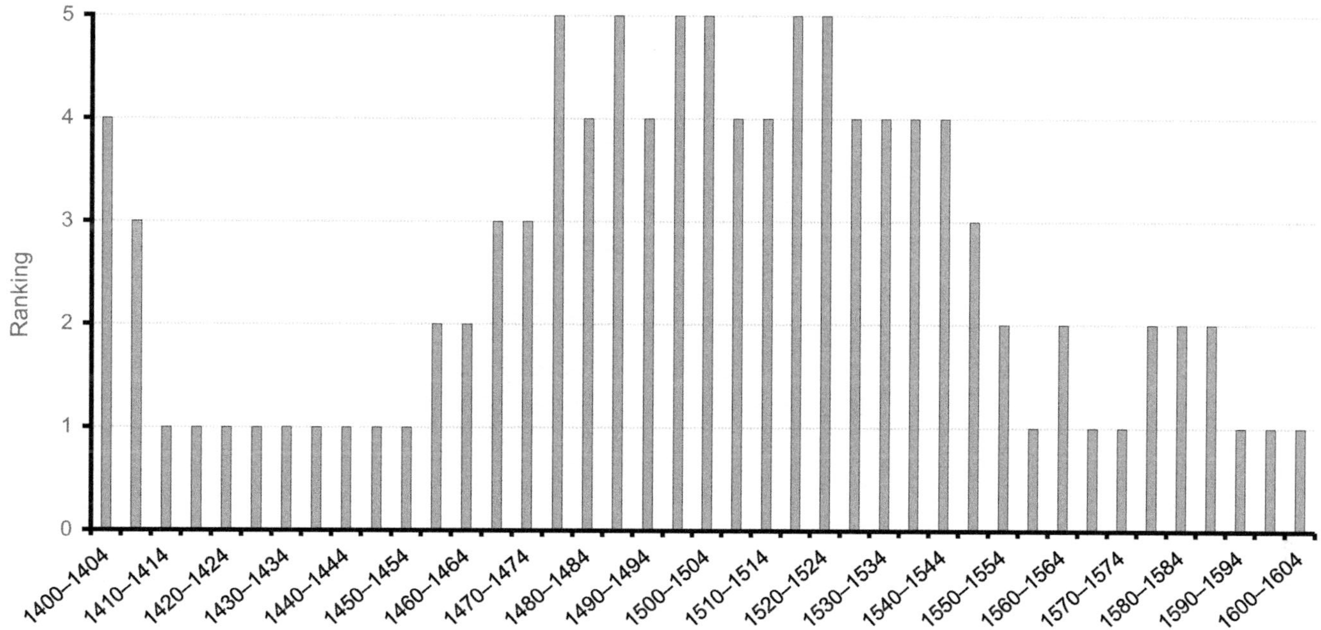

Figure 6.1. Trend of Irish marine fisheries, 1400–1604.

these ten, the closest in proximity to Ireland is Loch Sunart on the east coast of Scotland. While less specific to Ireland, the broader series is possibly the more accurate, as the combining of multiple series should emphasise common trends while reducing individual errors.[6] All three were tested against our estimated fishing trend.

The effect of SST on fish abundance depends on geographic area and species.[7] The relationship between temperature and herring abundance in the North Sea is well understood.[8] Lower temperatures generally result in higher plankton abundance, which in turn leads to greater herring populations. In the historical period, this increase in herring also led to higher herring catches.[9] In our case, herring was the most important marine fish, followed by hake. Understanding how these two species respond to changing temperatures should allow us to interpret the relationship between SST and our fishing trend.

Herring are particularly sensitive to changes in ocean conditions and temperature.[10] On the basis of historical catch data, a study from 1978 showed that herring abundance on the west coast of Ireland was negatively correlated to SST from 1887 to 1974. Lower sea temperatures increase herring abundance, which can result in better catches by fisheries. Again, this could be due to increased plankton production, as in the North Sea, or the impact of temperature on the distribution of herring.[11] Herring and pilchards can alternate on the Irish and English south coasts in response to changing climate and temperature, with herring preferring cooler water.[12]

There are no specific studies on the impact of temperature changes on Irish hake, but a general study indicates that they increase abundance in response to warming.[13] However, the response of hake and all species is also subject to much regional

[6] Cunningham et al., 'Reconstructions of Surface Ocean Conditions', p. 926.

[7] William W. L. Cheung et al., 'Review of Climate Change Impacts on Marine Fisheries in the UK and Ireland', *Aquatic Conservation: Marine and Freshwater Ecosystems*, 22 (2012), pp. 368–88.

[8] Martin Edwards et al., 'Ocean Climate Anomalies and the Ecology of the North Sea', *Marine Ecology Progress Series*, 239 (2002), pp. 1–10; R. D. M. Nash and M. Dickey-Collas, 'The Influence of Life History Dynamics and Environment on the Determination of Year Class Strength in North Sea Herring (Clupea harengus L.)', *Fisheries Oceanography*, 14 (2005), pp. 279–91.

[9] Scherer et al., 'A Historical Plankton Index'.

[10] Adrianus Corten, 'Herring and Climate: Changes in the Distribution of North Sea Herring Due to Climate Fluctuations' (PhD thesis, University of Groningen, 1994); Cheung et al., 'Climate Change Impacts on Marine Fisheries'.

[11] R. J. R. Grainger, 'Herring Abundance off the West of Ireland in Relation to Oceanographic Variation', *ICES Journal of Marine Science*, 38 (1978), pp. 180–88.

[12] Grainger, 'Herring Abundance', p. 186; A. J. Southward et al., 'Fluctuations in the Herring and Pilchard Fisheries of Devon and Cornwall Linked to Change in Climate since the 16th Century', *Journal of the Marine Biological Association of the United Kingdom*, 68 (1988), pp. 423–45.

[13] Cheung et al., 'Climate Change Impacts on Marine Fisheries', p. 383.

variation.[14] So, on the whole, given that herring is very sensitive to temperature changes and was by far the most important catch for the fifteenth- and sixteenth-century fisheries, we can hypothesise that lower average SSTs in our five-year periods should have resulted in a greater abundance of herring and therefore a better score on our fishing trend. Figures 6.2, 6.3 and 6.4 show the variation between the fishing trend and AMV, the Loch and north-east Atlantic SST.

At first glance, there does seem to be a clear correlation between the three series and the fishing trend, particularly for the north-east Atlantic and Loch Sunart. In all series, temperatures dipped at the beginning of the fifteenth century, particularly so in the north-east series, and this coincides with the first recorded large export of fish from Ireland.[15] In all series temperatures then rose for the first four decades of the fifteenth century, coinciding with a period of limited evidence, and thus a conservatively low score. The unfavourable climate conditions may have suppressed fisheries in this period, leading to fewer records. But this is impossible to say for certain.

In all three, there was also a significant dip in temperatures in the 1450s or 1460s, coming slightly later in the north-east compared to the other two. Again, this would seem to correspond to increased exports of fish from Ireland to Bristol, producing a high score in the fishing trend. Temperatures stayed at their lowest levels in all three for several decades in the late fifteenth and early sixteenth century, rising most quickly in the AMV but staying low until the 1550s in the other two series. These differences are probably owing to the fact that the AMV covers a far larger geographical area. The period of suppressed temperatures again seems to correlate with what we know to have been the height of the fisheries in this period, when exports were highest, and fishers from across Europe visited Irish waters.

Finally, the rise in temperatures toward the middle part of the sixteenth century occurred at the same time that, according to export figures, the industry was declining. In the case of Loch Sunart, temperatures stayed high for the remainder of our period, but in the other two series, temperatures dropped again over the 1560s, 1570s and 1580s. By the 1590s, we know that fisheries were in sharp decline. The AMV would seem to suggest that temperature conditions were favourable, as the series shows a significant drop. In contrast, the north-east Atlantic series shows a rapid rise in temperatures, while Loch Sunart remained high. We know from at least one account that the herring fisheries failed in both

[14] Cheung et al., 'Climate Change Impacts on Marine Fisheries', p. 383; Valentina Lauria et al., 'Regional Variation in the Impact of Climate Change: Evidence That Bottom-up Regulation from Plankton to Seabirds Is Weak in Parts of the Northeast Atlantic', *Marine Ecology Progress Series*, 488 (2013), pp. 11–22.

[15] Wendy R. Childs, 'Ireland's Trade with England in the Later Middle Ages', *Irish Economic and Social History*, 9 (1982), pp. 5–33.

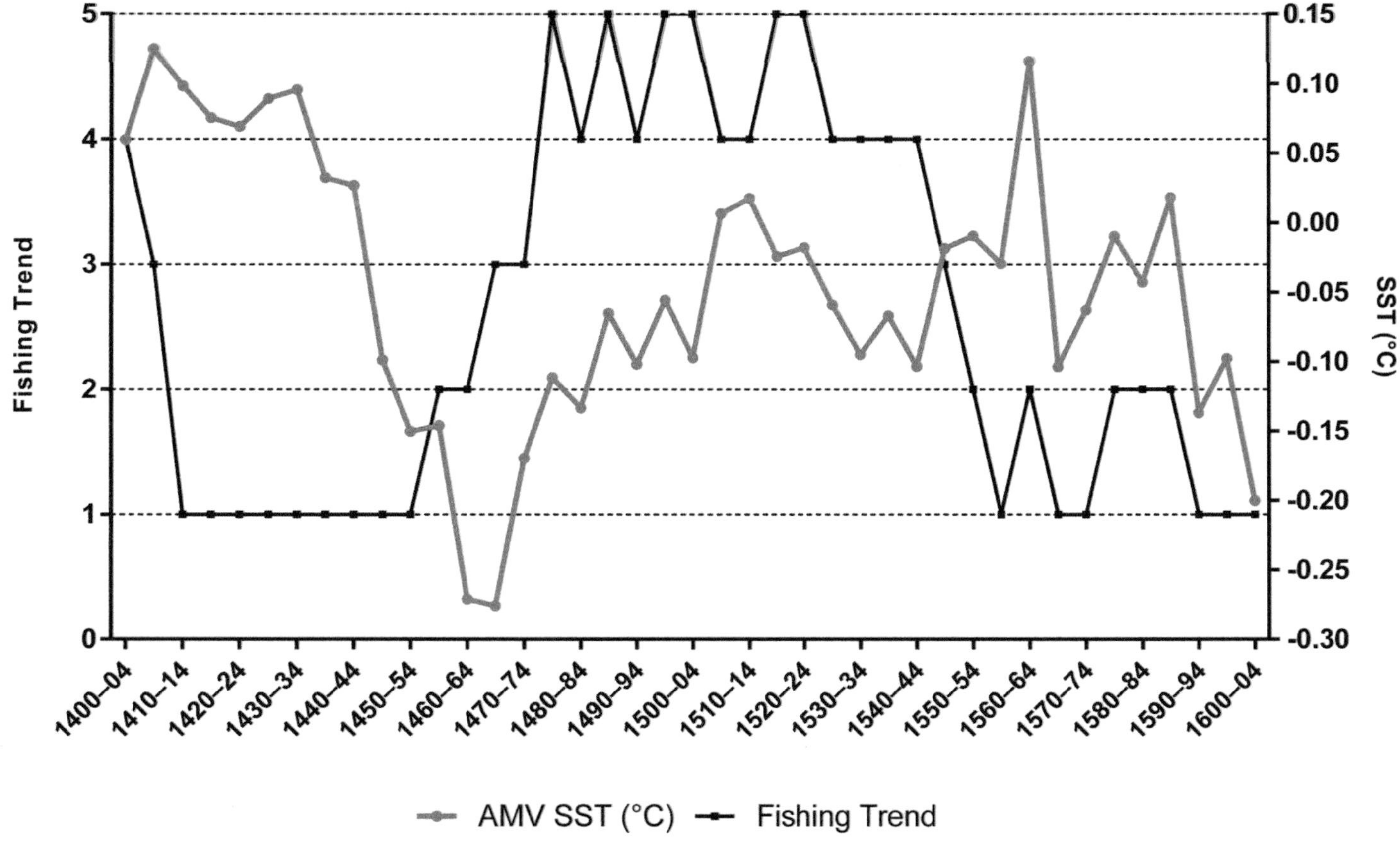

Figure 6.2. AMV compared to fishing trend, 1400–1604.

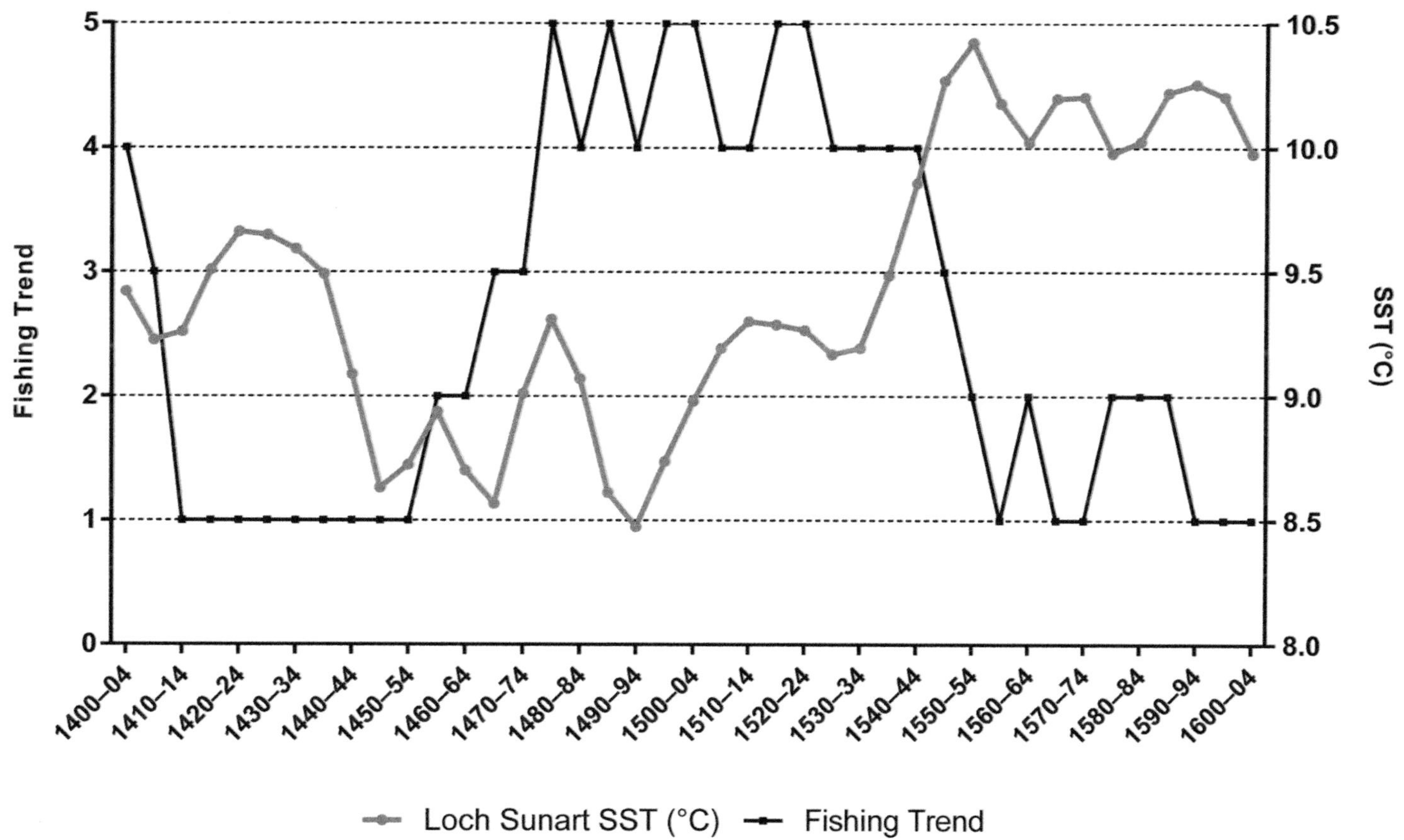

Figure 6.3. Loch Sunart SST compared to fishing trend, 1400–1604. It should be noted that Loch Sunart is considerably warmer than the other two study regions due to its limited geographic extent, ranging from 8.4 to 10.4°C.

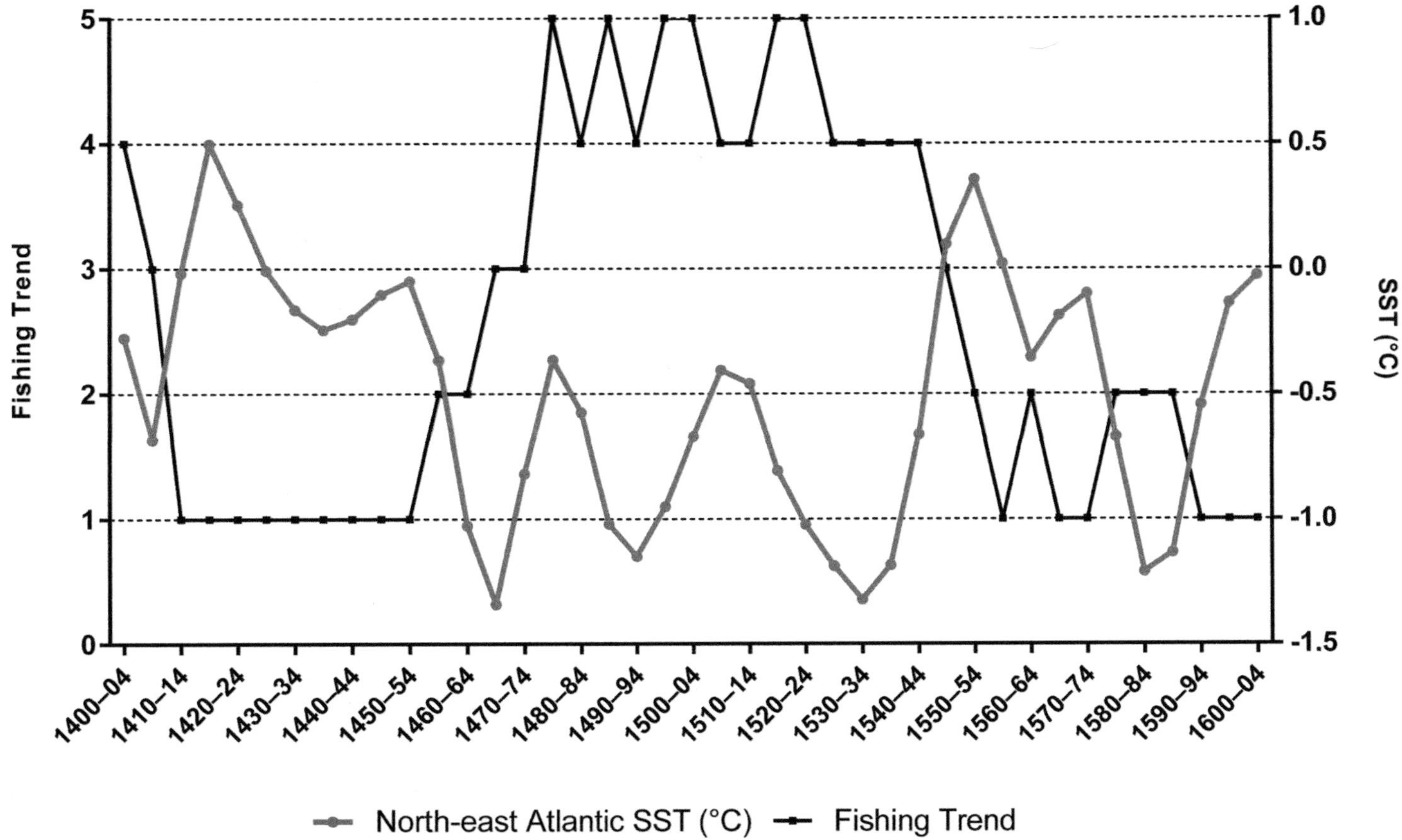

Figure 6.4. North-east Atlantic SST compared to fishing trend, 1400–1604.

Ireland and Scotland in 1597.[16] Scotland was not in the same turmoil as Ireland, so while warfare and piracy were evidently having a negative impact on fisheries in Ireland, two of these historical data series suggest ocean conditions were also not ideal.

While there seems to be a superficial connection between these temperature series and our fishing trend, we can also test any potential relationship statistically. Table 6.1 shows the results of a Pearson linear correlation between each of the temperature series and the fishing trend. This test was chosen because previous studies have indicated a linear relationship between herring abundance and SST in other contexts.[17] In his study of the relationship between SST and herring on the Irish west coast, R. J. R. Grainger found the strongest correlation was present, with a lag of 3–6 between temperature observations and herring catches.[18] This is likely to have been because favourable ocean conditions led to better recruitment and stronger year classes of herring. These juvenile herring reached maturity after several years and then became targets for fishers, stimulating better catches. With annually resolved data for the fifteenth and sixteenth centuries, we could have performed a cross-correlation analysis and possibly have seen stronger correlations, but the data is unavailable. As it stands, the fishing trend is too limited in its temporal scope for such an analysis, but it is also true that the five-year averages will capture some of the influence of a delayed response to changes in SST.

Table 6.1. Temperature series correlated with the fishing trend, 1400–1604.

Temperature series	**Pearson r-value**	**r-squared**	**p-value**
AMV (°C)	-0.14	2%	0.383
Loch Sunart SST (°C)	-0.43	18.62%	0.005
North-east Atlantic SST (°C)	-0.61	36.90%	<0.001

The results indicate a small negative correlation between the fishing trend and the AMV, with an r-value of only -0.14 and an r-squared value of 2 per cent. However, the p-value suggests the results are not statistically significant. We can see a negative correlation, but the results are too weak to confirm it. The AMV does still show synergy with our fishing trend in some key periods, as in the early and late fifteenth century and again later in the sixteenth century, but over the

[16] The Lords Justices Loftus and Gardener and Council of Ireland to the Privy Council, 29 November 1597 (TNA, SP 63/201, f.175).

[17] Grainger, 'Herring Abundance off the West of Ireland'; Scherer et al., 'A Historical Plankton Index'.

[18] Nash and Dickey-Collas, 'Influence of Life History Dynamics'; Scherer et al., 'A Historical Plankton Index'.

whole period, there is no clear statistical connection. This may be due to the broader area covered by the AMV, which may not be specific enough to capture changes around Ireland alone. Temperature values in the AMV also relate only to the summer season (May to September), which does not include the key winter fishing season for herring.[19] There is still a connection between summer and winter sea temperatures, which means the results remain valid, but the lack of winter values weakens the correlation. This result could also be due to inaccuracies in the fishing trend and our overall small sample size of five-year periods.

In the case of the correlation between the Loch Sunart SST and the fishing trend, the results are stronger. Again, there is a negative correlation and a p-value suggesting statistically significant results. The r-value implies that 18.62 per cent of the variability in our fishing trend is driven by the SST at Loch Sunart, not an extremely strong result, but nonetheless suggesting that temperature played a role in the trajectory of Irish fisheries. Again, the fact Loch Sunart is only a small area and not on the Irish coast weakens the correlation, and like the AMV, key winter months are missing from the data. However, both the Loch Sunart and north-east Atlantic series do include one extra month compared to the AMV (October).[20]

The best correlation was found between the fishing trend and temperatures across the north-east Atlantic, with another negative r value of -0.61 and a highly significant p-value. While this is the strongest result, the r-squared value still only suggests that 36.90 per cent of the variability in our fishing trend was caused by north-east Atlantic SST. The same caveats discussed for the previous series apply here too. The stronger result may suggest that the north-east Atlantic temperature series is a better reflection of sea temperatures all around the Irish coast. It is not as specific as Loch Sunart but not as broad as the AMV. Figure 6.5 shows a scatterplot of the fishing trend compared to north-east Atlantic SST (plots for the other two series can be found in Appendix 3). The chart clearly shows the negative relationship between north-east Atlantic SST and the fishing trend, something we observed to a lesser degree for Loch Sunart. While there was also a small negative correlation with the AMV, those results have to be discarded because of the high p-value. Overall, the results are mixed, but two of the three series support the hypothesis that lower temperatures did contribute to some extent to a better score on the fishing trend.

However, this evidence for a link between SST and better fishing outcomes is far from conclusive. These results depend entirely on the construction of the fishing trend, which is not based on completely reliable data. The results should be taken as an indicator of a possible connection rather than solid proof. There

19 Wang et al., 'Multidecadal Atlantic Climate Variability'.

20 Cunningham et al., 'Reconstructions of Surface Ocean Conditions', p. 922; David O'Sullivan et al., 'An Inventory of Irish Herring Spawning Grounds', *Marine Institute, Irish Fisheries Bulletin*, 42 (2013), pp. 1–31, at p. 26.

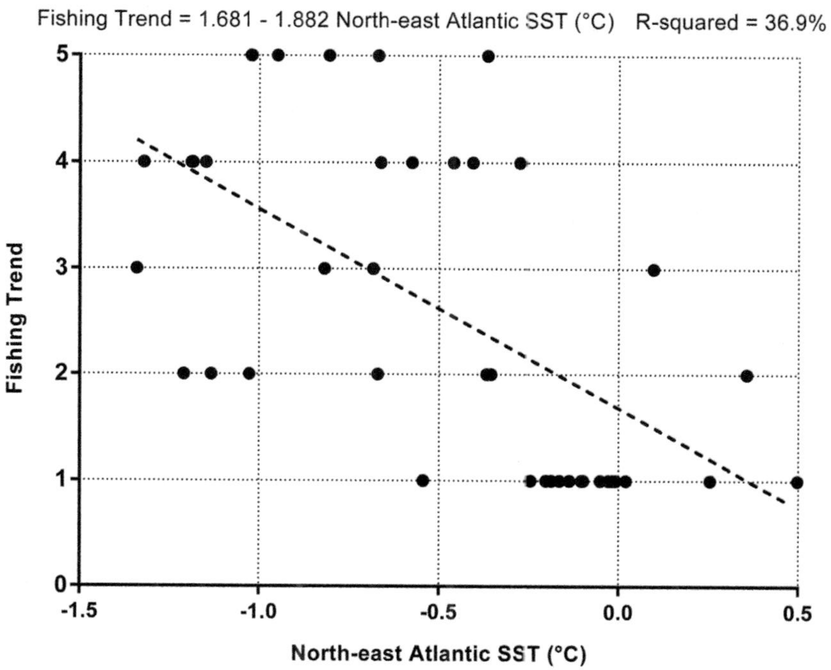

Figure 6.5. Scatterplot of fishing trend vs north-east Atlantic SST.

were of course other influences on the fisheries; piracy and warfare caused serious disruption in the late sixteenth century, as did the shift of European fisheries to Newfoundland. Other climatic and oceanographic changes, not all of which are quantifiable, would also have had an influence. Nevertheless, the SST data do suggest that ocean temperatures were favourable at key points across our study period.

The north-east Atlantic series, in particular, suggests that low temperatures in the mid- to late fifteenth century provided ideal conditions for herring. The period from 1450 to 1550 experienced the lowest sustained SSTs hitherto recorded. A hypothesised period of lower than usual solar activity from the mid-fifteenth century to 1550, known as the Spörer minimum, may have contributed to this cooling.[21] Temperatures in later centuries would get lower for longer, but not before

[21] There is disagreement concerning the exact period covered by the Spörer minimum; see C. Camenisch et al., 'The 1430s: A Cold Period of Extraordinary Internal Climate Variability during the Early Spörer Minimum with Social and Economic Impacts in North-Western and

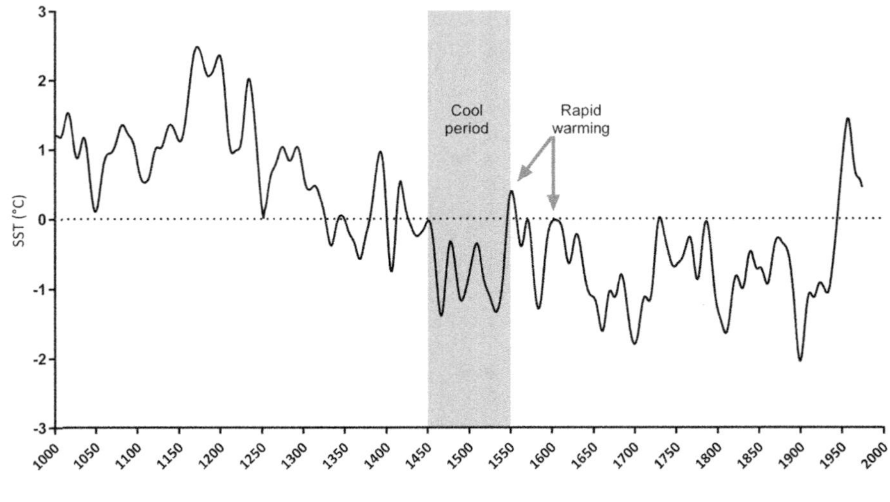

Figure 6.6. Reconstruction of north-east Atlantic SST, 1000–1974.

two periods of rapid warming in the mid- and late sixteenth century.[22] Figure 6.6 shows the whole temperature series for the north-east Atlantic, with these key periods highlighted. It does not seem coincidental that the most prosperous period in the fisheries, from the 1460s to the 1550s, coincides with this long period of cool temperatures. Likewise, periods of rapid warming happened in the late sixteenth century, when we know fisheries were in decline.

In their review of the historical climate in Ireland, Campbell and Ludlow found that the decades from 1440 to 1460 were one of the 'most distinctive episodes on paleoclimatic record'.[23] Those years saw Arctic winters and cooler than usual summers, leading to poor harvests and significant cattle losses that in turn contributed to hunger and social strife. It seems possible that these abnormal conditions were caused by a series of volcanic eruptions, first in 1452 and 1457 and then in 1470, 1477 and 1480. Such volcanic events have the potential to induce severe short-term climate alterations, which may have been compounded by the Spörer minimum.[24]

Central Europe', *Climate of the Past*, 12 (2016), pp. 2107–26, at p. 2108; for the potential impact on SST, see Cunningham et al., 'Reconstructions of Surface Ocean Conditions', p. 929.

22 Cunningham et al., 'Reconstructions of Surface Ocean Conditions', p. 929.

23 Campbell and Ludlow, 'Climate, Disease and Society', p. 228.

24 Campbell and Ludlow, 'Climate, Disease and Society', pp. 228–29.

These eruptions may also have contributed to the cooling represented in the SST data. Unpublished research from Ludlow and others (including the current author) has found that past volcanic eruptions led to a short-term lowering of ocean temperatures in the North Atlantic.[25] Modelling that simulates a large volcanic eruption predicts that much of the North Atlantic Ocean, including Irish waters, would experience significant cooling for five years after such an event.[26] These periods of lower temperatures would in theory be beneficial for herring. This research also shows that catches in the early modern Dutch herring fisheries increased significantly in the years following eruptions. Our data is not robust enough to enable a similar statistical test for Irish fisheries, but Scottish herring export data from Rorke over the period 1473–1599 shows a similar trend, with exports peaking markedly four and five years after eruptions.[27]

It is difficult to conclude with certainty that volcanic eruptions did cause cooling and, in turn, led to better fisheries in the mid-fifteenth century. But the evidence, including the SST data and our fishing trend, would support the idea. Additionally, famines caused by poor harvests and high cattle mortality from 1440 to 1460 may have motivated people in Ireland to turn to the sea for food. If so, they would have found very favourable ocean conditions. The Gaelic annalist Dubhaltach Mac Fhirbhishigh hints at this possibility, observing that the extremely wet summers of 1443 and 1444 were 'good for fish but bad for bees and sheep'.[28]

The time of the greatest decline in fisheries, the 1590s, was also a particularly turbulent time for terrestrial climates.[29] This decade saw persistent food shortages and harvest failures, in part due to war but also possibly due to bad weather. Other parts of Europe also experienced similar political upheaval and famine conditions during the decade.[30] There are many accounts in the State Papers of the extreme weather experienced by the English army in Ireland during the Nine Years War.[31]

[25] Francis Ludlow et al., 'Explosive Volcanism as a Driver of North Atlantic Marine Productivity, Fisheries and Markets, 1400–1850 CE', paper in progress (2023), pp. 1–27.

[26] Ludlow et al., 'Explosive Volcanism as a Driver'; Francesco S. R. Pausata et al., 'Impacts of High-Latitude Volcanic Eruptions on ENSO and AMOC', *Proceedings of the National Academy of Sciences*, 112 (2015), pp. 13784–88.

[27] Ludlow et al., 'Explosive Volcanism as a Driver'; Martin Rorke, 'Scottish Overseas Trade, 1275/86–1597' (PhD thesis, 2 vols, University of Edinburgh, 2001).

[28] Campbell and Ludlow, 'Climate, Disease and Society', p. 225.

[29] Geoffrey Parker, 'History and Climate: The Crisis of the 1590s Reconsidered', in Claus Leggewie and Franz Mauelshagen (eds), *Climate Change and Cultural Transition in Europe* (Leiden, 2018), pp. 119–55, at p. 122.

[30] Parker, 'History and Climate', pp. 153–55; Guido Alfani, 'The Famine of the 1590s in Northern Italy. An Analysis of the Greatest "System Shock" of Sixteenth Century', *Histoire & Mesure*, 26 (2011), pp. 17–50; Peter Clark (ed.), *The European Crisis of the 1590's: Essays in Comparative History* (London, 1985).

[31] Hubert Lamb has also traced several major storm events that struck northern Europe in

In September 1601, Captain Charles Plessington reported from Lough Swilly 'the fiercest storm that ever man saw', lasting '3 or 4 days'.[32] At Kinsale in November, lord deputy Mountjoy wrote, 'God send us better weather than this; for it exceedingly hinders us and kills our men.'[33] The direct impact of the extreme weather on shipping is detailed in a letter from Sir George Carew the following May: 'I have not dared to cross the Bay of Bantry, and we have had such storms of rain and wind that our ships have been in danger even riding in the harbour.' Carew wrote that he 'partly believed' a Catholic priest called Archer was conjuring the weather because 'the old men here never saw the like in May'.[34]

The extremity of conditions in this period would have also impacted fisheries, by sinking vessels and thus dissuading fishers from venturing out, and by cutting off trade routes. Additionally, increased storm activity could also have altered ocean conditions in Europe, leading to the warmer water seen in the SST data.[35] People in the fifteenth century even recognised that storms could drive herring away from certain areas. A statute concerning fisheries around Dublin in 1470 opened with the statement that 'the herring fishery is always casual and removable by storms and winds from one place to another'.[36] Thus the climatically abnormal period of the late sixteenth and early seventeenth centuries probably contributed to and exacerbated the decline of Irish fisheries.

The evidence presented here does not prove that changes in SST, whatever the cause, resulted in the rise and fall of late medieval and early modern Irish fisheries. A corresponding range of human factors were also involved. Unpicking the exact relationship between all the competing influences is virtually impossible. But all three SST series agree that ocean conditions were favourable in the mid-fifteenth and early sixteenth centuries when other anthropogenic factors aligned to allow fisheries to grow. There is also strong evidence for the presence of volcanic forcing in the fifteenth century. Additionally, two of the series (excluding the AMV) agree that ocean conditions were unfavourable in the late sixteenth century when other factors were putting pressure on the industry. The evidence suggests a link between the fortunes of the fisheries and environmental fluctuations.

---

this period: Hubert Lamb, *Historical Storms of the North Sea, British Isles and Europe* (Cambridge, 1991). Francis Ludlow and Arlene Crampsie also comment on the environmental impact of the Nine Years War in *CHI vol. 2*, pp. 627–33.

32 *Cal. Car.* MSS, 1601–03, no. 54, p. 54.

33 *Cal. Car.* MSS, 1601–03, no. 209, p. 192.

34 *Cal. Car.* MSS, 1601–03, no. 386, p. 392. For more examples, see no. 33, p. 36; no. 35, p. 37; no. 209, p. 119; no. 145, p. 125; no. 150, p. 134; no. 214, p. 202; no. 238, p. 219; no. 267, pp. 261–63; no. 491, p. 499.

35 Corten, 'Herring and Climate', p. 195.

36 *Statute Rolls, Ireland*, vol. III, 10 Edward IV, A.D. 1470, p. 665.

Various Irish historians have made claims concerning the influence of fish migrations on fisheries. It is worth exploring these assertions and dispelling some common misconceptions that have been frequently repeated. One of the earliest such claims came from Ada Kathleen Longfield, who wrote that the decline of Irish fisheries at the end of the sixteenth century was caused by a 'natural movement of the best shoals of fish from the Irish to the Newfoundland banks'.[37] This assertion has since been repeated by Susan Flavin and Karin Schüller.[38] Tim O'Neill has also written that 'about 1450 the migrating herring from the Baltic appeared in increasing numbers in the North Sea, the Irish Sea and off the southern and western coasts of Ireland'.[39] He argued that this event caused a boom in the fishing industry and stimulated the Irish economy, writing that it was 'the greatest single economic event of the two [the fourteenth and fifteenth] centuries'.[40] Thomas McErlean et al. also wrote that 'in the late 14th century', 'the herring which traditionally shoaled in the Baltic changed to the North Sea'.[41] In her chapter in the new *Cambridge History of Ireland* (2018), Margaret Murphy cited O'Neill's claim without question, re-emphasising the economic importance of the herring's arrival on Irish shores.[42]

There is however no scientific basis for such migrations. Not only is the ecological basis for Longfield's claim flawed, as herring do not migrate across the Atlantic, the species targeted in Newfoundland was entirely different. Cod was the primary catch for fishers in Newfoundland rather than herring. Additionally, Baltic herring is a separate subspecies of the Atlantic herring that populate Irish waters and the North Sea, which are adapted to their environment and are not capable of migrating out of their natural habitat.[43]

Such claims matter. O'Neill and others have used them to explain the rise and fall of the fisheries and have connected fish migrations to broader themes of Irish economic history. The real reasons for the growth and decline of fisheries in this

37 Ada K. Longfield, *Anglo-Irish Trade in the Sixteenth Century* (London, 1929), p. 56.

38 Susan Flavin, 'The Development of Anglo-Irish Trade in the Sixteenth Century' (MA thesis, University of Bristol, 2004), p. 66; Karin Schüller, 'Special Conditions of the Irish-Iberian Trade during the Spanish-English War (1585–1604)', in Óscar Recio Morales et al. (eds), *Irlanda y la Monarquía Hispánica. Kinsale 1601–2001: Guerra, Política, Exilio y Religión* (Madrid, 2002), pp. 447–68, at p. 450.

39 Timothy O'Neill, *Merchants and Mariners in Medieval Ireland* (Dublin, 1987), p. 30.

40 O'Neill, *Merchants and Mariners*, p. 131.

41 Thomas McErlean et al. (eds), *Strangford Lough: An Archaeological Survey of the Maritime Culture Landscape* (Belfast, 2002), p. 137; Mary O'Dowd, *Power, Politics and Land: Early Modern Sligo 1568–1688* (Belfast, 1991), p. 149.

42 *CHI vol. 1*, p. 414.

43 Sangeet Lamichhaney et al., 'Population-Scale Sequencing Reveals Genetic Differentiation Due to Local Adaptation in Atlantic Herring', *Proceedings of the National Academy of Sciences*, 109 (2012), pp. 19345–50.

period were more complex. If there was an oceanographic reason for the increase in herring abundance and fisheries in the fifteenth century, then it likely lies in the low sea surface temperatures experienced at that time. Such conditions may have increased herring populations and also extended the species' range further south along the Irish coast.[44] At the other end of our period, warmer conditions in the late sixteenth century could have caused the herring to distribute north to cooler waters.

A study of the south of England has shown that, over the last 400 years, herring and pilchard have alternated on the coast in response to changes in ocean temperature, with pilchards becoming abundant on the coast when the temperature was warmer and herring booming when the temperature cooled. In 1590–1640 the pilchard fishing industry was particularly strong in England.[45] In the Irish case, pilchards were not an important catch during the sixteenth century, but in the trade data from 1621 to 1626 they suddenly became the most exported fish by far.[46] This spike suggests that Ireland experienced a similar boom in pilchard fishing in the early seventeenth century and that ocean temperatures in southern Ireland had probably increased. Hitherto, Irish historians have shown little interest in the topic, possibly because the pilchard fishery was very prominent in England in the early twentieth century and subsequently suffered a dramatic decline. No such boom-and-bust occurred in modern Ireland, and consequently pilchards have been given little attention by Irish fisheries scientists.[47] R. J. R. Grainger briefly discussed the link between SST and pilchards in modern times but did not go into detail, while in an article from 1946, Arthur Went had also observed that pilchards were not always present on the Irish coast and established a tentative timeline of those periods when they were abundant, going back to 1611.[48] Ingelise Stuijts and Liam Downey refined this chronology in 2016.[49] If the research was developed further, it might be possible to correlate pilchard abundance with changes in ocean temperature over a much longer period.

Contemporaries in the early seventeenth century did notice that things were shifting, reporting that 'fish begins to wander from our shores', referring to herring.[50] But this also indicates that there was not always a simple relationship

44 Grainger, 'Herring Abundance off the West of Ireland'.

45 Southward et al., 'Fluctuations in the Herring and Pilchard Fisheries'; Donald Woodward, 'Irish Trade and Customs Statistics, 1614–1641', *Irish Economic and Social History*, 26 (1999), pp. 54–80.

46 Woodward, 'Irish Trade and Customs Statistics'.

47 Southward et al., 'Fluctuations in the Herring and Pilchard Fisheries'.

48 Grainger, 'Herring Abundance off the West of Ireland', p. 186; A. E. J. Went, 'The Irish Pilchard Fishery', *Proceedings of the Royal Irish Academy*, 51B (1946), pp. 81–120, at p. 119.

49 Ingelise Stuijts and Liam Downey, *Historical Farm and Fish Food Products* (Carlow, 2016), p. 22.

50 *NHI, vol. III*, p. 181.

between SST and fisheries. In the case of a fishing industry dominated by herring, cooling was favourable, but if warming reached a critical level, combined with other favourable oceanographic conditions, pilchards could become abundant. As the seventeenth-century records show, pilchard fisheries could be extremely prosperous for some but leave others with lower catches. It seems evident that the south of Ireland was well positioned to exploit pilchard fisheries, particularly Munster, with places like Baltimore becoming major producers. In contrast, the north of Ireland may have experienced only a lower abundance of herring, with no pilchards present to provide new fishing opportunities. This pattern also seems to be confirmed by the seventeenth-century trade visualised in Map 3.2.

More localised migrations of herring around Ireland also took place. Spawning herring populations fluctuate on the Mourne coast near Carlingford and the Isle of Man.[51] This natural change in spawning behaviour influenced fisheries. In 1217 fishers from Ireland travelled in large numbers to the Isle of Man when herring became abundant there.[52] No records specifically mention a similar migration in the fifteenth or sixteenth century, but fisheries on the coast of Ulster did clearly decline toward the end of the sixteenth century, so it is possible that migration cycles were involved. Other research has shown that winter and autumn spawning populations of herring in the Celtic Sea alternate in abundance year on year, the variability being driven not only by local conditions such as SST but also by global changes in climate.[53]

Migrations of fish like herring do happen and can have a major influence on the strength of fisheries, but they are also not nearly as dramatic as O'Neill and others have suggested. Instead, they are far more local and connected with fundamental changes in oceanographic conditions. We can begin to understand how past conditions influenced the fishing industry, but more research is required to fully understand the relationship.

Human Impact on Aquatic Ecosystems

Humans also exert their own direct influence on the marine environment. Today we are familiar with problems of overfishing, pollution, and ecosystem damage, and these issues were present in the past. While the evidence is not concrete, we know enough to suggest how early modern peoples could have altered and damaged the aquatic environments on which they relied.

[51] Mark Dickey-Collas et al., 'The Location of Spawning of Irish Sea Herring (Clupea harengus)', *Journal of the Marine Biological Association of the UK*, 81 (2001), pp. 713–14.

[52] William M. Hennessy (ed.), *Annals of Loch Cé 1* (*CELT*, 2008), p. 257.

[53] Clémentine Harma et al., 'The Rise and Fall of Autumn Spawning Herring in the Celtic Sea between 1959 and 2009', *Fisheries Research*, 121 (2012), pp. 31–42, at p. 31.

There are few examples of sixteenth-century fishers showing any concern for the degradation of the marine species they harvested. However, there are indications that people were aware of their potential impact on riverine ecosystems. While river fisheries are not the focus of this work, the examples discussed in this section will highlight that some fisheries in the sixteenth century were vulnerable to human impacts. Furthermore, certain river fisheries, such as those on the Bann, were regularly visited by foreign fisheries and were economically significant, so it is essential to consider how anthropogenic factors influenced them.

In 1466, Dublin Corporation ordered 'that no tanner, glover nor any person use limed ware or leather work in the river Liffey on account of the destruction of the salmon'. Perpetrators had to pay a fine if they polluted the river.[54] The tanning of leather produces a range of pollutants potentially damaging to fish populations.[55] Additionally, a statute of 1538 ordered the removal of all 'weirs, purprestures, millponds, engines, and other obstacles on the river Boyne and other streams in Meath and Uriell'. In the words of the Act, the order was necessary because these structures were doing 'great hurt' to the 'salmon fry' in the rivers.[56] Moreover it was forbidden in Brehon law to erect a fishing weir on or near a neighbour's land.[57] Concern over the raising of weirs was also related to navigation, as additional obstacles in the river would create barriers to boats and migrating fish.

Recent research has linked the decline of salmon in northern European rivers during the medieval and early modern periods to the rise of water-powered mills.[58] Quantitative data shows a correlation between the increasing number of watermills and the decline of salmon in selected rivers. We cannot perform the same analysis for Irish rivers, but the qualitative evidence does suggest that sixteenth-century Irish people were concerned with the health of their rivers. In 1569, the Irish Parliament legislated to prevent the grazing of 'swine, hogge or pigge' near rivers from the fifth of March to the end of September each year. The law noted the 'great hurt and hindrance' livestock did to 'the frye of salmon, eel and other commodious fishes' that inhabited rivers.[59] Several more statutes were passed in 1569, including an 'act against the killing of young fry of eels and salmon', and

[54] A. E. J. Went, 'Fisheries of the River Liffey. Notes on the Corporation Fishery up to the Dissolution of the Monasteries', *JRSAI*, 83 (1953), pp. 163–73, at p. 171.

[55] Mwinyikione Mwinyihija, 'Main Pollutants and Environmental Impacts of the Tanning Industry', in Mwinyikione Mwinyihija (ed.), *Ecotoxicological Diagnosis in the Tanning Industry* (New York, 2010), pp. 17–35.

[56] *Cal. Car. MSS*, vi, f.79, p. 437.

[57] Jim Mac Laughlin, *Troubled Waters: A Social and Cultural History of Ireland's Sea Fisheries* (Dublin, 2010), p. 90.

[58] R. H. J. Lenders et al., 'Historical Rise of Waterpower Initiated the Collapse of Salmon Stocks', *Scientific Reports*, 6 (2016), pp. 1–9.

[59] Longfield, *Anglo-Irish Trade*, p. 56.

an 'act against laying hemp, flax or limed hides in freshwater or rivers'.[60] Similar restrictions were implemented against the flax industry by Dublin Corporation in 1585. They banned the 'retting' or soaking of flax in ditches on the grounds of All Hallows priory 'for avoiding hurt to the increase of fish for that it flows often to the said ditches'.[61] The processing of flax has been shown to lead to damaging pollution as the runoff from the process can easily seep into river systems.[62]

While these examples would suggest that people in the sixteenth century were concerned with conservation, we must remember that powerful and wealthy individuals generally controlled fishing rights on rivers. It was in their financial interest to protect the fisheries. However, in many ways, the impact of the laws highlighted here would be the same as current conservation efforts. Contemporaries clearly understood to some extent the impact industry, river infrastructure and agriculture could have on rivers and how sensitive freshwater ecosystems could be to change. The trade data discussed in Chapter 3 showed that salmon exports in the sixteenth century declined earlier than marine species like herring. It is possible that pollution and the degradation of river ecosystems damaged salmon stocks and contributed to this decline.

Arthur Went has also suggested that long-term overexploitation has damaged or destroyed many oyster colonies in Irish waters.[63] Even in 1600, Dublin Corporation was concerned about the sustainability of their oyster beds. In that year, the Corporation ordered that 'the dredgers for oysters in Poolbeg should bring up the gravel and stones which they find there and cast the same upon the banks for the better preserving of the Pool'.[64] These regulations suggest the oyster habitat was being degraded over time by the harvesting practices.

Legislators did not show the same concern for preserving sea fisheries in the sixteenth century, and there are several reasons why this might have been the case. Observing changes in river systems was more straightforward, as they were smaller and more enclosed habitats. Furthermore, the system of rights and ownership applied to river fisheries did not apply to marine fishing; it was far harder to claim ownership of catches in the sea. River systems were also more vulnerable to human influence than marine ecosystems. However, that is not to say that humans could not negatively impact marine ecosystems. There is also an ongoing debate between marine environmental historians as to the degree to which people impacted fish stocks before the industrialised fishing of the nineteenth century.

[60] *CSPI* (IMC), 1568–71, no. 353, p. 140, no. 354, p. 141, no. 355, p. 141.

[61] A. E. J. Went, 'Fisheries of the River Liffey: II. Notes on the Corporation Fishery from the Time of the Dissolution of the Monasteries', *JRSAI*, 84 (1954), pp. 41–58, at p. 42.

[62] E. F. Howard et al., 'The Pollutional Character of Flax Retting Wastes', *Sewage Works Journal*, 6 (1934), pp. 596–600, at p. 598.

[63] A. E. J. Went, 'Oyster Fisheries', *Dublin Historical Record*, 18 (1963), pp. 56–63, at p. 56.

[64] Went, 'Oyster Fisheries', p. 57.

Some, such as Jeffrey Bolster, have argued that fishers in the past, using primitive technologies, had a substantial impact on marine ecosystems and altered them significantly.[65] According to Bolster, while the impact of human manipulation has become visible in recent years, 'the process itself – little understood and, until recently, generally ignored – has been underway for centuries'.[66]

Consideration of the longer-term history of fish extractions has also filtered into marine science. In 1995 Daniel Pauly called on fisheries managers to take into account the long history of marine systems when determining quotas.[67] Pauly's ideas are important for historians as well because they mean the study of historical fisheries could, in some instances, help to inform the current management of fisheries. Extending the 'baseline' back to the sixteenth century is almost impossible in the Irish case, but the theory does invite us to consider the idea that the ecosystems we view today are far from pristine and have undergone generations of human interference and exploitation. On the other hand, other studies have concluded that past fishers had no substantial impact on stocks. In his work on the Dutch North Sea herring fishery, Bo Poulsen compared the modern total allowable catch (TAC) to historical extractions. Across Europe, TAC is agreed by member states through mechanisms set down in the Common Fisheries Policy. TAC amounts are intended to set extractions at limits deemed sustainable.[68] Poulsen argued that historical herring fishing could not have impacted stocks because the catches were far below the modern TAC.[69] In the case of marine mammals, however, research has shown that humans significantly impacted past populations.[70] Studies focused on Spain, Brazil, and the north-eastern USA have demonstrated that hunters decimated whale populations with rudimentary technology.[71] That said, whales are fundamentally different from superabundant

65 Jeffrey W. Bolster, *The Mortal Sea: Fishing the Atlantic in the Age of Sail* (Cambridge, MA, 2012).

66 Jeffrey W. Bolster, 'Opportunities in Marine Environmental History', *Environmental History*, 11 (2006), pp. 567–97.

67 Daniel Pauly, 'Anecdotes and the Shifting Baseline Syndrome of Fisheries', *Trends in Ecology and Evolution*, 10 (1995), p. 430.

68 European Commission, 'The Common Fisheries Policy (CFP)' (https://ec.europa.eu/fisheries/cfp_en) (accessed 5 April 2020).

69 Bo Poulsen, *Dutch Herring: An Environmental History, c. 1600–1860* (Amsterdam, 2008), p. 80.

70 Poul Holm et al., 'The North Atlantic Fish Revolution (ca. AD 1500)', *Quaternary Research*, 108 (2019), pp. 1–15, at p. 4.

71 There are many studies on this topic, for examples see Randall R. Reeves et al., 'History of Whaling and Estimated Kill of Right Whales, Balaena glacialis, in the Northeastern United States, 1620–1924', *Marine Fisheries Review*, 61 (1999), pp. 1–36; Michael M. Barkham, 'French Basque "New Found Land" Entrepreneurs and the Import of Codfish and Whale Oil to Northern Spain, c. 1580 to c. 1620', *Newfoundland Studies*, 10 (1994), pp. 1–43; Alex Aguilar, 'A Review of Old Basque Whaling and Its Effect on the Right Whales (Eubalaena glacialis) of the North Atlantic', *Report – International Whaling Commission*, 10 (1986), pp. 191–99.

species such as herring, since there are far fewer individuals in whale populations, and they are more vulnerable to human exploitation.

The data collected for the present book suggests that fisheries in Ireland declined towards the end of the sixteenth century. As we have seen, there were shortages of herring in Ireland during the 1590s, and exports to England and Wales completely collapsed. We know that war and piracy damaged fisheries, and that oceanographic conditions were probably unfavourable, but it is also important to ask whether overexploitation contributed to this decline. We can approach the question by comparing Ireland's modern TAC with historical catches. As herring was the most important and most consistent catch during the sixteenth century, it makes sense to focus on that species. In 2019, the TAC for herring in waters surrounding Ireland was 13,268 TLW.[72] Not all these catches go to Irish fisheries but are instead shared among EU member states.

Our best estimate for how much herring was extracted in the sixteenth century derives from the trade data discussed in Chapter 3. That showed the highest level of exports occurring in 1525/6 when 1,092 TLW was exported from Ireland. The highest figure from the seventeenth-century data was 3,679 in 1640/1.[73] However, these are only export figures. Domestic consumption would have to be included in order to calculate the total level of extraction. Unfortunately, this is almost impossible to estimate. Nonetheless, overall the evidence suggests that it is unlikely that sixteenth-century fishers were capable of exceeding or reaching the current TAC of 13,268.

Given the disparity between historical extractions and the current sustainable limits, it would seem unlikely that people in the early modern period were capable of overfishing. However, there are some complicating factors. The Irish coastal region, as defined by the EU and ICES, is far more extensive than the region that would have been accessible to early modern fisheries. For this reason, the stocks available to fishers in the past were more restricted compared to today. Furthermore, the current TACs represent a system that has already undergone massive shifts since the sixteenth century. Additionally, herring were caught during the spawning season when they came close to shore in large numbers. This meant that fishers heavily targeted specific herring populations that were accessible and suited their boats and fishing methods. Modern fisheries scientists have noted that high levels of mortality can be exerted on spawning herring populations with very little fishing effort.[74] That is to say, it does not take many fishers

[72] 'COUNCIL REGULATION (EU) 2019/124 of 30 January 2019 fixing for 2019 the fishing opportunities for certain fish stocks and groups of fish stocks, applicable in Union waters and, for Union fishing vessels, in certain non-Union waters' (The Council of the European Union, 2019).

[73] Woodward, 'Irish Trade and Customs Statistics'.

[74] O'Sullivan et al., 'Irish Herring Spawning Grounds', p. 2.

or many fishing hours to catch or kill a large number of herring during their spawning period. Targeting herring during this time could also negatively impact recruitment and lead to weak class groups in subsequent years, as many fish do not get the chance to reproduce.[75] Fishers in the fifteenth and sixteenth centuries targeted the same herring populations year after year; it is feasible, then, that they could have reduced stock levels of specific spawning populations in the most popular fishing spots.[76] Unfortunately, the evidence is not strong enough to prove this with certainty, but it is possible that unsustainable fishing practices damaged the long-term productivity of herring fisheries in the early modern period.

In the case of every other species of fish, the TAC allocated to Irish fishers exceeds the amount of fish that we can confirm was extracted during the fifteenth and sixteenth centuries. The Irish TAC for hake was 3,493 TLW in 2020, while the highest export of hake on record from our period was 350 TLW in 1492.[77] However, it is notable that the largest catches today are very different from those in the sixteenth century. Today the most popular species of fish caught by Irish fishers are mackerel, horse mackerel, and blue whiting. None of these species featured prominently in medieval or early modern fisheries. This change in target species is down to several factors, including changing tastes and consumption patterns, the overexploitation of traditional catches like herring over the centuries, and the development of new fishing technology allowing for the targeting of deeper-water species much further off the coast. Although a historical comparison illustrates that fisheries in the past were very different from those today, many of the same issues, including pollution, habitat degradation and overfishing, were faced by early modern fishers. It is particularly interesting to note that the pollution and the degradation of river systems have a far longer history than might be expected, and that these were critical issues even in the sixteenth century.

Fundamentally, comparing modern TAC figures to historical catches presents difficulties, especially the assumption that the ecosystem in the sixteenth century was the same as it is today.[78] Many aspects of the modern ecosystem have changed, not least the physical structure of many parts of the coastline. As well as being vulnerable to overexploitation, herring spawning habitats are incredibly fragile and easily damaged by natural or human disturbances.[79] Coastlines have undergone both natural and anthropogenic changes since the sixteenth century,

[75] Corten, 'Herring and Climate', p. 17.

[76] A similar observation is made in Alison Mary Locker, 'The Role of Stored Fish in England 900–1750AD; the Evidence from Historical and Archaeological Data' (PhD thesis, University of Southampton, 2000), p. 34.

[77] Sea Fisheries Protection Authority, '2020 Quota Uptake – Landings' (https://www.sfpa.ie/Statistics/Quotas) (accessed 15 June 2020).

[78] Poulsen, *Dutch Herring*, p. 80.

[79] O'Sullivan et al., 'Irish Herring Spawning Grounds', pp. 2–3.

and in some cases, these changes have altered the habitats of marine species and rendered areas no longer suitable for certain species.

Coastal change occurs all the time through natural physical processes. Sediment transported by waves and currents, as well as coastal erosion, are both major drivers of change.[80] Additionally, high-energy events like storms can have a significant effect over short periods, especially in areas like the Irish west coast, which is exposed to intense storms from the North Atlantic.[81] Coastal change is also human-made: over centuries, land has been reclaimed in many parts of the shoreline, artificial harbours and sea walls have been constructed, and channels have been dredged in bays.[82] Most fifteenth- and sixteenth-century fisheries took place within inshore waters and close to settlements, and these waters have undergone the most noticeable shifts, as towns and industry have grown and begun to reclaim land. Invisible changes have taken place as well, with problems such as nutrient enrichment altering the chemistry of coastal waters and damaging fisheries.[83]

Tracing coastal change from the sixteenth century onwards is difficult. Since most maps made in our period are impressionistic, they are not practical for tracing coastal change. However, for certain urban areas, such as Dublin Bay, we can use more recent historical maps to gain a better understanding of what the area, and its fisheries, would have looked like. Captain Greenville Collins' 1696 map of Dublin Bay was drawn long after our period ended, but can still tell us something about Dublin Bay in pre-modern times: before the construction of the north and south sea walls, the city could only be reached through a narrow channel in the sand. The construction of the north wall significantly altered the structure of the bay, as it caused Bull Island to form behind it.[84] Previously this area would have been a sizeable shallow sandbank. Large amounts of land have also been reclaimed close to Dublin city: the areas of Rings End, the North Strand and the modern Docklands lie almost entirely on reclaimed land. Collins' map also gives us clues as to the marine species inhabiting this area. Just south of modern Clontarf, a

80 J. R. French and H. Burningham, 'Coastal Geomorphology: Trends and Challenges', *Progress in Physical Geography*, 33 (2009), pp. 117–29, p. 117.

81 J. A. G. Cooper et al., 'Identifying Storm Impacts on an Embayed, High-Energy Coastline: Examples from Western Ireland', *Marine Geology*, 210 (2004), pp. 261–80.

82 Aidan O'Sullivan and Colin Breen, *Maritime Ireland: An Archaeology of Coastal Communities* (Stroud, 2007), p. 17.

83 R. J. Gowen et al., 'Production and Its Fate in Two Coastal Regions of the Irish Sea: The Influence of Anthropogenic Nutrients', *Marine Ecology Progress Series*, 208 (2000), pp. 51–64; Jesper H. Andersen et al., 'Long-Term Temporal and Spatial Trends in Eutrophication Status of the Baltic Sea', *Biological Reviews*, 92 (2017), pp. 135–49.

84 Donal T. Flood, 'The Birth of the Bull Island', *Dublin Historical Record*, 28 (1975), pp. 142–53.

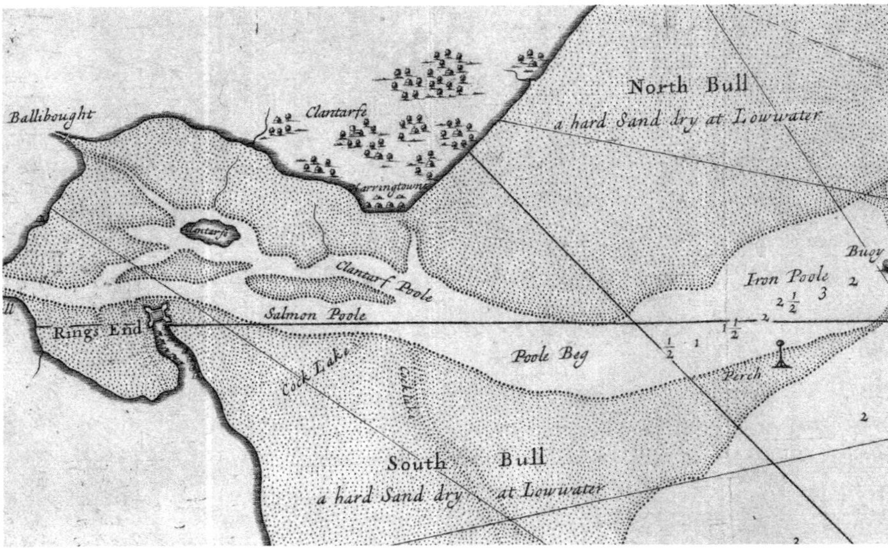

Plate 6.1. Extract from 'Dublin Bay' by Greenville Collins 1696.

small village called 'Herringstowne' is marked on the map.[85] The name of the village suggests a site where herring fisheries were based, and the abundant sand and gravel beds in the bay would also have provided potential habitat for spawning herring.[86] South of Herringstowne is a channel marked 'Salmon Poole', suggesting that salmon were caught not just in the river Liffey but also further out in the bay as the fish approached in the narrow channels. Further south again is an area marked 'Cock Lake', probably indicating a habitat for cockles and oysters. Thus coastal change has clearly altered the physical structure of Dublin Bay and the species of marine life that fishing communities were able to harvest.

Collins' map highlights the radical changes that have occurred in Dublin Bay in more recent times. Similar developments have occurred in many other parts of the Irish coast. And while we cannot quantify or map all these changes, it is still important to recognise that they occurred and probably altered the nature of marine life and fisheries that could operate there.

Ultimately, the natural environment had a range of powerful impacts on fisheries in the sixteenth century, and, conversely, people from the past shaped and influenced the environments in which they lived. We might imagine that environmental degradation and industrial pollution are only aspects of our modern

[85] The name can be found on several other maps from earlier and later periods; Donal T. Flood, 'Dublin Bay in the 18th Century', *Dublin Historical Record*, 31 (1978), pp. 129–41, at pp. 129 and 131.

[86] Henk J. L. Heessen, 'ICES FishMap Species Factsheet: Herring' (ICES, 2005).

world, but these issues were present in the medieval and early modern periods. The science and philosophy that underlie the current environmental movement did not exist then, but contemporaries cared about protecting aquatic environments and the fish that inhabited them. On a broader scale, natural changes in oceanographic conditions influenced the rise and fall of fisheries. In the fifteenth century, when other human factors aligned, the fishing industry experienced conditions conducive to high fish populations. Conversely, when war and famine were disrupting the fisheries in the late sixteenth century, the situation was made worse by warmer than usual waters that reduced the abundance of herring and other key species.

Understanding the impact humans have had on sea fisheries is more complicated. Current quotas are not always the best metric to understand the sustainability of past fish extractions. Catches in the fifteenth and sixteenth centuries were concentrated in a few key areas, while the modern industry covers a far larger area and operates much further out to sea. In the case of herring, it is conceivable that intense fishing pressure on a few key spawning populations of the fish did have an impact on stocks.

Conclusion

> Their fishful bays, were taken from the Irish of the province of Ulster, and given in their presence to foreign tribes.
>
> *Annals of the Four Masters, Volume Six*, 1608, discussing the Flight of the Earls

This quotation encapsulates the profound changes the seventeenth century brought to the fisheries of Ulster and Ireland as a whole. But without the context of the previous two centuries, we could not possibly understand why it was significant that these fisheries were 'taken from the Irish'. It is to be hoped that this book has provided that background and shown that fisheries were one of the most important and most productive industries in fifteenth- and sixteenth-century Ireland. Nor is the growth, decline and dispossession of the fisheries an obscure historical curiosity, only of interest to the specialist; these changes were fundamental elements of Ireland's economic, environmental, and cultural history.

Earlier I quoted Tim O'Neill as saying that the arrival of herring shoals and the subsequent herring fisheries was the 'greatest single economic event' of the fifteenth century.[1] Others have made similar claims without substantiation or explanation.[2] In some ways, O'Neill was correct; the growth of fisheries in the fifteenth and sixteenth centuries was highly significant to the development of the Irish economy, but until now the scale and reasons for this growth have not been explored in detail.

In reality, it was not just one factor, such as the 'arrival' of herring shoals, that led to the growth of fisheries in the fifteenth century. Rather, it was a confluence of events that created conditions ripe for fisheries to prosper. Raymond Gillespie has written that regarding the economy as 'socially embedded relationships, rather than mechanistic markets for exchange' can allow us to 'bridge cultural worlds and use economic life as an organising principle for describing early modern Ireland'.[3] The same can also apply to fisheries. We have considered the more mechanistic elements of the fish trade, including export volumes and prices, and these are valuable data points for understanding just how much fish was extracted from the ocean. But ultimately, the most important characteristic of fisheries during the period was the relationships built between landowners, fishers, and merchants

[1] Timothy O'Neill, *Merchants and Mariners in Medieval Ireland* (Dublin, 1987), p. 131.

[2] *CHI vol. 1*, p. 414; Raymond Gillespie, *The Transformation of the Irish Economy 1550–1700* (Dublin, 1991), p. 7.

[3] *CHI vol. 2*, p. 532.

from across different communities in Ireland and Europe. These facilitated and strengthened fisheries and propelled them to prosperity. The breakdown of these relationships seriously harmed the industry in the later sixteenth century.

Among the most critical of these relationships was that between Ireland's coastal communities and visiting fishers. While many Gaelic Irish families looked to cattle and land for wealth, others looked out to the sea. We know that English, Spanish, and French fishers regularly visited Ireland – and maritime communities developed complex systems to welcome, facilitate, protect, and tax these visiting fishers. Often these communities operated in areas with unproductive soil and grazing. The sea offered wealth, food and a solid connection to the rest of Europe. These families were not too primitive to fish for themselves; rather, they formed a symbiotic relationship with visitors who could do the fishing for them while also bringing in highly desirable goods like salt and wine. Places like Baltimore, Berehaven, and Donegal became key locations for foreign fishers to visit, and developed slightly different fishing systems to suit their local geographic and political needs.

Fishers from Europe found Ireland a welcoming place in the fifteenth and early sixteenth centuries because there was no one dominant power on the island. While the English, French and Spanish states regularly fought each other, the coast of Ireland remained, for the most part, neutral ground in which all fishers could operate. Gaelic and Old English lords were free to operate and profit from their own fisheries without interference from a central government in Dublin or London which could tax and control these fisheries. Attempts were certainly made; by 1465, legislation was already introduced to require licences for foreign fishing vessels in Ireland.[4] But the English government was in no position to enforce these rules. Instead, the Old English and Irish even co-managed and oversaw fisheries in some places along the coast. Before relations broke down in the mid-sixteenth century, Baltimore and Kinsale jointly elected a fishing admiral each year to police the fisheries on their coast.[5]

The English also actively participated in Irish fisheries in the fifteenth and early sixteenth centuries. They had complete control of some key fishing locations, such as Carlingford, but also visited Gaelic Irish areas just like the Spanish or French. English fishers, especially from the West Country, formed a vital part of Ireland's fishing economy, and before the 1530s, seem to have peacefully coexisted with the other fishing groups.

Of course, Ireland had domestic fisheries too. Dublin, the largest city in Ireland, had multiple fishing towns that funnelled fish into the metropolis to feed the population. Old English towns such as Waterford, Wexford, Cork and

4 *CSPI* (IMC), 1566–67, no. 117, p. 62.

5 John O'Donovan (ed.), *The Genealogy of Corca Laidhe* (*CELT*, 2011).

Youghal also operated their own fleets, which fed local populations and exported much of their catch to England.

By amalgamating data from multiple sources, this study could illuminate the fifteenth- and sixteenth-century fish trade to Britain in far more detail than has been previously possible. Fish was by far Ireland's most significant export commodity in the late fifteenth and early sixteenth centuries. Dublin and other ports in the Pale mainly traded with Chester, but the largest sector of the fish trade operated between Munster ports and Bristol. We know considerably less about the fish trade to continental Europe. The evidence suggests that merchants did not regularly export Irish fish to Europe, but instead, European fishers exchanged goods for the right to fish on the Irish coast. Trade still took place but was not recorded in formal customs accounts. Ultimately, we cannot accurately quantify the scale of these foreign fisheries every year, so we may never know the true extent of fish extractions in late medieval or early modern Ireland. Nevertheless, the evidence suggests these operations were significant and perhaps large enough to contribute to general economic growth and a building boom in fifteenth-century Ireland.[6]

Despite limitations, the figures for the fish trade to Britain still provide great insight into the health and trajectory of the fishing industry in the fifteenth and sixteenth centuries. Herring, hake, and salmon were the primary species exploited. Of these, herring was by far the largest catch over the two centuries and peaked in 1525/6 with a total export of 1,271 tonnes of live weight herring. This alone would not place Ireland among Europe's largest herring fisheries at the time, but if we factor in domestic consumption and extractions from visiting fishers, the total would be much larger.[7] Overall, fish exports peaked around the same time as herring in the 1520s, but then declined over the rest of the century. The trade data is interrupted by tax changes in the 1560s, and when available again in the 1590s shows that the fish trade had all but collapsed. It would recover and adjust in the following century, but fisheries would never regain the economic status previously held.

The reasons for this crash were manifold, but chief among them was the disrupting and corrosive influence of war and conflict. The fifteenth and early sixteenth centuries were a relatively peaceful period in Ireland, and fishers could operate mostly undisturbed. This began to change from the 1530s. Much subsequent conflict related to the Tudor reconquest.

Fisheries also played a role in the justification for English conquest and plantation. English writers and pamphleteers expounded on the mismanagement

6 O'Neill, *Merchants and Mariners*, p. 36; *CHI vol. 2*, p. 538; *CHI vol. 1*, p. 411.

7 Bo Poulsen, *Dutch Herring: An Environmental History, c. 1600–1860* (Amsterdam, 2008), pp. 69–70.

and underuse of fishing resources by the native Irish while emphasising the benefits and profits the same could bring to their own people. The river Bann, for example, was one of the most valuable economic assets in Ulster and was crucial to the profitability of the English plantation.[8] The presence of Catholic fisheries from continental Europe was a cause of particular concern for English writers and politicians, and as religious divisions increased across Europe, English policy only became more hostile toward foreign fisheries.

Ultimately, all the fisheries in Ireland were damaged and disrupted to some degree by conflict. Successive wars destroyed infrastructure, depleted the population, and deterred fishers from operating. Towns like Kinsale, Cork and Youghal were themselves involved in conflict and suffered direct damage but were largely able to recover in successive decades and into the next century. In contrast, Gaelic Irish fisheries across Ireland, and especially in Munster and Ulster, had their ownership and access to fishing permanently transformed by English intervention.

This change of ownership and the related breaking of relationships had the longest-lasting impact on fisheries. It seems to have happened at all levels, not just between Gaelic Irish and English. Already in 1537, a fleet from Waterford attacked Baltimore.[9] This is just one example, but in the years before this, there is no record of a major assault by an Old English town on a Gaelic Irish port. Both places were major fishing spots and seem to have worked together previously. The situation would only deteriorate in the late decades of the sixteenth century, with the Desmond rebellions and Nine Years War seriously harming coastal communities across the country.

The outcome of the Nine Years War would see the end of Gaelic power in Ireland and complete English hegemony. In the years that followed, the English would take control over some of the most prosperous former Gaelic Irish and Old English fisheries, including those at Baltimore, Berehaven, and the Bann. Fisheries would emerge again in these places during the seventeenth century, but would be fundamentally different in ownership and character. Foreign fishers from Europe were gone, replaced by English settlers who made their own catches. By the beginning of the seventeenth century, trading and fishing relationships that had developed over centuries and allowed many Irish fisheries to operate had been severed, and all the most prominent Gaelic Irish and many of the Old English fishing locations were in English hands.

Furthermore, the English presence in Ireland drew the island into the broad geopolitical conflicts of the sixteenth century. In turn, these conflicts brought pirates of all nationalities to Irish shores. Piracy in Irish and adjacent waters

8 *CSPI* (IMC), 1571–75, no. 930.1, p. 566.

9 *CSPI* (IMC), 1509–47, no. 627, pp. 349–50.

spiked markedly at the end of the sixteenth century, in parallel with the Anglo-Spanish war of 1585–1604. Pirates captured fishing boats and merchants, and fishers were deterred from operating as the waters around Ireland became increasingly dangerous and lawless toward the end of the century.

But conflict was not the only factor operating. It is likely that even without English intervention, fisheries in Ireland would have changed in line with broader European trends. The most important was the movement of European fishers to Newfoundland, including many fishers who had once come to Ireland. The Spanish were the first to exploit the cod fishery in Newfoundland, but later the English, especially West Country fisheries, would dominate,[10] having presumably developed skills in Irish waters that allowed them to prosper further afield. A limited amount of evidence also suggests that Irish vessels were engaging in fishing in Newfoundland from the 1540s. In this respect, Ireland was not an isolated island outside the broader development of European fisheries but was central to the general westward expansion of European fisheries in the sixteenth century. However, it is harder to understand if poor conditions in Ireland drove fishers to shift their focus to Newfoundland, or if this would have happened even if Ireland had remained safe.

Changing patterns of consumption and tradition also influenced fisheries. Fish was a crucial dietary item in sixteenth-century Ireland and held special religious and cultural significance. It served an essential role in provisioning the English army. Herring was the fish most commonly eaten by soldiers and citizens, but other species, such as cod, whiting and river fish, featured in diets. Catholic tradition involved eating fish on certain days of the week and during Lent. Towns like Dublin had a 'Lent store' of herring to supply the population during the religious season.[11] With the Protestant Reformation the justification for fish fasting changed for some communities. While most of Protestant Europe abandoned the practice, in England it was retained but reframed as a secular activity, resulting in a divergence in fish-eating culture in Ireland. While the English army and towns like Dublin observed secular fish days, Catholics continued to consume fish for religious reasons. Gaelic poetry from the sixteenth and early seventeenth centuries illustrates the important and complex role held by fish in Irish culture. The poem *Mo-chean do theacht, a sgadáin*, written in 1630, praises the herring and its arrival to feed the people during Lent.[12] Other poems speak of the health benefits of eating fish and the mythological past of salmon. Additionally, the Gaelic Irish lords of

10 Poul Holm et al., 'The North Atlantic Fish Revolution (ca. AD 1500)', *Quaternary Research*, 108 (2019), pp. 1–15.

11 Thomas Might to Burghley, 28 September 1580 (TNA, SP 63/76, f.146); Thomas Might to Burghley, 18 October 1580 (TNA, SP 63/77, f.104).

12 *Mo-chean do theacht, a sgadáin*, poem 1, in Thomas F. O'Rahilly, *Measgra Dánta: Miscellaneous Irish Poems, Part I* (Cork, 1970).

Baltimore prized species like halibut, which must have featured prominently on their tables when available. While we cannot say exactly how much fish the Gaelic Irish consumed, the literary evidence suggests that fish was held in high esteem and formed an essential part of the Irish diet and tradition.

Army ration records also illustrated that by the end of the sixteenth century, local supplies of fish in Ireland were becoming critically short. In response, large amounts of Newfoundland cod began to be imported, much of it from England, though enterprising Irish merchants like Nicholas Weston engaged in voyages to Newfoundland and kept the English army supplied when other foodstuffs were failing. Newfoundland fish essentially saved the English war effort at the end of the Nine Years War by keeping the army from starvation, but, because of this over-reliance, English soldiers came to loathe it. Soldiers and officers complained of its taste, quality, the difficulty of preparation and its poor effect on health. As a result, perhaps, Newfoundland cod never became a popular item of mass consumption in Ireland as it did elsewhere.[13]

In later centuries fish consumption in Ireland would fall well behind other European nations, with parity only achieved by the 1990s.[14] We do not fully understand why this is the case, because, as we have seen, Ireland had abundant supplies of fish. While there are probably many reasons, it is feasible that events in the sixteenth century contributed. The poor response to Newfoundland cod could have had a lasting impact. Additionally, many Irish communities were divorced from sea fisheries during the late sixteenth and early seventeenth centuries. Control of these fisheries went to settlers who were more concerned with exports than domestic supply. If the Irish had remained in control of marine fisheries, a stronger fish-eating culture might have developed on the island.

Finally, changes in ocean conditions had an important influence on the rise and fall of fisheries over the fifteenth and sixteenth centuries. In the past, historians have attributed the growth of Irish fisheries in the mid-fifteenth century to unfounded claims about the migration of herring to Irish shores.[15] While this was not the case, it seems that lower than usual sea surface temperatures (SST) provided ideal conditions for Irish herring fisheries to grow in the fifteenth century. These conditions may have been promoted by a combination of lower solar activity and volcanic eruptions.[16] Likewise, warmer waters in the late sixteenth century may

[13] Terje Inderhaug, 'Stockfish Production, Cultural and Culinary Values', *Food Ethics*, 5 (2020), pp. 1–40, at p. 11; Regina Grafe, *Distant Tyranny: Markets, Power, and Backwardness in Spain, 1650–1800* (Princeton, NJ, 2012), p. 52.

[14] FAO, 'Food Balances (–2013, Old Methodology and Population)' (2021) (https://www.fao.org/faostat/en/#data/FBSH) (accessed 21 May 2021).

[15] O'Neill, *Merchants and Mariners*, p. 131.

[16] Laura K. Cunningham et al., 'Reconstructions of Surface Ocean Conditions from the Northeast Atlantic and Nordic Seas during the Last Millennium', *The Holocene*, 23 (2013), pp.

have lowered fish abundance and catches during a time when the fishing industry was facing pressure from other, human factors. By the early seventeenth century, warming waters seem to have led to the appearance of pilchards on the Irish south coast, fundamentally altering fisheries for several decades.

Overall, the fishing trend established for this study showed a negative correlation with three reconstructions of SST from the North Atlantic and Scotland, although only two of these were statistically significant. However, the evidence for the connection between SST and fisheries is inconclusive, because of the limitations of the data. Evidence suggests that oceanographic conditions played an important auxiliary role in the growth and decline of fisheries, but human factors were ultimately more important. Overfishing could have also damaged the longer-term sustainability of fisheries. This is particularly the case for local spawning populations of herring, which fishers may have overexploited in the fifteenth and sixteenth centuries. The anthropogenic effect on fisheries did not end with herring; we know that people in the sixteenth century were aware that their industrial and agricultural activities could harm river fisheries. While this impact cannot be quantified, increased levels of pollution and the construction of riverine infrastructure could have contributed to the decline of salmon fisheries in sixteenth-century Ireland, as in other parts of Europe around the same period.[17]

Proving these points is difficult, but it is essential that they be raised because they force us to consider our longer-term relationship with and impact on aquatic environments. Fisheries were crucial to the economy and life of fifteenth- and sixteenth-century Ireland – but the communities, relationships, and environment on which fisheries were built were fragile. No coastal community in Ireland emerged unchanged, while others did not re-emerge at all. Ultimately, cross-cultural cooperation, peace and a favourable climate brought fisheries to their highest point in the late fifteenth and early sixteenth centuries. In contrast, violence, competition, conquest, and the failure to adapt to broader changes in markets and climate then led to collapse. In a modern world where ocean conditions are changing at unprecedented rates and where conflict and competition for marine resources are increasing, there may be lessons to be learnt from fisheries in the fifteenth and sixteenth centuries.

921–35, at p. 929; Francis Ludlow et al., 'Explosive Volcanism as a Driver of North Atlantic Marine Productivity, Fisheries and Markets, 1400–1850 CE', paper in progress (2023), pp. 1–27.

[17] R. H. J. Lenders et al., 'Historical Rise of Waterpower Initiated the Collapse of Salmon Stocks', *Scientific Reports*, 6 (2016), pp. 1–9.

Appendices

They fish with hooks of gold and we but with speech.
Sir William Cecil to Sir Henry Sidney, 13 July 1567[1]

1. Qualitative Methods

There have been few comprehensive studies of Ireland's late medieval or early modern fisheries, and most of what has been written tends to rely on anecdotes rather than methodically collected evidence. The current study aimed to change that by utilising a range of digital methods designed to collect and analyse evidence systematically from a variety of qualitative and quantitative sources. Information collected from these sources was then amalgamated into two central relational databases: the QFD for qualitative information and the FTD for quantitative. From these databases, three datasets have been released alongside this study. One dataset contains qualitative information related to fishing from the QFD, and another relates to piracy events from 1564 to 1603.[2] The third dataset is a compilation of fish trade data from the FTD.[3] All three are hosted on the file-sharing service Figshare. The rest of this appendix will detail how these databases and their related datasets were constructed and explore some of the methods used to analyse data from them.

No sources from early modern Ireland deal specifically with fisheries; the evidence is scattered throughout a range of primary sources and can be found in many different contexts. Three main categories of evidence have been employed in this study: qualitative, quantitative, and visual. Each of them demands a different method and approach. In addition, archaeological and ecological observations from existing studies were interwoven with the historical data.

The corpus of qualitative primary sources collected for this study contains ninety documents and over 11 million words. Given this scale, it would be nearly impossible to locate all entries relevant to fisheries using manual means. Instead, Qualitative Data Analysis (QDA) methods were used to search sources.[4]

[1] *CSPI*, 1509–73, no. 59, p. 342.

[2] Patrick Hayes, 'Qualitative Fishing Dataset' (2022) (10.6084/m9.figshare.20367951) (25 July 2022); 'Piracy Events, 1564 to 1603' (2022) (10.6084/m9.figshare.20367972) (25 July 2022).

[3] Patrick Hayes, 'Fish Trade Dataset' (2022) (10.6084/m9.figshare.20367882) (25 July 2022).

[4] Paul Mihas, *Qualitative Data Analysis* (Oxford, 2019), p. 2.

MaxQDA was the software chosen; it allows for the large-scale searching and coding of textual and multimedia sources. MaxQDA, and similar software packages, use 'codes' to collect and apply meaning systematically to text segments. In the a priori QDA approach, codes are determined prior to data collection, while in a more exploratory approach, codes are 'continually refined as inductive codes are developed and applied to more data'.[5] The latter approach was taken.

The methodology used was inspired by Robert M. Schwartz's analysis of the British parliamentary reports on sea fisheries. Schwartz used a variety of digital methods to analyse and visualise the reports and provided a broad picture of how fisheries fared in late nineteenth-century Britain. For example, he performed a QDA analysis on interviews with fishers and applied codes to reflect how positively or pessimistically the fishers perceived fisheries in their area. The results were then mapped to provide a spatial overview of fishers' perceptions.[6]

The first step of the QDA process is using optical character recognition software to make texts machine-readable. This process allows us to search texts using keywords, then apply codes to relevant segments. The keywords used in this study were developed and refined to ensure all entries relating to fishing activity were found within the sources. The records were also searched for records relating to piracy. Records discussing the impact of extreme weather on fisheries and shipping were also looked for, but this data was not ultimately used, although Chapter 6 does have a brief discussion of extreme weather events and their impact on fisheries in the early seventeenth century. Keywords for this search were chosen through a close reading of the primary sources and improved over time to ensure all relevant entries were found. For example, common alternative spellings of words were included to increase the accuracy of the search. The final search string of fifty-seven keywords is listed in the introduction to the QFD dataset online.[7] The accuracy and effectiveness of these keywords were verified using spot checks and existing indexes.

After relevant entries were found in the sources, they were coded into different categories using MaxQDA. This software does contain its own tools for analysing and visualising the codes, but these were too limited. Instead, external software was used to organise and analyse the entries found using MaxQDA. Microsoft Access was the software chosen to build the two databases for its user-friendly and intuitive approach. A relational database structure was adopted because it allows for the linking of records between multiple sheets, thus offering a much more flexible structure than if the data were to be entered into a single spreadsheet.

[5] Mihas, *Qualitative Data Analysis*, p. 2.

[6] Robert M. Schwartz, 'Digital Partnership: Combining Text Mining and GIS in a Spatial History of Sea Fishing in the United Kingdom, 1860 to 1900', *International Journal of Humanities and Arts Computing*, 9 (2015), pp. 36–56, at p. 48.

[7] Hayes, 'Qualitative Fishing Dataset'.

The first version of the QFD was designed to record only instances of piracy found in the primary sources. The database documented vital details related to each piracy event, such as date, attacker, victim, and location.[8] Next, extreme weather events were added, with different details, such as strength and damage caused.[9] Finally, the database was extended to include a range of details about fishing operations; this aspect is the most complex and extensive. The QFD works hierarchically, with the *Voyage* sheet being the highest record level. Each voyage record connects to an 'event sheet', such as *Piracy Events*, *Extreme Weather Events* and *Fishing Mentions*. All the sheets are linked together through a central QFD identification number.

Figure A.1 provides details on the full procedure for processing and entering information into the database. The first four steps, to locate texts, translate them into a machine-readable format, search, and code them in MaxQDA, have already been described. The next step was to enter information found in the sources into the database using a custom data entry form. Information entered using this form creates a new record that relates to a specific event or type of data. Different types of information are sorted using a system of categorisation. Following data entry, the records can then be queried for analysis. A data query is a process by which relevant information is requested from the database for further analysis in other software, such as Microsoft Excel.

Ensuring the accuracy and reliability of data were key concerns in this work. Turning qualitative information into definite data points can be challenging as historical data is often not definite, can be open to interpretation, or is vague and incomplete. This level of inaccuracy is particularly true in relation to dates, with some dates in the sources being more exact than others. A three-point scale was assigned to each date to rank them based on reliability. This method also allowed the filtering out of those dates with the lowest accuracy rating and the rapid sorting of events based on the reliability of dates. A score of one was applied to dates with low accuracy; this might be a record generated from a letter without an exact date and only the year it was sent. A score of two was applied to records with middling accuracy; these might be letters listing the exact day the letter was sent without stating when events took place. The most accurate rating of three was reserved for records that discuss the specific day events happened.

When information from the State Papers and other sources was entered into the database using the entry form, it was placed into several different categories and subcategories for further analysis. For example, a letter in the State Papers might mention herring fishing taking place at Waterford and a barrel of herring costing 14*s*. in the town. In this case, a record would be generated under the *Fishing*

[8] This piracy information has a separate dataset: Hayes, 'Piracy Events, 1564 to 1603'.

[9] As mentioned, the data collected on extreme weather events was not used in the final study.

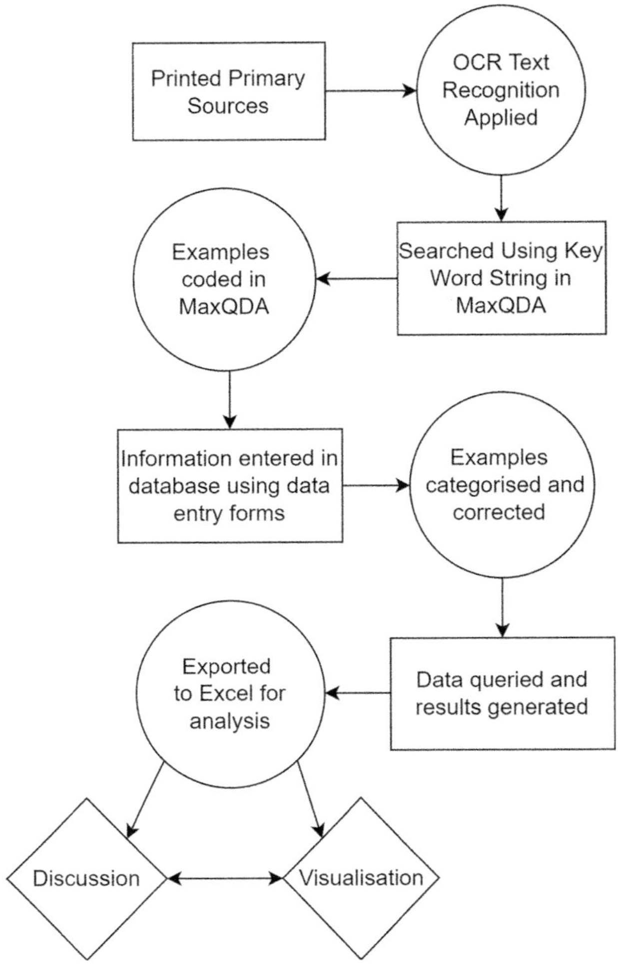

Figure A.1. QFD data entry process.

Location category, with a *Fishing Towns* subcategory. The species (herring), location (Waterford), and date the letter was sent would also be recorded. Additionally, a second entry would be generated under the *Fish Price* category, listing much of the same information, such as species and date, and also recording the unit of fish used (barrel) and the cost of that unit (14*s*.). This example illustrates how one letter can generate several records in the database. Following the categorisation of all the records, a full list of data points relating to topics like fishing locations or prices could be generated quickly for further analysis. The full list of categories and subcategories may be found in the introduction to the online dataset.

Where available, a 'Fishing Party' tag was also assigned to entries, denoting the nationality or background of the fishers or other actors in the record. For example, in the case of a record stating that Spanish fishers were seen fishing in Baltimore, the main category would be *Fishing Location* and the *Fishing Party* would be Spanish. Ultimately, the categorisation of records allows for many types of analysis and quick recall. Based on the assembled data, we can pull out all records relating to Spanish fishing locations between the years 1500 to 1575. The data generated can then be used to plot a map of locations. Similar queries can be written to extract records such as fishing voyages or fish consumption by the English army. In this way, the database can be used to generate quantitative data, like maps of fishing locations or lists of prices, and to find examples for qualitative discussions.

The methods described here could be criticised as being reductive, and, in some ways, they are. However, these methods have not replaced close readings of primary texts or traditional qualitative analysis, which remain a vital element of the research. The context around individual entries and the broader historical trends at play have been considered in detail. If anything, databases allow for comprehensive note-keeping and for organising records that enhance qualitative discussions. Moreover, the thorough quantification of qualitative information can allow us to identify and understand trends in sources that would not be visible through qualitative analysis alone.

2. Quantitative Methods

The other main type of information used in this study is numeric, quantitative information. This type of information provides us with the best chance of understanding the extent of ocean extractions in the past and gauging the scale of fisheries. Before discussing the methods used to process historical data, it is useful to discuss why quantitative information is vital to modern fisheries. Today fishery statistics are essential to understand the health and sustainability of fisheries over time, as well as to gauge the impact of humans on fish populations. Knowing how much fish is being extracted allows fishery managers to set quotas and track changes in the biomass of different stocks. Additionally, comprehending the level of extractions allows regulating bodies like the European Union to share fishing resources between nations. Like any natural resource, fish has economic and political significance, and can thus create conflict between competing parties. The sharing of fisheries in Ireland and the United Kingdom, for example, remains a contentious issue, especially after the United Kingdom's vote to leave the EU in 2016.[10]

[10] Simon Carswell, 'State on Watch to Protect Sea Borders amid Brexit Fears for Fishing', *The Irish Times*, 30 September 2019; 'Fisheries Dispute Breaks out between Ireland and Scotland', *RTE News*, 8 June 2019.

Historic quantitative information is important to the understanding of man's impact on aquatic environments and the economic value of fisheries. The research on which this book is based has combined data from a variety of quantitative sources into one central database, the FTD, which uses a standard set of field names to amalgamate different sources and render them comparable. While the quality of the sources varies, most contain common core pieces of information, such as name and home port of the ship, shipmaster's name, and sometimes origin and destination. Often, they only record what port the ship is entering or leaving. Commonly, the goods carried are listed, with the name of the merchant responsible. The hometown of the master and merchants is also frequently reported. In the absence of other information, these details can be used to estimate the origin or destination of the ship.

The initial framework of the FTD was based on the sixteenth-century Bristol trade dataset of Jones and Flavin.[11] Many of the same data field names were initially employed in the FTD, but over time, additional fields were added to accommodate different sources and pieces of information. Like the QFD, the FTD is a relational database and was built within Microsoft Access. The database comprises two central sheets: the *Main Reference* sheet and the *Commodities and Merchants* sheet. The *Main Reference* sheet records details about each cargo ship: tonnage, master, home port, and origin. This sheet also clearly indicates the original authors and source of the data that has been combined. The *Commodities and Merchants* sheet records all details about the cargo carried by ships and the merchants responsible. The *Sources* sheet provides detailed information regarding the original sources. Finally, the *Fish Conversions* sheet contains the details necessary to convert every unit of fish mentioned in the original sources into tonnes of live weight (TLW).

This conversion of historical units to TLW is a core element of the database. The reason for converting fish in this way is to establish a single standard measure, which will allow us to compare volumes of traded fish between sources and time periods, and to compare historical records across periods, up to the present day. Modern fisheries' statistics are calculated by weighing fish on board ships or upon its landing at port. Fish is generally weighed in a processed state, so to calculate live weight, a conversion factor is applied.[12] The conversion factors used are specific to the type of fish, but the common goal of conversions is to estimate the weight of fish while it was alive. Within the EU, conversion factors are regulated through the Common Fisheries Policy.[13] Other institutions, such as the Food

[11] Evan Jones and Susan Flavin (eds), *Bristol's Trade with Ireland and the Continent, 1503–1601: The Evidence of the Exchequer Customs Accounts* (Dublin, 2009).

[12] Eurostat, European Commission, 'Aquaculture Production in Tonnes and Value' (https://ec.europa.eu/eurostat/web/products-datasets/-/tag00075) (accessed 12 May 2019).

[13] European Commission, 'The Common Fisheries Policy (CFP)' (https://ec.europa.eu/

and Agriculture Organisation of the United Nations (FAO), provide their own conversion factors.[14]

To work out the TLW, we first need to know the live weight of each individual unit of fish. Then to reach the final TLW figures, we multiply the number of each unit by its calculated live weight. These live weight calculations work by either knowing the volume of a historical unit or the number of individual fish contained in that unit. This book builds on existing work done by scholars, such as Jones and Flavin, on historical units, but further research was conducted to attain the most accurate estimates for units of fish.[15] For example, in the case of the Bristol records, seven different units were used to transport herring from Ireland (different units were used for other fish). The most common unit used for both white and red herring was the barrel. It is important to note these are the same species; white herring were salted, while red herring were salted and smoked.[16] However, the difference in processing methods requires unique live weight conversion methods. We know from a letter sent by Henry Wallop to William Cecil in 1583 that an Irish white herring barrel contained 900 fish.[17] We can combine this with data from the Marine Institute of Ireland relating to the average weight of herring in Irish water. Since fishers in the sixteenth century targeted herring on their spawning grounds and would have caught a greater portion of sexually mature fish, herring under the age of three years were excluded. The average weight of herring older than three years of age is 175.36 grams (g).[18] Given these details, the final calculation works as such,

1 barrel white herring = 900 fish × 175.36g = 157.83kg live weight herring

This method gives us the live weight of a white herring barrel. However, the alternative method to reach live weight is to use the volume of the unit. In the case of herring, a statute from 1451 mentions that a herring barrel in Bristol was meant

fisheries/cfp_en) (accessed 5 April 2020).

14 FAO, 'Indicative Factors for Converting Product Weight to Live Weight for a Selection of Major Fishery Commodities' (Coordinating Working Party on Fishery Statistics, 2020).

15 See, for example, Susan Flavin and Evan Jones, *A Glossary of Commodities, Weights and Measures Found in the Sixteenth-Century Bristol Customs Accounts* (Bristol, 2009); Evan Jones, *Tonnage Estimates of Goods Recorded in the Bristol 'Particular' Customs Accounts: 1465/6, 1485/6, 1503/4* (Bristol, 2017).

16 Alison Mary Locker, 'The Role of Stored Fish in England 900–1750AD; the Evidence from Historical and Archaeological Data' (PhD thesis, University of Southampton, 2000), pp. 43–44; Charles L. Cutting, *Fish Saving* (London, 1955), p. 76.

17 Wallop to Burghley, 13 January 1583 (TNA, SP 63/99, f.40).

18 Herring live weight data from 'Species Dashboard, Herring' (https://shiny.marine.ie/speciesdash/) (accessed 30 September 2020).

to be 30 gallons or 113.56 litres (L) in volume.[19] For the purpose of our calculations, we may assume that herring had the same density as water, so a barrel with a volume of 113.56L would contain 113.56 kilograms (kg) of herring. To convert this processed weight to live weight, we then use the modern conversion factors from the FAO. For salted white herring, the conversion factor is 1.39.[20] Again, given these details, the final calculation in this example is as follows,

1 barrel white herring = 113.56kg × 1.39 = 157.85kg live weight herring

Encouragingly, in the case of white herring, calculations using the number of fish and the calculation using the volume both come to 157.8kg, suggesting that the information used in both is accurate. Similar calculations were done for every species and cure of fish found in the sources. Evan Jones has estimated that a barrel of red herring contained 800 fish, so to determine the live weight, we multiply 800 by the average weight of a herring from the Marine Institute (.). In the case of hake, the fish were commonly shipped by the long hundred (C) (containing 120 fish).[21] So, to establish the live weight, we multiply 120 by the average weight of a hake, which the Marine Institute data records as 1.363kg.[22] The calculation works like this,

1 C Hake = 120 × 1.363kg = 163.56kg live weight hake

The estimates discussed so far have all been based on modern data, but there is an argument to be made against using modern measurements when studying historical fisheries. Researchers have shown that many species of fish were on average larger in pre-industrial times. James Barrett has noted that the archaeological remains of cod found on Orkney from the eleventh and twelfth centuries indicated an average fish size of > 80 cm. Remains from eleventh- to thirteenth-century York indicated an average size of between 50 and 80 cm. In contrast, modern estimates of cod from the North Sea range between 15 and 30 cm, far below the historical estimates. However, as Barrett observes, 'one must exercise caution when interpreting archaeological fish size evidence', because fish size may also relate to other social and economic factors unrelated to the characteristics of the fish population itself. The most convincing evidence for a reduction in fish size over time comes from freshwater species, while the evidence from marine

19 Statute made in Bristol in 1451; R. B. Mowat and E. E. Butcher (eds), *The Great Red Book of Bristol*, vol. IV (part I) (Bristol, 1933), p. 134.

20 FAO, 'Indicative Factors for Converting Product Weight to Live Weight'.

21 Flavin and Jones, *A Glossary of Commodities*, p. 32.

22 Hake live weight data, from 'Species Dashboard, Hake (Merluccius merluccius)' (https://shiny.marine.ie/speciesdash/) (accessed 30 September 2020).

species is more ambiguous and open to interpretation.[23] Additionally, within marine species, the evidence for a reduction in the size of large fish, such as cod, is more extensive than that for smaller species such as herring. Finally, there are currently no studies of the historical size of fish in Irish waters.

For these reasons, I have used modern data from the Marine Institute of Ireland to estimate the size of herring and several other species. However, the Marine Institute provides no modern data for cod, and the evidence for a change in the average size of cod over time is far more robust than for other species. Consequently, resort has been made to Barbara Harvey's historical estimate for the weight of cod in medieval England. Harvey studied the diets of medieval English monks and was able to estimate that cod weighed 2.996kg per fish at the beginning of the sixteenth century. Harvey also provides an estimate for herring that is only marginally higher than the estimate for herring used in this study.[24]

Moving on from the live weight calculations, existing printed and digitised sources were chosen for this study of the fish trade to maximise the amount of relevant data that could be collected and analysed. It is important to note that the data assembled is far from a perfect time series due to its many temporal and spatial gaps. There are undoubtedly other records that could help fill these holes; some have been printed, and much else remains in manuscript form. However, this work has assembled data from a variety of ports trading heavily with Ireland and has cohesively combined this disparate data. Moreover, the live weight conversions performed here allow us to comprehend the number of fish extracted from Irish waters in this period and to compare volumes between sources. Overall, the combination of information from multiple sources in a database like this allows us to see and understand patterns that would be otherwise indiscernible if examining these sources in isolation.

A large portion of the FTD, including all the fish trade data, has been released online as a single dataset in spreadsheet format.[25] This includes a tab with all the live weight conversion factors used to generate figures for this book. There is also an introductory document outlining the sources used, together with an explanation for each column heading.

[23] James H. Barrett, 'An Environmental (Pre)History of European Fishing: Past and Future Archaeological Contributions to Sustainable Fisheries', *Journal of Fish Biology*, 94 (2019), pp. 1033–44, at pp. 1037–38.

[24] Barbara F. Harvey, *Living and Dying in England, 1100–1540: The Monastic Experience* (Oxford, 2002), 'Table A: Messes of Fish in the Refectory at Westminster Abbey, c. 1495–1525'. This conversion is listed beside 'Fish, Salted' in the data appendices, as cod was rarely called 'cod' in the trade sources.

[25] Hayes, 'Fish Trade Dataset'.

3. Fishing Trend Methods

This section will briefly outline how the fishing trend discussed in Chapter 6 was designed. First, the trade data discussed in Chapter 3 were combined into a single time series covering the period from 1400 to 1604, with TLW and value both represented. This data is not a perfect reflection of the scale of the fishing industry but does provide a reasonable representation of how well the fisheries were doing over time. The trajectory indicated by the trade data is also supported by a great deal of qualitative evidence.

As there are many gaps in the trade data, the period was split into five-year sections to give the best chance of having data for each. The total value and TLW of fish exports were then aggregated, and each section was assigned a score of 1 to 5 based on the relative level of exports, with 1 indicating a poor period and 5 a good period for fishing.

The rankings of 1–5 are based on breakpoints generated using the Jenks natural breaks classification method. The breakpoints used to assign the scores for both nominal value and TLW are indicated in Tables A.1 and A.2.

Table A.1. Natural Jenks ranking for nominal value of fish exports.

Natural Jenks ranking	High	Low
1	5	88
2	89	628
3	629	1,238
4	1,239	2,094
5	2,095	3,006

Table A.2. Natural Jenks ranking for TLW of fish exports.

Natural Jenks ranking	High	Low
1	1	25
2	26	286
3	287	587
4	588	974
5	975	1,272

These breakpoints essentially mean that if a five-year period had a total fish export of between 975 and 1,272 TLW, then it would be assigned a score of 5. Likewise, if it had only 20 TLW, it would get a score of 1. The averages of the values and TLW scores were then calculated to assign each five-year period with trade data a final score. If there was no trade data or other quantitative evidence, the score was based on a careful assessment of the available qualitative evidence. The final scores, and the basis behind them, are listed in Table A.3.

Table A.3. Fishing trend scores per five-year period.

Period	Score	Basis	Period	Score	Basis
1400–04	4	Trade data	1505–09	4	Qualitative
1405–09	3	Qualitative	1510–14	4	Qualitative
1410–14	1	Qualitative	1515–19	5	Trade data
1415–19	1	Qualitative	1520–24	5	Qualitative
1420–24	1	Qualitative	1525–29	4	Trade data
1425–29	1	Qualitative	1530–34	4	Qualitative
1430–34	1	Qualitative	1535–39	4	Qualitative
1435–39	1	Qualitative	1540–44	4	Trade data
1440–44	1	Qualitative	1545–49	3	Trade data
1445–49	1	Qualitative	1550–54	2	Trade data
1450–54	1	Qualitative	1555–59	1	Trade data
1455–59	2	Qualitative	1560–64	2	Trade data
1460–64	2	Trade data	1565–69	1	Trade data
1465–69	3	Trade data	1570–74	1	Qualitative
1470–74	3	Trade data	1575–79	2	Qualitative
1475–79	5	Trade data	1580–84	2	Trade data
1480–84	4	Qualitative	1585–89	2	Trade data
1485–89	5	Trade data	1590–94	1	Trade data
1490–94	4	Trade data	1595–99	1	Trade data
1495–99	5	Qualitative	1600–04	1	Qualitative
1500–04	5	Trade data			

Bibliography

Manuscript Sources

The National Archives [of the UK]

E 190/1133/1: Port: Bristol, Official: Controller Overseas, Mich. 1601, - Mich. 1602.

HCA 13, The High Court of Admiralty: Instance and Prize Courts: Examinations and Answers, 1536–1826.

SP 60–65: State Papers, Ireland, 1509–1714.

Printed Primary Sources

Ainsworth, John (ed.), 'Corporation Book of the Irishtown of Kilkenny, 1537–1628', *Analecta Hibernica*, 28 (1978), pp. 1, 3–78.

Berry, Henry F. (ed.), 'Minute Book of the Corporation of Dublin, Known as the "Friday Book", 1567–1611', *Proceedings of the Royal Irish Academy: Archaeology, Culture, History, Literature*, 30 (1912/13), pp. 477–514.

Berry, Henry F., and James F. Morrissey, *Statute Rolls of the Parliament of Ireland*, vols II–IV (Dublin, 1910–39).

Caulfield, Richard (ed.), *The Council Book of the Corporation of Kinsale* (Guildford, 1870).

—. (ed.), 'The Council Book of the Corporation of the City of Cork: from 1609 to 1643, and from 1690 to 1800' (Cork, 1876).

—. (ed.), *The Council Book of the Corporation of Youghal* (Guildford, 1878).

Chichester, Arthur, and Dudley R. Edwards, 'Letter-Book of Sir Arthur Chichester', *Analecta Hibernica*, 8 (1938), pp. 3, 5–177.

Connolly, Philomena (ed.), *Statute Rolls of the Irish Parliament, Richard III – Henry VIII* (Dublin, 2002).

de Santa Cruz, Alonso, *Circa 1540. Alonso de Santa Cruz on the geography of Northeastern North America after Gomes. In his Islario general de todas las islas del mundo*, ed. Antonio Blásquez (Madrid, 1922).

Dutton, Hely, *Statistical Survey of the County of Clare: With Observations on the Means of Improvement* (Dublin, 1808).

Gillespie, Raymond (ed.), *The Proctor's Accounts of Peter Lewis 1564–1565* (Dublin, 1996).

Griffith, Margaret (ed.), *Calendar of Inquisitions Formerly in the Office of the Chief Remembrancer of the Exchequer Prepared from the MSS of the Irish Record Commission* (Dublin, 1991).

Hunter, R. J., and Brendan Scott (eds), *The Ulster Port Books, 1612–15* (Belfast, 2012).

Jeninges, Edward, *A briefe discouery of the damages that happen to this realme by disordered and vnlawfull diet* (London, 1590).

Jones, Evan, and Susan Flavin (eds), *Bristol's Trade with Ireland and the Continent, 1503–1601: The Evidence of the Exchequer Customs Accounts* (Dublin, 2009).

Lewis, Arthur Edward (ed.), *The Welsh Port Books, 1550–1603* (Cymmrodorion Record Series, xii, London, 1927).

Longfield, Ada Kathleen (ed.), *Fitzwilliam Accounts 1560–64* (Dublin, 1960).

Mac Niocaill, Gearóid (ed.), *Crown Survey of Lands 1540–41* (Dublin, 1992).

McGrath, Bríd (ed.), *The Minute Book of the Corporation of Clonmel, 1608–1649* (Dublin, 2006).

McManus, Damian, and Eoghan Ó Raghallaigh (eds), *A Bardic Miscellany: Five Hundred Bardic Poems from Manuscripts in Irish and British Libraries* (Dublin, 2010).

Millett, Benignus, 'A Report from the President of Munster, Sir William Drury, 24 March 1577 (1578 new style)', *Collectanea Hibernica*, 46/47 (2004/2005), pp. 7–15.

Mowat, R. B., and E. E. Butcher (eds), *The Great Red Book of Bristol*, vol. iv (part I) (Bristol, 1933).

Nicholls, K. W. (ed.), *The Irish Fiants of the Tudor Sovereigns: During the Reigns of Henry VIII, Edward VI, Philip & Mary, and Elizabeth I* (Dublin, 1994).

Ó Dálaigh, Brian, *Corporation Book of Ennis* (Dublin, 1990).

O'Donovan, John (ed.), *Miscellany of the Celtic Society* (Dublin, 1849).

O'Rahilly, Thomas F., *Measgra Dánta: Miscellaneous Irish Poems, Part I* (Cork, 1970).

Quinn, David B., *New American World. A Documentary History of North America to 1612* (London, 1979).

Sharpe, Edward, *Britaines Busse. Or A Computation Aswell of the Charge of a Busse or Herring-Fishing Ship* (London, 1615).

Warner, Sir George (ed.), *The Libelle of Englyshe Polycye: A Poem on the Use of Sea-power, 1436* (Oxford, 1926).

White, Newport Benjamin, *Extents of Irish Monastic Possessions, 1540–1541, from Manuscripts in the Public Record Office* (London and Dublin, 1943).

Wilson, K. P. (ed.), *Chester Customs Accounts, 1301–1566* (Record Society of Lancashire and Cheshire, vol. CXI, Liverpool, 1969).

Young, Robert M. (ed.), *The Town Book of the Corporation of Belfast, 1613–1816* (Belfast, 1892).

Printed Official Documents and Calendars

Appleby, John C. (ed.), *A Calendar of Material Relating to Ireland from The High Court of the Admiralty Examinations 1536–1641* (Dublin, 1992).

Brewer, J. S., and William Bullen (eds), *Calendar of the Carew Manuscripts Preserved in the Archiepiscopal Library at Lambeth*, 6 vols (London, 1867–73).

Calendar of State Papers, Ireland, 1509–1606, 12 vols (London, 1860–1912).

Calendar of State Papers, Ireland, Tudor Period, 1509–1575, 5 vols (Dublin, 2000–17).

M'Neill, George Powell (ed.), *Rotuli Scaccarii Regum Scotorum (The Exchequer Rolls of Scotland)*, vol. XXIII, *A.D. 1595–1600* (Edinburgh, 1908).

Morrin, James (ed.), *Calendar of the patent and close Rolls of Chancery in Ireland, of the reigns of Henry VIII., Edward VI., Mary, and Elizabeth*, vol. i (Dublin, 1861).

—. *Calendar of the patent and close Rolls of Chancery in Ireland, from the 18th to the 45th of Queen Elizabeth* (Dublin, 1862).

Murphy, Elaine (ed.), *A Calendar of Material Relating to Ireland from the High Court of the Admiralty, 1641–1660* (Dublin, 2011).

Murray, James (ed.), *State Papers, King Henry the Eight, Part III – Continued, 1538–1546* (London, 1834).

Tresham, E. (ed.), *Rotulorum Patentium et Clausorum Cancellariae Hiberniae Calendarium* (Dublin, 1828).

Electronic Primary Sources

Allen, Robert C., and Richard W. Unger (eds), *London Prices, Allen-Unger Global Commodity Prices Database* (http://www.gcpdb.info/) (accessed 29 July 2016).

Condon, M., and Evan Jones (eds), *Bristol 'Particular' Customs Account Transcriptions* (http://www.bristol.ac.uk/history/research/cabot/publications/) (accessed 9 September 2020).

Crooks, Peter (ed.), '*CIRCLE* – A Calendar of Irish Chancery Letters c. 1244–1509' (Dublin, 2019) (https://chancery.tcd.ie) (accessed 13 April 2020).

Gadd, Stephen (ed.), *Portfolio: Exchequer Port Book Project* (Winchester, 2017) (http://portfolio.winchester.ac.uk/).

Hussey, D. et al. (eds), *Gloucester Port Books, 1575–1765*, SN: 3218 vols (http://doi.org/10.5255/UKDA-SN-3218-1) (accessed 9 October 2017).

Jones, Evan, and Susan Flavin (eds), *Ireland-Bristol Trade in the Sixteenth Century* (www.bris.ac.uk/Depts/History/Ireland/datasets.htm) (accessed 23 April 2020).

Simms, Katharine, and Mícheál Hoyne (eds), 'Bardic Poetry Database' (https://bardic.celt.dias.ie/) (accessed 25 August 2020).

University College Cork – Corpus of Electronic Texts (CELT)

'CELT: Corpus of Electronic Texts' (University College Cork, 1992–2020) (https://celt.ucc.ie//index.htm).

Balé, Marcos, and Emer Purcell (eds), *The Annals of Ulster AD 1379–1541* (*CELT*, 2010) (https://celt.ucc.ie//published/T100001C/index.html).

Bambury, Pádraig (ed.), *Annals of Connacht* (*CELT*, 2008) (https://celt.ucc.ie//published/T100011/index.html).

Bambury, Pádraig, and Stephen Beechinor (eds), *The Annals of Ulster AD 1202–1378* (*CELT*, 2003) (https://celt.ucc.ie//published/T100001B/index.html).

Hennessy, William M. (ed.), *Annals of Loch Cé 1* (*CELT*, 2008) (https://celt.ucc.ie//published/T100010A/index.html).

—. (ed.), *Annals of Loch Cé 2* (*CELT*, 2008) (https://celt.ucc.ie//published/T100010B/index.html).

Hondelink, Floortje (ed.), *Annals of the Four Masters, Volume 4* (*CELT*, 2008) (https://celt.ucc.ie//published/T100005D/index.html).

Knott, Eleanor (ed.), *The Bardic Poems of Tadhg Dall Ó Huiginn (1550–1591)* (*CELT*, 2008) (https://celt.ucc.ie/published/T402563/index.html).

Mac Airt, Seán, and Beatrix Färber (eds), *Annals of Inisfallen* (*CELT*, 2008) (https://celt.ucc.ie//published/T100004/index.html).

McDonald, Orla (ed.), *Annals of the Four Masters, Volume 5* (*CELT*, 2002) (https://celt.ucc.ie//published/T100005E/index.html).

Murphy, John A. (ed.), *The Desmond Survey* (*CELT*, 2013) (https://celt.ucc.ie//published/E580000-001/index.html).

O'Brien, Karen (ed.), *Annals of the Four Masters, Volume 6* (*CELT*, 2008) (https://celt.ucc.ie//published/T100005F/index.html).

Ó Corráin, Donnchadh (ed.), *The Annals of Tigernach (Gearóid Mac Niocaill)* (*CELT*, 2010) (https://celt.ucc.ie//published/T100002A/index.html).

Ó Corráin, Donnchadh, and Mavis Cournane (eds), *The Annals of Ulster AD 431–1201* (*CELT*, 2020) (https://celt.ucc.ie//published/T100001A/index.html).

O'Donovan, John (ed.), *The Genealogy of Corca Laidhe* (*CELT*, 2011) (https://celt.ucc.ie//published/T105009/header.html).

O'Donovan, John, and Emma Ryan (eds), *Annals of the Four Masters, Volume 1* (*CELT*, 2002) (https://celt.ucc.ie//published/T100005A/index.html).

Priour, Myriam (ed.), *Annals of the Four Masters, Volume 3* (*CELT*, 2002) (https://celt.ucc.ie//published/T100005C/index.html).

Priour, Myriam, and Stephen Beechinor (eds), *Annals of the Four Masters, Volume 2* (*CELT*, 2008) (https://celt.ucc.ie//published/T100005B/index.html).

Smith, Sir Thomas, 'A Letter sent by I. B. Gentleman vnto his very frende Mayster

R. C. Esquire, 1572', ed. Beatrix Färber (*CELT*, 2019) (https://celt.ucc.ie//published/E570001-001/index.html).

Spenser, Edmund, 'A View of the Present State of Ireland, 1596', ed. William M. Hennessy (*CELT*, 2008) (https://celt.ucc.ie//published/E500000-001).

Photographs

Wilse, Anders Beer, 'Klippfisken Stacks in Piles, Kristiansund, 11 August 1922' (Norsk Folkemuseum, Gallery Nor, 1922).

Maps

Miscellaneous Maps

'Baltimore, Ireland', c. 1620 (Wentworth Woodhouse Muniments, Sheffield Archives: WWM/StrP/20/100).

Collins, Greenville, 'Dublin Bay', 1696 (Bibliothèque nationale de France, département Cartes et plans, GE DD-2987 (2653).

'The New Description of the Cittie of Waterford with the New Fortifications There...' (Hardiman atlas) (The Board of TCD, IE TCD MS 1209/64, 1591).

Irish Maps c.1558–c.1610, The National Archives, London

Jobson, Francis, 'The Plotte of the Citie of Limbrick', 1587 (TNA, MPF 1/96).

'Map of area in County Cork and County Kerry, Ireland, around Bantry Bay. Shows Kenmare', 1558 (TNA, MPF 1/94).

'Map of Belfast Lough, showing Belfast, Carrickfergus, Larne Lough, and Island Magee', 1569 (TNA, MPF 1/77).

'Map of the Coast of County Antrim, Ireland', 1580 (TNA, MPF 1/88).

'Map of the Coast of County Down, Ireland', 1568 (TNA, MPF 1/89).

'Map of the coast of Ireland from Carrickfergus (County Antrim) to Dundrum Bay (County Down)', 1580 (TNA, MPF 1/87).

'Map of the coast of Ireland from Dublin to Carrickfergus', 1580 (TNA, MPF 1/86).

'Map of the coast of Munster from Kinsale (County Cork) and Dingle (County Kerry)', 1587 (TNA, MPF 1/100).

'Plat of the fort of Enniskillen', 1594 (TNA, MPF 1/80).

Research Projects

Flavin, Susan, 'FoodCult: Food, Culture and Identity in Ireland c. 1550–1650' (TCD, 2020) (https://foodcult.eu/).

'Food Smart Dublin' (TCD, 2022) (https://www.tcd.ie/tceh/projects/foodsmartdublin/).

'The North Atlantic Fish Revolution: An Environmental History of the North Atlantic 1400–1700 (NorFish)', European Research Council Advanced Grant ERC-2014-ADG (TCD, 2016–20) (http://cehresearch.org/norfish.html).

'4-Oceans Project' (TCD, CHAM Centro de Humanidades, Norwegian University of Science and Technology, 2023) (https://www.tcd.ie/tceh/4-oceans/).

Reports and Regulations

Bjarnason, B. A., and M. Carlesi, *Handlining and Squid Jigging* (FAO, Rome, 1992).

'COUNCIL REGULATION (EU) 2019/124 of 30 January 2019 fixing for 2019 the fishing opportunities for certain fish stocks and groups of fish stocks, applicable in Union waters and, for Union fishing vessels, in certain non-Union waters' (The Council of the European Union, 2019).

Dransfeld, L. et al., *North Western Waters Atlas* (3rd edn, Oranmore, 2014).

European Commission, 'The Common Fisheries Policy (CFP)' (https://ec.europa.eu/fisheries/cfp_en).

FAO, 'Indicative Factors for Converting Product Weight to Live Weight for a Selection of Major Fishery Commodities' (Coordinating Working Party on Fishery Statistics, 2020).

'Herring in the Southern Irish Sea, Celtic Sea and South West of Ireland, Pelagic Trawl' (SEAFISH, 2017).

ICES, *Report of the Working Group on the Assessment of Southern Shelf Stocks of Hake, Monk and Megrim* (Copenhagen, 2013).

The Ocean and Cryosphere in a Changing Climate: Special Report of the Intergovernmental Panel on Climate Change (Cambridge, 2022).

'Overview on Dietary Reference Values for the EU Population as Derived by the EFSA Panel on Dietetic Products, Nutrition and Allergies' (European Food Safety Authority, 2017).

Simpson, Linzi, and Margaret Gowen, '0529 - 161–168 Church Street/3–15 Hammond Lane, Dublin' (Dublin, 2002).

UN General Assembly, 'United Nations Convention on the Law of the Sea' (1982).

Datasets

Cunningham, Laura K. et al., 'NOAA/WDS Paleoclimatology – 1000 Year Composite Sea Surface Temperature Record from the North Atlantic Ocean, NENA Comp & S1', (2013) (https://doi.org/10.25921/p44h-7t71) (accessed 27 July 2022).

FAO, 'Food Balances (2010–)' (2022) (https://www.fao.org/faostat/en/#data/FBS) (accessed 20 January 2022).

FAO, 'Food Balances (–2013, Old Methodology and Population)' (2021) (https://www.fao.org/faostat/en/#data/FBSH) (accessed 21 May 2021).

Hayes, Patrick, 'Fish Trade Dataset' (2022) (https://doi.org/10.6084/m9.figshare.20367882) (25 July 2022).

Hayes, Patrick, 'Piracy Events, 1564 to 1603' (2022) (https://doi.org/10.6084/m9.figshare.20367972) (25 July 2022).

Hayes, Patrick, 'Qualitative Fishing Dataset' (2022) (https://doi.org/10.6084/m9.figshare.20367951) (25 July 2022).

Holm, Poul et al., 'Accelerated Extractions of North Atlantic Cod and Herring Dataset' (Dublin, 2021) (https://doi.org/10.6084/m9.figshare.13614452.v14) (accessed 25 July 2022).

Secondary Sources

Aaronson, Sheldon, 'Algae', in K. F. Kiple and K. C. Ornelas (eds), *The Cambridge World History of Food* (New York, 2000), pp. 231–48.

Aguilar, Alex, 'A Review of Old Basque Whaling and Its Effect on the Right Whales (Eubalaena glacialis) of the North Atlantic', *Report – International Whaling Commission*, 10 (1986), pp. 191–99.

Albala, Ken, *Eating Right in the Renaissance* (Berkeley, CA, 2002).

Alfani, Guido, 'The Famine of the 1590s in Northern Italy. An Analysis of the Greatest "System Shock" of Sixteenth Century', *Histoire & Mesure*, 26 (2011), pp. 17–50.

Alheit, Jürgen, and Eberhand Hagen, 'The Bohuslän Herring Periods: Are They Controlled by Climate Variations or Local Phenomena?', *ICES Theme Session O*, 6 (1996), pp. 1–21.

Alvarez, Paula et al., 'Distribution and Abundance of European Hake Merluccius merluccius (L.), Eggs and Larvae in the North East Atlantic Waters in 1995 and 1998 in Relation to Hydrographic Conditions', *Journal of Plankton Research*, 26 (2004), pp. 811–26.

Andersen, Jesper H. et al., 'Long-Term Temporal and Spatial Trends in Eutrophication Status of the Baltic Sea', *Biological Reviews*, 92 (2017), pp. 135–49.

Anderson, J. T., 'Piracy and World History - An Economic Perspective on Maritime Predation', *Journal of World History*, 6 (1995), pp. 175–99.

Anderson, K., 'Does History Count?', *Endeavour*, 30 (December 2006), pp. 150–55.

Andrews, J. H., *Shapes of Ireland, Maps and Their Makers 1564–1839* (Dublin, 1997).

—. *Maps in Those Days: Cartographic Methods before 1850* (Dublin, 2009).

Andrews, Kenneth R., *Elizabethan Privateering: English Privateering during the Spanish War, 1585–1603* (London, 1964).

—. 'The Elizabethan Seaman', *The Mariner's Mirror*, 68 (1982), pp. 245–62.

—. *Trade, Plunder and Settlement: Maritime Enterprise and the Genesis of the British Empire, 1480–1630* (Cambridge, 1984).

Appleby, John C., 'The Fishing Ventures of Nicholas Weston of Dublin', *Dublin Historical Record*, 39 (1986), pp. 150–55.

—. 'A Nursery of Pirates: The English Community in Ireland in the Early Seventeenth Century', *International Journal of Maritime History*, 2 (1990), pp. 1–27.

—. 'Conflict, Cooperation and Competition: The Rise and Fall of the Hull Whaling Trade during the Seventeenth Century', *The Northern Mariner*, 18 (2008), pp. 23–59.

—. *Under the Bloody Flag: Pirates of the Tudor Age* (Stroud, 2009).

—. *Women and English Piracy: 1540–1720: Partners and Victims of Crime* (Woodbridge, 2013).

Appleby, John C., and Michael McCaughan (eds), *Irish Sea: Aspects of Maritime History* (Belfast, 1989).

Appleby, John C., and Mary O'Dowd, 'The Irish Admiralty: Organisation and Development 1570–1640', *Irish Historical Studies*, 24 (1985–86), pp. 299–326.

Barkham, Michael M., 'French Basque "New Found Land" Entrepreneurs and the Import of Codfish and Whale Oil to Northern Spain, c. 1580 to c. 1620', *Newfoundland Studies*, 10 (1994), pp. 1–43.

—. 'La industria pesquera en el País Vasco peninsular al principio de la Edad Moderna: ¿una edad de oro?', *Itsas Memoria. Revista de Estudios Marítimos del País Vasco*, 3 (2000), pp. 29–75.

—. 'The Spanish Basque Irish Fishery and Trade in the Sixteenth-Century', *History Ireland*, 9, no. 3 (2001), pp. 12–15.

—. 'El comercio marítimo vizcaíno y guipuzcoana con el Atlántico peninsular (Asturias, Galicia, Portugal y Andalucía) y con los archipiélagos de Canarias y Madeira al principio de la Edad Moderna', *Itsas Memoria. Revista de Estudios Marítimos del País Vasco*, 4 (2003), pp. 147–64.

—. 'The Offshore and Distant-Water Fisheries of the Spanish Basques, c.1500-1650', in D. J. Starkey et al. (eds), *A History of the North Atlantic Fisheries 1* (Bremerhaven, 2009), pp. 229–49.

Barkham, Selma Huxley, 'The Basque Whaling Establishments in Labrador 1536–1632 - A Summary', *Arctic*, 37 (1984), pp. 515–19.

Barnett, Eleanor, 'Reforming Food and Eating in Protestant England, c. 1560–c. 1640', *Historical Journal*, 63 (2020), pp. 507–27.

Barrett, James H., 'The Pirate Fishermen: The Political Economy of a Medieval Maritime Society', in Ballin B. Smith et al. (eds), *West Over Sea: Studies in Scandinavian Sea-Borne Expansion and Settlement before 1300* (Leiden, 2007), pp. 299–340.

—. 'An Environmental (Pre)History of European Fishing: Past and Future Archaeological Contributions to Sustainable Fisheries', *Journal of Fish Biology*, 94 (2019), pp. 1033–44.

Barrett, James H., and David C. Orton (eds), *Cod and Herring: The Archaeology and History of Medieval Sea Fishing* (Oxford, 2016).

Barrett, James H. et al., 'The Origins of Intensive Marine Fishing in Medieval Europe: The English Evidence', *Proceedings of the Royal Society B: Biological Sciences*, 271 (2004), pp. 2417–21.

—. 'Detecting the Medieval Cod Trade: A New Method and First Results', *JAS*, 35 (2008), pp. 850–61.

—. 'Interpreting the Expansion of Sea Fishing in Medieval Europe Using Stable Isotope Analysis of Archaeological Cod Bones', *JAS*, 38 (2011), pp. 1516–24.

Bernard, Jacques, 'The Maritime Intercourse between Bordeaux and Ireland, c. 1450–1520', *Irish Economic and Social History*, 7 (1980), pp. 7–21.

Blake, John, *Sea Charts of the British Isles* (London, 2005).

—. *The Sea Chart: The Illustrated History of Nautical Maps and Navigational Charts* (London, 2016).

Blanchard, Ian, 'The Continental European Cattle Trades, 1400–1600', *Economic History Review*, 39 (1986), pp. 427–60.

Bolster, Jeffrey, W., 'Opportunities in Marine Environmental History', *Environmental History*, 11 (2006), pp. 567–97.

—. *The Mortal Sea: Fishing the Atlantic in the Age of Sail* (Cambridge, MA, 2012).

Bradshaw, Brendan, *The Dissolution of the Religious Orders in Ireland under Henry VIII* (Cambridge, 1974).

—. 'The Edwardian Reformation in Ireland, 1547–53', *Archivium Hibernicum*, 34 (1977), pp. 83–99.

Brady, Ciaran, 'Faction and the Origins of the Desmond Rebellion of 1579', *Irish Historical Studies*, 22 (1981), pp. 289–312.

—. *The Chief Governors: The Rise and Fall of Reform Government in Tudor Ireland 1536–1588* (Cambridge, 1994).

Brady, Ciaran, and Raymond Gillespie (eds), *Natives and Newcomers: Making of Irish Colonial Society, 1534–1641* (Dublin, 1986).

Breen, Colin, 'A Medieval Fishery at Rossglass, Co. Down', *Ulster Journal of Archaeology*, 58 (1999), pp. 153–55.

—. *The Gaelic Lordship of the O'Sullivan Beare* (Dublin, 2005).

—. 'Marine Fisheries and Society in Medieval Ireland', in *C&H*, pp. 91–98.

—. 'Maritime Connections: Landscape and Lordship along the Gaelic Atlantic Seaboard of Scotland and the North of Ireland during the Middle Ages', *Maritime Communities of the North Atlantic Arc in the Early Modern Period, Journal of the North Atlantic*, Special Volume 12 (2019), pp. 3–15.

Breen, Colin, and John Raven, 'Maritime Lordship in Late-Medieval Gaelic Ireland', *Medieval Archaeology*, 61 (2017), pp. 149–82.

Brenner, Y. S., 'The Inflation of Prices in Early Sixteenth Century England', *Economic History Review*, 14 (1961), pp. 225–39.

Bridbury, Anthony Randolph, *England and the Salt Trade in the Later Middle Ages* (Oxford, 1955).

Brito, Cristina, 'On Mermaids and Manatees: A First Approach to the Evolution of Natural History Images in Early Modern Times', Paper presented at the Report from the European Cetacean Society Conference Workshop (Setúbal, 2013).

Brooks, F. W., 'A Wage Scale for Seamen, 1546', *English Historical Review*, 60 (1945), pp. 234–46.

Brophy, D. et al., 'Temporal Trends in Stock Origin and Abundance of Juvenile Herring (Clupea harengus) in the Irish Sea', *ICES Journal of Marine Science*, 66 (2008), pp. 1749–53.

Buchanan, R. H., 'The Irish Sea: The Geographical Framework', in Michael McCaughan and John C. Appleby (eds), *The Irish Sea: Aspects of Maritime History* (Belfast, 1989), pp. 1–12.

Caddy, J. F., 'Zoogeographical Categories Relevant to Fisheries Management in the Mediterranean', *Studies and Reviews – General Fisheries Council for the Mediterranean* (FAO), 70 (1998), pp. 1–30.

Camenisch, C. et al., 'The 1430s: A Cold Period of Extraordinary Internal Climate Variability during the Early Spörer Minimum with Social and Economic Impacts in North-Western and Central Europe', *Climate of the Past*, 12 (2016), pp. 2107–26.

Campbell, Bruce M. S., 'Benchmarking Medieval Economic Development: England, Wales, Scotland, and Ireland, c.1290', *Economic History Review*, 61 (2008), pp. 896–945.

—. 'Nature as Historical Protagonist: Environment and Society in Pre-Industrial England', *Economic History Review*, 63 (2010), pp. 281–314.

—. *The Great Transition: Climate, Disease and Society in the Late-Medieval World* (Cambridge, 2013).

Campbell, Bruce M. S., and Francis Ludlow, 'Climate, Disease and Society in

Late-Medieval Ireland', *Proceedings of the Royal Irish Academy*, 120C (2020), pp. 159–252.

Canning, Ruth A., 'Profits and Patriotism: Nicholas Weston, Old English Merchants, and Ireland's Nine Years' War, 1594–1603', *Irish Economic and Social History*, 43 (2016), pp. 85–112.

Canny, Nicholas, *Making Ireland British, 1580–1650* (Oxford, 2001).

—. 'Writing Early Modern History Ireland, Britain, and the Wider World', *Historical Journal*, 46, (2003), pp. 723–47.

Carswell, Simon, 'State on Watch to Protect Sea Borders amid Brexit Fears for Fishing', *Irish Times*, 30 September 2019.

Carus-Wilson, E. M. (ed.), *The Overseas Trade of Bristol in the Later Middle Ages* (Bristol Record Society, vii, Bristol, 1937).

Cell, Gillian T., *English Enterprise in Newfoundland, 1577–1660* (Toronto, 1969).

Challis, C. E., 'The Debasement of the Coinage, 1542–1551', *Economic History Review*, 20 (1967), pp. 441–66.

Cheung, William W. L. et al., 'Review of Climate Change Impacts on Marine Fisheries in the UK and Ireland', *Aquatic Conservation: Marine and Freshwater Ecosystems*, 22 (2012), pp. 368–88.

Childs, Wendy R., 'Ireland's Trade with England in the Later Middle Ages', *Irish Economic and Social History*, 9 (1982), pp. 5–33.

—. 'England's Icelandic Trade in the Fifteenth Century: The Role of the Port of Hull', in Poul Holm et al. (eds), *Northern Seas Yearbook* (Esbjerg, 1995), pp. 11-31.

—. 'The Commercial Shipping of South-Western England in the Later Fifteenth Century', *The Mariner's Mirror*, 83 (1997), pp. 272–92.

—. 'Medieval English Trade in the North Sea: Denmark and Jutland, c1300-1500', in Ellen Damgaard et al. (eds), *A North Sea Region. West Jutland and the World II* (Esbjerg, 1998), pp. 47–74.

—. 'Irish Merchants and Seamen in Late Medieval England', *Irish Historical Studies*, 32 (2000–01), pp. 22–43.

—. 'Commercial Relations between Basque Provinces and England in the Later Middle Ages, c.1200–c.1500', *Itsas Memoria. Revista de Estudios Marítimos del País Vasco*, 4 (2003), pp. 55–64.

Clark, Gregory, 'The Price History of English Agriculture, 1209–1914', *Research in Economic History*, 22 (2003), pp. 41–123.

—. 'The Long March of History: Farm Wages, Population, and Economic Growth, England 1209–1869', *Economic History Review*, 60 (2007), pp. 97–135.

Clark, Peter (ed.), *The European Crisis of the 1590's: Essays in Comparative History* (London, 1985).

Clarke, H. B., 'Dublin, Part I, to 1610', *IHTA*, 11 (2002), pp. 1–36.

Clarkson, L. A., and Margaret E. Crawford, *Feast and Famine – History of Food and Nutrition in Ireland, 1500–1920* (Oxford, 2001).

Collins, John, *Salt and Fishery* (London, 1682).

Collins, Timothy, 'From Hoekers to Hookers: A Survey of the Literature and Annotated Bibliography on the Origins of the Galway Hooker', *Journal of the Galway Archaeological and Historical Society*, 54 (2001), pp. 66–83.

Conbhuí, C. Ó., 'The Lands of St. Mary's Abbey, Dublin', *Proceedings of the Royal Irish Academy*, 62C (1961-63), pp. 21–86.

Cooper, Ernest R., 'The Dunwich Iceland Ships', *The Mariner's Mirror*, 25 (1939), pp. 170–77.

Cooper, J. A. G. et al., 'Identifying Storm Impacts on an Embayed, High-Energy Coastline: Examples from Western Ireland', *Marine Geology*, 210 (2004), pp. 261–80.

Costello, Kevin, *The Court of Admiralty of Ireland, 1575–1893* (Dublin, 2011).

Crawford, Margaret E. (ed.), *Famine: The Irish Experience 900–1900: Subsistence Crises and Famines in Ireland* (Edinburgh, 1989).

Cullen, L. M., *Anglo-Irish Trade, 1660–1800* (Manchester, 1968).

—. 'Population Trends in Seventeenth-Century Ireland', *Economic and Social Review*, 6 (1975), pp. 149–65.

—. 'Comparative Aspects of Irish Diet, 1550–1850', in Hans J. Teuteberg (ed.), *European Food History, a Research Review* (Leicester, 1992), pp. 45–55.

—. *Economy, Trade and Irish Merchants at Home and Abroad, 1600–1988* (Dublin, 2012).

Cunningham, Laura K. et al., 'Reconstructions of Surface Ocean Conditions from the Northeast Atlantic and Nordic Seas during the Last Millennium', *The Holocene*, 23 (2013), pp. 921–35.

Curtis, Edmund, 'Norse Dublin', *Dublin Historical Record*, 41 (1998), pp. 86–97.

Cushing, D. H., *The Provident Sea* (Cambridge, 1988).

Cutting, Charles L., *Fish Saving* (London, 1955).

Davies, C. S. L., 'Provisions for Armies, 1509–50; A Study in the Effectiveness of Early Tudor Government', *Economic History Review*, 16 (1964), pp. 234–48.

Dean, David M., 'Parliament, Privy Council, and Local Politics in Elizabethan England: The Yarmouth-Lowestoft Fishing Dispute', *Albion*, 22 (1990), pp. 39–64.

de Courcy, John, *Ireland's Sea Fisheries: A History* (Dublin, 1981).

—. 'A Survey of Early Irish Maritime Trade and Ships', in Michael McCaughan and John C. Appleby (eds), *The Irish Sea: Aspects of Maritime History* (Belfast, 1989), pp. 21–26.

Dickey-Collas, Mark et al., 'The Location of Spawning of Irish Sea Herring (Clupea harengus)', *Journal of the Marine Biological Association of the UK*, 81 (2001), pp. 713–14.

Dickson, David, *Old World Colony: Cork and South Munster 1630–1830* (Cork, 2005).

Dollinger, Philippe, *The German Hansa* (London, 1970).

Donovan, Brian C., and David Edwards (eds), *British Sources for Irish History, 1485–1641: A Guide to Manuscripts in Local, Regional and Specialised Repositories in England, Scotland and Wales* (Dublin, 1997).

Douglas, K. S. et al., 'A Meteorological Study of July to October 1588: The Spanish Armada Storms', *Climatic Research Unit Research Publication*, 6 (1978), pp. 1-83.

Downey, Liam, and Ingelise Stuijts, 'Historical Fisheries: Irish Pilchard Fishery and Historical Fish Harvesting and Processing Methods, Fish Products and Consumption', *Kerry Archaeological & Historical Journal*, 2 (2014), pp. 1–35.

Duffy, A. J. F., 'The Early History of the Rye Fishing Industry', *Sussex Archaeological Collections*, 107 (1969), pp. 37–64.

Dunn, Shannon, and Chuck Meide, 'The "Wretched Poor" and the Sea: Contest and Exploitation of Achill Island's Historic Maritime Landscape', *Internet Archaeology*, 37 (2014), pp. 1–35.

Edwards, Martin et al., 'Ocean Climate Anomalies and the Ecology of the North Sea', *Marine Ecology Progress Series*, 239 (2002), pp. 1–10.

Edwards, R. Dudley, and Mary O'Dowd, *Sources for Early Modern Irish History, 1534–1641* (Cambridge, 1985).

Egan, Simon, 'By Land and by Sea: The Role of the Maritime Sphere in the Expansion of O'Donnell Power ca. 1380–1500', *Maritime Communities of the North Atlantic Arc in the Early Modern Period, Journal of the North Atlantic*, Special Volume 12 (2019), pp. 16–31.

Ellis, Steven G., 'Thomas Cromwell and Ireland, 1532-1540', *Historical Journal*, 23 (1980), pp. 497–519.

Elton, G. R., 'Piscatorial Politics in the Early Parliaments of Elizabeth I', in Neil McKendrick and R. B. Outhwaite (eds), *Business Life and Public Policy* (Cambridge, 1986), pp. 1–20.

Fasting, Kåre, *En industri av verdensformat. Utviklingen av sildolje og sildemelindustrien i Norge* (Bergen, 1954).

'Fisheries Dispute Breaks out between Ireland and Scotland', *RTE News*, 8 June 2019.

Fitzsimons, Fiona, 'Tudor Fiants', *History Ireland*, 23, no. 4 (2015), p. 25.

'Fitzwilliam Manuscripts at Milton, England', *Analecta Hibernica*, 4 (1932), pp. 287–326.

Flanagan, Laurence, 'The Irish Legacy of the Spanish Armada', *Archaeology Ireland*, 2 (1998), pp. 145–49.

Flavin, Susan, 'Consumption and Material Culture in Sixteenth-Century Ireland', *Economic History Review*, 64 (2011), pp. 1144–74.

—. Review of 'A Calendar of Material Relating to Ireland from the High Court of Admiralty, 1641–1660', *Irish Studies Review*, 20 (2012), pp. 493–94.

—. *Consumption and Culture in Sixteenth-Century Ireland: Saffron, Stockings and Silk* (Woodbridge, 2014).

—. 'Food, Drink and Society in Sixteenth-Century Ireland: Cultures of Consumption', in Eve Campbell et al. (eds), *Becoming and Belonging in Ireland c.1200–1600* (Cork, 2018), pp. 212–43.

—. 'Food and Social Politics in Early Modern Ireland: Representing the Peasant in The Parliament of Clan Tomas', *Food and History*, 20 (2022), pp. 9–40.

Flavin, Susan, and Evan Jones, *A Glossary of Commodities, Weights and Measures Found in the Sixteenth-Century Bristol Customs Accounts* (Bristol, 2009).

Flavin, Susan et al., 'Understanding Early Modern Beer: An Interdisciplinary Case-Study', *Historical Journal* (2023), pp. 1–34.

Flood, Donal T., 'The Birth of the Bull Island', *Dublin Historical Record*, 28 (1975), pp. 142–53.

—. 'Dublin Bay in the 18th Century', *Dublin Historical Record*, 31 (1978), pp. 129–41.

Forsythe, Wes at al., 'Persistence and Risk: Salt Production in Post-Medieval Ireland', *World Archaeology*, 50 (2018), pp. 603–19.

Fox, H. S. A., *The Evolution of the Fishing Village: Landscape and Society along the South Devon Coast 1086–1550* (Oxford, 2001).

French, J. R., and H. Burningham, 'Coastal Geomorphology: Trends and Challenges', *Progress in Physical Geography*, 33 (2009), pp. 117–29.

Fury, Cheryl A., 'Health and Care at Sea', in Cheryl A. Fury (ed.), *The Social History of English Seamen, 1485–1649* (Woodbridge, 2012).

Gallagher, Fióna, and Marie-Louise Legg, 'Sligo', *IHTA*, 24 (2012), pp. 1–27.

Gardiner, Mark, 'The Late Medieval Port Buildings of Ardglass, County Down, in the Context of Northern European Trade', *Ulster Journal of Archaeology*, 74 (2017–18), pp. 202–12.

Gardiner, Mark, and T. E. McNeill, 'Seaborne Trade and the Commercialisation of Fifteenth- and Sixteenth-Century Gaelic Ulster', *Proceedings of the Royal Irish Academy*, 116C (2016), pp. 229–62.

Gerritsen, H. D., and E. Kelly (eds), *Atlas of Commercial Fisheries around Ireland* (3rd edn, Oranmore, 2019).

Gillespie, Raymond, 'Harvest Crises in Early Seventeenth-Century Ireland', *Irish Economic and Social History*, 11 (1984), pp. 5–18.

—. *The Transformation of the Irish Economy 1550–1700* (Dublin, 1991).

Goldman, Mara et al. (eds), *Knowing Nature: Conversations at the Intersection of Political Ecology and Science Studies* (Chicago, 2011).

Gowen, R. J. et al., 'Production and Its Fate in Two Coastal Regions of the Irish

Sea: The Influence of Anthropogenic Nutrients', *Marine Ecology Progress Series*, 208 (2000), pp. 51–64.

Grafe, Regina, 'The Globalisation of Codfish and Wool: Spanish-English-North American Triangular Trade in the Early Modern Period', *Economic History Working Papers*, 71 (2003), pp. 1–37.

—. *Distant Tyranny: Markets, Power, and Backwardness in Spain, 1650–1800* (Princeton, NJ, 2012).

Grainger, R. J. R., 'Herring Abundance off the West of Ireland in Relation to Oceanographic Variation', *ICES Journal of Marine Science*, 38 (1978), pp. 180–88.

Gregory, Ian N., and Richard G. Healey, 'Historical GIS: Structuring, Mapping and Analysing Geographies of the Past', *Progress in Human Geography*, 31 (2016), pp. 638–53.

Guénette, Sylvie, and Didier Gascuel, 'Shifting Baselines in European Fisheries: The Case of the Celtic Sea and Bay of Biscay', *Ocean & Coastal Management*, 70 (2012), pp. 10–21.

Halpern, Benjamin S. et al., 'Spatial and Temporal Changes in Cumulative Human Impacts on the World's Ocean', *Nature Communications*, 6 (2015), pp. 1–7.

Hamilton-Dyer, Sheila, 'Exploring the Contrast: Fish-Bone Assemblages from Medieval Ireland', in *C&H*, pp. 231–38.

Hampton, Stephen, '"Welcome Dear Feast of Lent": River Understandings of the Forty-Day Fast in Early Stuart England', *Journal of Theological Studies*, 63 (2012), pp. 608–48.

Harma, Clémentine et al., 'The Rise and Fall of Autumn Spawning Herring in the Celtic Sea between 1959 and 2009', *Fisheries Research*, 121 (2012), pp. 31–42.

Harvey, Barbara F., *Living and Dying in England, 1100–1540: The Monastic Experience* (Oxford, 2002).

Hayes, Patrick W. et al., 'European Naval Diets in the Sixteenth Century: A Quantitative Method for Comparative and Nutritional Analysis', *Historical Methods*, 52 (2019), pp. 195–212.

Heessen, Henk J. L., 'ICES FishMap Species Factsheet: Herring' (ICES, 2005).

Heessen, Henk J. L. et al. (eds), *Fish Atlas of the Celtic Sea, North Sea, and Baltic Sea* (Wageningen, 2015).

Herbruck, Wendell, 'Forestalling, Regrating and Engrossing', *Michigan Law Review*, 27 (1929), pp. 365–88.

Hickling, C. F., *The Hake and the Hake Fishery* (London, 1934).

Hitchcock, Richard, 'Dingle in the Sixteenth Century. With an Introduction and Notes', *Transactions of the Kilkenny Archaeological Society*, 2 (1852), pp. 133–43.

Hodkinson, B. J., *Index of Placenames of the Desmond Survey* (Limerick, 2013).

Hoffmann, Richard C., 'A Brief History of Aquatic Resource Use in Medieval Europe', *Helgoland Marine Research*, 59 (2004), pp. 22–30.

Holm, Poul, 'The Danish Fisheries, c. 1450–1800: Medieval and Early Modern Sources and Their Potential for Marine Environmental History', in P. Holm et al. (eds), *The Exploited Seas: New Directions for Marine Environmental History* (St. John's, Nfld, 2001), pp. 97–122.

—. 'Human Impacts on Fisheries Resources and Abundance of the Danish Wadden Sea, c. 1520 to the Present', *Helgoland Marine Research*, 59 (2005), pp. 39–44.

—. Review of Jim Mac Laughlin, *Troubled Waters*, *The Mariner's Mirror*, 97 (2011), pp. 366–68.

—. Review of W. Jeffrey Bolster, *The Mortal Sea: Fishing the Atlantic in the Age of Sail*, *H-Environment Roundtable Review*, 4 (2014), pp. 1–4.

Holm, Poul et al., *Proteiner til verden. Fiskemelsindustrien i Esbjerg 1948–1998* (Esbjerg, 1998).

—. 'Marine Animal Populations: A New Look Back in Time', in A. D. MacIntyre (ed.), *Life in the World's Oceans: Diversity, Distribution, and Abundance* (Oxford, 2010), pp. 1–24.

—. 'The North Atlantic Fish Revolution (ca. AD 1500)', *Quaternary Research*, 108 (2019), pp. 1–15.

—. 'Accelerated Extractions of North Atlantic Cod and Herring, 1520–1790', *Fish and Fisheries*, 23 (2022), pp. 54–72.

Horner, Arnold, 'Through Fractured Lens of Civil Survey – Buildings across Mid-Seventeenth Century Dublin Region', in Patrick J. Duffy and William Nolan (eds), *At the Anvil: Essays in Honour of William J. Smyth* (Dublin, 2012).

Horning, Audrey J., 'On the Banks of the Bann – the Riverine Economy of an Ulster Plantation Village', *Historical Archaeology*, 41 (2007), pp. 94–114.

Howard, E. F. et al., 'The Pollutional Character of Flax Retting Wastes', *Sewage Works Journal*, 6 (1934), pp. 596–600.

Hughes, Edward, 'England's Monopoly of Salt, 1563–1571', *English Historical Review*, 40 (1925), pp. 334–50.

Hulme, M., and E. M. Barrow (eds), *Climates of the British Isles: Present, Past and Future* (London, 1997).

Hurrel, James W., and Harry Van Loon, 'Decadal Variations in Climate Associated with the North Atlantic Oscillation', *Climatic Change*, 36 (1997), pp. 301–26.

Hutchinson, W. F. et al., 'The Globalization of Naval Provisioning: Ancient DNA and Stable Isotope Analyses of Stored Cod from the Wreck of the Mary Rose, AD 1545', *Royal Society Open Science*, 2 (2015), pp. 1–13.

Inderhaug, Terje, 'Stockfish Production, Cultural and Culinary Values', *Food Ethics*, 5 (2020), pp. 1–40.

Innis, Harold Adams, *The Cod Fisheries: The History of an International Economy* (Toronto, 1978).

Jackson, Jeremy B. C. et al., 'Historical Overfishing and the Recent Collapse of Coastal Ecosystems', *Science*, 293 (2001), pp. 629–37.

Janzen, Olaf Uwe, 'The Problem of Piracy in the Newfoundland Fishery in the Aftermath of the War of the Spanish Succession', in Poul Holm and Olaf Janzen (eds), *Northern Seas Yearbook* (Esbjerg, 1998), pp. 57–75.

Jenkins, James Travis, *The Herring and the Herring Fisheries* (London, 1927).

Johnston, Don, 'Two Pre-Ordnance Survey Maps of Carlingford', *Journal of the County Louth Archaeological and Historical Society*, 27 (2009), pp. 8–25.

Jones, Evan, 'Illicit Business: Accounting for Smuggling in Mid-Sixteenth-Century Bristol', *Economic History Review*, 54 (2001), pp. 17–38.

—. *Tonnage Estimates of Goods Recorded in the Bristol 'Particular' Customs Accounts: 1465/6, 1485/6, 1503/4* (Bristol, 2017).

Jónsson, J., 'Fisheries off Iceland, 1600–1900', *ICES Marine Science Symposia*, 198 (1992), pp. 3–16.

Kearney, H. F., 'The Irish Wine Trade, 1614–15', *Irish Historical Studies*, 9 (1954–55), pp. 400–42.

Kelleher, Connie, 'The Gaelic O'Driscoll Lords of Baltimore', in Linda Doran and James Lyttleton (eds), *Lordship in Medieval Ireland, Image and Reality* (Dublin, 2007), pp. 130–59.

—. *The Alliance of Pirates: Ireland and Atlantic Piracy in the Early Seventeenth Century* (Cork, 2020).

Kelly, Maria, '"Unheard-of Mortality": The Black Death in Ireland', *History Ireland*, 9, no. 4 (2001), pp. 12–17.

Kissane, Christopher et al., *Food, Religion and Communities in Early Modern Europe* (London, 2018).

Kjellström, Anna et al., 'Dietary Patterns and Social Structures in Medieval Sigtuna, Sweden, as Reflected in Stable Isotope Values in Human Skeletal Remains', *JAS*, 36 (2009), pp. 2689–99.

Korta, Maria et al., 'European Hake (Merluccius merluccius) in the Northeast Atlantic Ocean', in Hugo Arancibia (ed.), *Hakes: Biology and Exploitation* (Chichester, 2015), pp. 1–37.

Kowaleski, M., 'The Expansion of the South-Western Fisheries in Late Medieval England', *Economic History Review*, 53 (2000), pp. 429–54.

Lamb, Hubert, *Historical Storms of the North Sea, British Isles and Europe* (Cambridge, 1991).

Lamichhaney, Sangeet et al., 'Population-Scale Sequencing Reveals Genetic Differentiation Due to Local Adaptation in Atlantic Herring', *Proceedings of the National Academy of Sciences*, 109 (2012), pp. 19345–50.

Lankford, Éamon, *O'Driscolls Past and Present* (Cork, 2005).

Lauria, Valentina et al., 'Regional Variation in the Impact of Climate Change: Evidence that Bottom-up Regulation from Plankton to Seabirds is Weak in Parts of the Northeast Atlantic', *Marine Ecology Progress Series*, 488 (2013), pp. 11–22.

Lenders, R. H. J. et al., 'Historical Rise of Waterpower Initiated the Collapse of Salmon Stocks', *Scientific Reports*, 6 (2016), pp. 1–9.

Lennon, Colm, *The Lords of Dublin in the Age of Reformation* (Dublin, 1989).

—. *Sixteenth Century Ireland: The Incomplete Conquest* (Dublin, 2005).

Loingsigh, Máire Ní, 'An Assessment of Castles and Landownership in Late Medieval North Donegal', *Ulster Journal of Archaeology*, 57 (1994), pp. 145–58.

Longfield, Ada Kathleen, 'Anglo-Irish Trade in the Sixteenth Century as Illustrated by the English Customs Accounts and Port Books', *Proceedings of the Royal Irish Academy*, 36C (1923), pp. 317–32.

—. *Anglo-Irish Trade in the Sixteenth Century* (London, 1929).

Lotze, Heike K. et al., 'Ecosystem-Based Management of Seaweed Harvesting', *Botanica Marina*, 62 (2019), pp. 395–409.

Lotze, Heike K., and Loren McClenachan, 'Historical Ecology: Informing the Future by Learning from the Past', in M. D. Bertness et al. (eds), *Marine Community Ecology and Conservation* (Sunderland, MA, 2014), pp. 165–200.

Ludlow, Francis, 'Interpretation of Celestial and Environmental Phenomena in the Irish Annals', *Atlas*, 10 (2005), pp. 48–60.

—. 'Three Hundred Year of Weather Extremes from the Annals of Connacht', *Journal of Postgraduate Research*, 5 (2006), pp. 46–65.

—. 'Tree Ring Chronology of Meteorological Extremes for Ireland, AD425–1650', *Irish Meteorological Society Newsletter*, 5 (2011), pp. 54–59.

Ludlow, Francis et al., 'Medieval Irish Chronicles Reveal Persistent Volcanic Forcing of Severe Winter Cold Events, 431–1649 CE', *Environmental Research Letters*, 8 (2013), pp. 1–10.

Luetić, Josip, 'English Mariners and Ships in Seventeenth-Century Dubrovnik', *The Mariner's Mirror*, 64 (1978), pp. 276–84.

Lydon, James, 'The Middle Nation', in Peter Crooks (ed.), *Government, War and Society in Medieval Ireland* (Dublin, 2019), pp. 332–52.

Lyons, Mary Ann, 'Maritime Relations between Ireland and France, c. 1480–c. 1630', *Irish Economic and Social History*, 27 (2000), pp. 1–24.

MacCarthy-Morrogh, Michael, *The Munster Plantation: English Migration to Southern Ireland 1583–1641* (Oxford, 1986).

Mac Con Iomaire, Máirtín, 'The History of Seafood in Irish Cuisine and Culture', *History Studies*, 5 (2004), pp. 61–76.

Mac Laughlin, Jim, *Troubled Waters: A Social and Cultural History of Ireland's Sea Fisheries* (Dublin, 2010).

Marsden, Reginald G. (ed.), *Documents Relating to Law and Custom of the Sea* (Navy Records Society, xlix–l, London, 1915).

Mathers, R. G. et al., 'A Review of the Potential Effect of Irish Hydroelectric Installations on Atlantic Salmon', *Biology and Environment: Proceedings of the Royal Irish Academy*, 102B (2002), pp. 69–79.

McCarty, T. K. et al., 'Long-Term Effects of Hydropower Installations and Associated River Regulation on River Shannon Eel Populations: Mitigation and Management', *Hydrobiologia*, 609 (2008), pp. 109–24.

McClenachan, Loren, and John N. Kittinger, 'Multicentury Trends and the Sustainability of Coral Reef Fisheries in Hawai'i and Florida', *Fish and Fisheries*, 14 (2013), pp. 239–55.

McCormack, Anthony M., 'The Social and Economic Consequences of the Desmond Rebellion of 1579–83', *Irish Historical Studies*, 34 (2004–05), pp. 1–15.

McErlean, Thomas et al. (eds), *Strangford Lough: An Archaeological Survey of the Maritime Culture Landscape* (Belfast, 2002).

McGowan-Doyle, Valerie, *The Book of Howth: The Elizabethan Re-Conquest of Ireland and the Old English* (Cork, 2011).

McKenna, John et al., 'Accurate Mental Maps as an Aspect of Local Ecological Knowledge (LEK): A Case Study from Lough Neagh, Northern Ireland', *Ecology and Society*, 13 (2008), pp. 1–24.

McInerney, Luke, 'A Sixteenth Century Bardic Poem Composed for Seán Mac Conmara, Lord of West "Clann Chuiléin"', *Seanchas Ardmhacha*, 23 (2010), pp. 33–56.

Michell, A. R., 'The European Fisheries in Early Modern History', in E. E. Rich and C. H. Wilson (eds), *The Cambridge Economic History of Europe* (Cambridge, 1977), pp. 133–84.

Mihas, Paul, *Qualitative Data Analysis* (Oxford, 2019).

Mikalsen, Gaute et al., 'Late-Holocene Changes in Ocean Circulation and Climate: Foraminiferal and Isotopic Evidence from Sulafjord, Western Norway', *The Holocene*, 11 (2001), pp. 437–46.

Moody, T. W. et al. (eds), *A New History of Ireland III: Early Modern Ireland 1534–1691* (Oxford, 2009).

Morgan, Hiram, 'The Colonial Venture of Sir Thomas Smith in Ulster, 1571–75', *Historical Journal*, 28 (1985), pp. 261–78.

—. 'Hugh O'Neill and the Nine Years War in Tudor Ireland', *Historical Journal*, 36 (1993), pp. 21–37.

—. *Tyrone's Rebellion: The Outbreak of the Nine Years War in Tudor Ireland* (Dublin, 1993).

Moriarty, Christopher, 'Arthur E. J. Went, 1910–1980', *The Irish Naturalists' Journal*, 20 (1981), pp. 309–12.

—. 'Fish and Fisheries', in Wilson John Foster and Helena C. G. Chesney (eds), *Nature in Ireland: A Scientific and Cultural History* (Montreal, 1997), pp. 283–98.

Morris, Marc, *A Great and Terrible King* (London, 2009).

Müldner, Gundula, 'Marine Fish Consumption in Medieval Britain: The Isotope Perspective from Human Skeletal Remains', in *C&H*, pp. 239–49.

Müldner, Gundula, and Michael P. Richards, 'Fast or Feast: Reconstructing Diet in Later Medieval England by Stable Isotope Analysis', *JAS*, 32 (2005), pp. 39–48.

Muldrew, Craig, *Food, Energy and the Creation of Industriousness: Work and Material Culture in Agrarian England, 1550–1780* (Cambridge, 2011).

Murphy, Margaret, and Michael Potterton, *The Dublin Region in the Middle Ages* (Dublin, 2010).

Murray, Theresa Denise, 'Baltimore to Barbary: The 1631 Sack of Baltimore', *History Ireland*, 14, no. 4 (2006), pp. 14–18.

Mwinyihija, Mwinyikione, 'Main Pollutants and Environmental Impacts of the Tanning Industry', in Mwinyikione Mwinyihija (ed.), *Ecotoxicological Diagnosis in the Tanning Industry* (New York, 2010), pp. 17–35.

Nash, R. D. M., and M. Dickley-Collas, 'The Influence of Life History Dynamics and Environment on the Determination of Year Class Strength in North Sea Herring (Clupea harengus L.)', *Fisheries Oceanography*, 14 (2005), pp. 279–91.

Nedkvitne, Arnved, 'The Development of the Norwegian Long-Distance Stockfish Trade', in *C&H*, pp. 50–59.

Nilelssen, Alf Ragnar, 'Norwegian Fisheries, c. 1100–1850', in D. J. Starkey and James E. Candow (eds), *The North Atlantic Fisheries: Supply, Marketing and Consumption, 1560–1990* (Hull, 2006), pp. 83–112.

Odland, John, 'The Allocation of Merchant Capital in Early Tudor London', *Economic History Review*, 63 (2010), pp. 1058–80.

O'Dowd, Mary, *Power, Politics and Land: Early Modern Sligo 1568–1688* (Belfast, 1991).

O'Flaherty, Eamon, 'Limerick', *IHTA*, 21 (2010), pp. 1–59.

Ohlmeyer, Jane, 'The Dunkirk of Ireland: Wexford Privateers during the 1640's', *Journal of the Wexford Historical Society*, 12 (1989), pp. 23–49.

—. (ed.), *The Cambridge History of Ireland, Volume 2, 1550–1730* (Cambridge, 2018).

O'Mahony, Edward, 'The O'Driscolls and Their Revenues from Fishing – the 1609 Inquisition', *Mizen Journal*, 8 (2000), pp. 128–30.

O'Neill, Timothy, *Merchants and Mariners in Medieval Ireland* (Dublin, 1987).

O'Rourke, Kevin H., and Jeffrey G. Williamson, 'After Columbus: Explaining Europe's Overseas Trade Boom, 1500–1800', *Journal of Economic History*, 62 (2002), pp. 417–56.

Orton, D. C. et al., 'Stable Isotope Evidence for Late Medieval (14th–15th C)

Origins of the Eastern Baltic Cod (Gadus morhua) Fishery', *PLoS One*, 6 (2011), pp. 1–11.

Orton, David C. et al., 'Fish for the City: Meta-Analysis of Archaeological Cod Remains and the Growth of London's Northern Trade', *Antiquity*, 88 (2015), pp. 516–30.

O'Sullivan, Aidan, *Foragers, Farmers and Fishers in a Coastal Landscape: An Intertidal Archaeological Survey of the Shannon Estuary* (Dublin, 2001).

—.'Medieval Fish Traps on the Shannon Estuary, Ireland: Interpreting People, Place and Identity in Estuarine Landscapes', *Journal of Wetland Archaeology*, 5 (2013), pp. 65–77.

O'Sullivan, Aidan, and Colin Breen, *Maritime Ireland: An Archaeology of Coastal Communities* (Stroud, 2007).

O'Sullivan, David et al., 'An Inventory of Irish Herring Spawning Grounds', *Marine Institute, Irish Fisheries Bulletin*, 42 (2013), pp. 1–31.

O'Sullivan, M. D. M., *Old Galway* (Cambridge, 1942).

O'Sullivan, Muiris, and Liam Downey, 'Medieval Fisheries', *Archaeology Ireland*, 23 (2009), pp. 30–33.

Palmer, William, 'The Problem of Ideology in the Elizabethan Conquest of Ireland: A Comparative Approach?', *New Hibernia Review*, 19 (2015), pp. 128–42.

Parker, Geoffrey, 'History and Climate: The Crisis of the 1590s Reconsidered', in Claus Leggewie and Franz Mauelshagen (eds), *Climate Change and Cultural Transition in Europe* (Leiden, 2018), pp. 119–55.

Pauly, Daniel, 'Anecdotes and the Shifting Baseline Syndrome of Fisheries', *Trends in Ecology and Evolution*, 10 (1995), p. 430.

Pausata, Francesco S. R. et al., 'Impacts of High-Latitude Volcanic Eruptions on ENSO and AMOC', *Proceedings of the National Academy of Sciences*, 112 (2015), pp. 13784–88.

Pennell, C. R. (ed.), *Bandits at Sea: A Pirates Reader* (New York, 2001).

—. 'The Geography of Piracy: Northern Morocco in the Mid-Nineteenth Century', in C. R. Pennell (ed.), *Bandits at Sea: A Pirate Reader* (New York, 2001), pp. 55–68.

Pereira, Leonel, and João Cotas, 'Historical Use of Seaweed as an Agricultural Fertilizer in the European Atlantic Area', in Leonel Pereira et al. (eds), *Seaweeds as Plant Fertilizer, Agricultural Biostimulants and Animal Fodder* (Boca Raton, FL, 2019), pp. 1–22.

Phadraig, B., 'Speed's Plan of Dublin', *Dublin Historical Record*, 10 (1949), pp. 97–105.

Pluymers, K., '"Pirates" and the Problems of Plantation in Seventeenth-Century Ireland', in Peter C. Mancall and Carole Shammas (eds), *Governing the Sea in the Early Modern Era* (San Marino, CA, 2015), pp. 79–107.

Pope, Peter, *Fish into Wine: The Newfoundland Plantation in the Seventeenth Century* (Chapel Hill, NC, 2004).

Poulsen, Bo, *Dutch Herring: An Environmental History, c. 1600–1860* (Amsterdam, 2008).

—. 'The Variability of Fisheries and Fish Populations Prior to Industrialized Fishing: An Appraisal of the Historical Evidence', *Journal of Marine Systems*, 79 (2010), pp. 327–32.

—. 'Marine Environmental History', in Mauro Agnoletti et al. (eds), *World Environmental History* (Oxford, 2012), pp. 1–35.

Proulx, Jean-Pierre, *Basque Whaling in Labrador in the 16th Century* (Ottawa, 1993).

Quinn, D. B., 'Anglo-Irish in Ulster in the Early Sixteenth Century', in Arthur Deane (ed.), *Proceedings and Reports of the Belfast Natural History and Philosophical Society* (Belfast, 1935), pp. 56–78.

—. 'Sir Thomas Smith (1513–1577) and the Beginnings of English Colonial Theory', *Proceedings of the American Philosophical Society*, 89 (1945), pp. 543–60.

—. *Explorers and Colonies: America, 1500–1625* (London, 1990).

Raicevich, S. et al., 'Community-Based Management of the Mediterranean Coastal Fisheries: Historical Reminiscence or the Root for New Fisheries Governance?', *Regional Studies in Marine Science*, 21 (2017), pp. 86–93.

Reeves, Randall R. et al., 'History of Whaling and Estimated Kill of Right Whales, Balaena glacialis, in the Northeastern United States, 1620–1924', *Marine Fisheries Review*, 61 (1999), pp. 1–36.

Riis, Thomas, *Crail and Its Fisheries 1550–1600* (Strathmartine, 2016).

Rixson, Denis, *The West Highland Galley* (Edinburgh, 1998).

Robinson, Philip, 'Carrickfergus', *IHTA*, 2 (1986), pp. 1–16.

Rodger, N.A.M., *The Wooden World: An Anatomy of the Georgian Navy* (Glasgow, 1986).

—. *The Safeguard of the Sea: A Naval History of Great Britain, Volume 1, 660–1649* (London, 1997).

Rogers, J. E. T., *A History of Agriculture and Prices in England* (7 vols, Oxford, 1866–92).

Rorke, Martin, 'The Scottish Herring Trade, 1470–1600', *Scottish Historical Review*, 84 (2005), pp. 149–65.

Rosenberg, Andrew A. at al., 'The History of Ocean Resources: Modeling Cod Biomass Using Historical Records', *Frontiers in Ecology and the Environment*, 3 (2005), pp. 78–84.

Ruddock, A. Alwyn, 'The Earliest Records of the High Court of Admiralty', *Bulletin of the Institute of Historical Research*, 22 (1949), pp. 139–51.

Samuel, Arthur Michael, *The Herring: Its Effect on the History of Britain* (London, 1918).

Schüller, Karin, 'Special Conditions of the Irish-Iberian Trade during the

Spanish-English War (1585–1604)', in Óscar Recio Morales et al. (eds), *Irlanda y la Monarquía Hispánica. Kinsale 1601–2001: Guerra, Política, Exilio y Religión* (Madrid, 2002), pp. 447–68.

Schwartz, Robert M., 'Digital Partnership: Combining Text Mining and GIS in a Spatial History of Sea Fishing in the United Kingdom, 1860 to 1900', *International Journal of Humanities and Arts Computing*, 9 (2015), pp. 36–56.

Schwerdtner Manez, K. et al., 'The Future of the Oceans Past: Towards a Global Marine Historical Research Initiative', *PLoS One*, 9 (2014), pp. 1–11.

Senior, C. M., *A Nation of Pirates: English Piracy in Its Heyday* (Newton Abbot, 1976).

Sgroi, R. C. L., 'Piscatorial Politics Revisited: The Language of Economic Debate and the Evolution of Fishing Policy in Elizabethan England', *Albion*, 35 (2003), pp. 1–24.

Sheehan, Anthony, 'Irish Towns in a Period of Change, 1558–1625', in Ciaran Brady and Raymond Gillespie (eds), *Natives and Newcomers: Making of Irish Colonial Society, 1534–1641* (Dublin, 1986), pp. 93–119.

Sicking, Louis, and Darlene Abreu-Ferreira (eds), *Beyond the Catch: Fisheries of the North Atlantic, the North Sea and the Baltic, 900–1850* (Leiden, 2009).

Simms, Anngret et al., 'Dundalk', *IHTA*, 16 (2006), pp. 1–24.

—. 'Carlingford', *IHTA*, 23 (2011), pp. 1–15.

Simms, Katharine, 'Guesting and Feasting in Gaelic Ireland', *JRSAI*, 108 (1978), pp. 67–100.

—. 'The O Hanlons, the O Neills and the Anglo-Normans in Thirteenth-Century Armagh', *Seanchas Ardmhacha*, 9 (1978), pp. 70–94.

—. 'References to Landscape and Economy in Irish Bardic Poetry', in H. B. Clark et al. (eds), *Surveying Ireland's Past: Multidisciplinary Essays in Honour of Anngret Simms* (Dublin, 2004), pp. 145–68.

Simpson, S. D. et al., 'Continental Shelf-Wide Response of a Fish Assemblage to Rapid Warming of the Sea', *Current Biology*, 21 (27 September 2011), pp. 1565–70.

Smith, Brendan (ed.), *The Cambridge History of Ireland, Volume 1, 600–1550* (Cambridge, 2018).

Smith, Hance D., *Shetland Life and Trade 1550–1914* (Glasgow, 1984).

Sorheim, Helge, 'The Birth of Commercial Fisheries and the Trade of Stockfish in the Borgundfjord, Norway', in *C&H*, pp. 60–70.

Southward, A. J. et al., 'Fluctuations in the Herring and Pilchard Fisheries of Devon and Cornwall Linked to Change in Climate since the 16th Century', *Journal of the Marine Biological Association of the United Kingdom*, 68 (1988), pp. 423–45.

Star, Bastiaan et al., 'Ancient DNA Reveals the Chronology of Walrus Ivory Trade from Norse Greenland', *Proceedings of the Royal Society B: Biological Sciences*, 285 (2018), pp. 1–9.

Starkey, D. J. et al. (eds), *A History of the North Atlantic Fisheries, Volume 1, From Early Times to the Mid-19th Century* (Bremerhaven, 2009).

Starkey, David J., 'Voluntaries and Sea Robbers: A Review of the Academic Literature on Privateering, Corsairing, Buccaneering and Piracy', *The Mariner's Mirror*, 97 (2011), pp. 127–47.

Stenseth, N. C. et al., *Marine Ecosystems and Climate Variation: The North Atlantic – A Comparative Perspective* (Oxford, 2005).

Stradling, R. A., *Armada of Flanders: Spanish Maritime Policy and European War, 1568–1668* (Cambridge, 1992).

Stuijts, Ingelise, and Liam Downey, *Historical Farm and Fish Food Products* (Carlow, 2016).

Taverner, Charlie, and Susan Flavin, 'Food and Power in Sixteenth-Century Ireland: Studying Household Accounts from Dublin Castle', *Historical Journal*, 66 (2023), pp. 1–26.

Thomas, Avril, 'Derry-Londonderry', *IHTA*, 15 (2005), pp. 1–47.

Tittler, R., 'The English Fishing Industry in the Sixteenth Century', *Albion*, 9 (1977), pp. 40–60.

Tracy, James D., 'Herring Wars: The Habsburg Netherlands and the Struggle for Control of the North Sea, ca. 1520–1560', *Sixteenth Century Journal*, 24 (1993), pp. 249–72.

Treadwell, Victor, 'The Establishment of the Farm of the Irish Customs 1603–13', *English Historical Review*, 93 (1978), pp. 580–602.

Trentmann, F. (ed.), *The Oxford Handbook of the History of Consumption* (Oxford, 2012).

Turgeon, Laurier, *The Era of the Far-Distant Fisheries: Permanence and Transformation (circa 1500–1850)* (St. John's, Nfld, 2005).

Tykot, R. H., 'Stable Isotopes and Diet: You Are What You Eat', in M. Martini et al. (eds), *Proceedings of the International School of Physics "Enrico Fermi"*, Course CLIV (Amsterdam, 2004), pp. 433–44.

Unger, Richard W., 'Dutch Herring, Technology, and International Trade in the Seventeenth Century', *Journal of Economic History*, 40 (1980), pp. 253–80.

Vallaux, Camille, *Penmarch, aux XVI et XVII siècles* (Société Nouvelle de Libraire et d'Edition, Paris, 1906).

van Vliet, A. P., 'The Influence of Dunkirk Privateering on the North Sea (Herring) Fishery during the Years 1580–1650', in J. Roding and Heerma L. van Voss (eds), *The North Sea and Culture* (Leiden, 1996), pp. 150–65.

Vesteinsson, Orri, 'Commercial Fishing and the Political Economy of Medieval Iceland', in *C&H*, pp. 71–79.

Wang, Jianglin et al., 'Internal and External Forcing of Multidecadal Atlantic Climate Variability over the Past 1,200 Years', *Nature Geoscience*, 10 (2017), pp. 512–18.

Watt, John Anthony, *The Church in Medieval Ireland* (Dublin, 1998).

Webb, John, 'Apprenticeship in the Maritime Occupations at Ipswich, 1596–1651', *The Mariner's Mirror*, 46 (1960), pp. 29–34.

Went, A. E. J., 'The Galway Fishery. An Account of the Ownership of the Fishery', *Proceedings of the Royal Irish Academy*, 48C (1942), pp. 233–53.

—. 'The Galway Fishery. An Account of the Modes of Fishing Together with Notes on Other Matters Connected with the Fishery', *Proceedings of the Royal Irish Academy*, 49C (1943), pp. 187–219.

—. 'Fishing Weirs of the River Erne', *JRSAI*, 75 (1945), pp. 213–23.

—. 'The Irish Hake Fishery, 1504–1824', *Journal of the Cork Historical and Archaeological Society*, 51 (1946), pp. 41–51.

—. 'The Irish Pilchard Fishery', *Proceedings of the Royal Irish Academy*, 51B (1946), pp. 81–120.

—. 'Pilchards in the South of Ireland', *Journal of the Cork Historical and Archaeological Society*, 51 (1946), pp. 137–57.

—. 'The Ling in Irish Commerce', *JRSAI*, 78 (1948), pp. 119–26.

—. 'Foreign Fishing Fleets along the Irish Coasts', *Journal of the Cork Historical and Archaeological Society*, 54 (1949), pp. 17–24.

—. 'Giraldus Cambrensis' Notes on Irish Fish', *Irish Naturalists' Journal*, 9 (1949), pp. 221–24.

—. 'Eel Fishing at Athlone: Past and Present', *JRSAI*, 80 (1950), pp. 146–54.

—. 'Fishing Scenes from Irish Topographical Prints', *JRSAI*, 81 (1951), pp. 156–60.

—. 'Fisheries of the River Liffey. Notes on the Corporation Fishery up to the Dissolution of the Monasteries', *JRSAI*, 83 (1953), pp. 163–73.

—. 'Material for a History of the Fisheries of the River Boyne', *Journal of the County Louth Archaeological Society*, 13 (1953), pp. 18–33.

—. 'Fisheries of the River Liffey: II. Notes on the Corporation Fishery from the Time of the Dissolution of the Monasteries', *JRSAI*, 84 (1954), pp. 41–58.

—. 'Fisheries of the Munster Blackwater', *JRSAI*, 90 (1960), pp. 97–131.

—. 'Fisheries of the Munster Blackwater (Continued)', *JRSAI*, 91 (1961), pp. 19–41.

—. 'Oyster Fisheries', *Dublin Historical Record*, 18 (1963), pp. 56–63.

—. 'The Pursuit of Salmon in Ireland', *Proceedings of the Royal Irish Academy*, 63C (1964), pp. 191–244.

—. '"Sweeper" Nets for Salmon in County Kerry', *JRSAI*, 102 (1972), pp. 178–80.

—. 'List of Some Historical Papers etc on Irish Fish, Fishing and Fisheries 1940–1974', *Department of Fisheries and Forestry Fishing Leaflet*, 97 (1978), pp. 1–7.

—. 'Historical Notes on the Fisheries of the Estuary of the River Shannon', *JRSAI*, 111 (1981), pp. 107–18.

Woodward, Donald, 'Irish Trade and Customs Statistics, 1614–1641', *Irish Economic and Social History*, 26 (1999), pp. 54–80.

Wubs-Mrozewicz, Justyna, and Stuart Jenks (eds), *The Hanse in Medieval and Early Modern Europe* (Leiden, 2013).

Yoder, Cassady, 'Diet in Medieval Denmark: A Regional and Temporal Comparison', *JAS*, 37 (2010), pp. 2224–36.

—. 'Let Them Eat Cake? Status-Based Differences in Diet in Medieval Denmark', *JAS*, 39 (2012), pp. 1183–93.

Theses

Armour, C., 'The Trade of Chester and the State of the Dee Navigation, 1600–1800' (PhD thesis, University of London, 1956).

Buldorini, Chiara, 'Drogheda as a Case Study of Anglo-Norman Town Foundation in Ireland, 1194–1412' (PhD thesis, TCD, 2009).

Corten, Adrianus, 'Herring and Climate: Changes in the Distribution of North Sea Herring Due to Climate Fluctuations' (PhD thesis, University of Groningen, 1994).

Flavin, Susan, 'The Development of Anglo-Irish Trade in the Sixteenth Century' (MA thesis, University of Bristol, 2004).

—. 'Consumption and Material Culture in Sixteenth-Century Ireland' (PhD thesis, University of Bristol, 2011).

Gray, T., 'Devon's Coastal and Overseas Fisheries 1597–1642' (PhD thesis, University of Exeter, 1988).

Jones, Evan, 'The Bristol Shipping Industry in the Sixteenth Century' (PhD thesis, University of Edinburgh, 1998).

Jones, Peter, 'Technological Innovation and Resource Management in the Fisheries of the British Isles, ca.1400–1900' (PhD thesis, University of Strathclyde, 2016).

Locker, Alison Mary, 'The Role of Stored Fish in England 900–1750AD; the Evidence from Historical and Archaeological Data' (PhD thesis, University of Southampton, 2000).

Matthews, Al, 'European Fish Market Dynamics from 1500 to 1800' (PhD thesis, TCD, 2021).

Murau, Hilario, 'Reproductive Fundamentals for the Estimation of Egg Production of the European Hake, Merluccius merluccius in the Bay of Biscay' (PhD thesis, Universidad del País Vasco, 2006).

Rorke, Martin, 'Scottish Overseas Trade, 1275/86–1597' (PhD thesis, 2 vols, University of Edinburgh, 2001).

Taylor, Duncan, 'The Maritime Trade of the Smaller Bristol Channel Ports in the Sixteenth Century' (PhD thesis, University of Bristol, 2009).

van Bochove, Christiaan, 'The Economic Consequences of the Dutch' (PhD thesis, Utrecht University, 1980).

Woodward, Donald, 'The Trade of Elizabethan Chester' (PhD thesis, University of Hull, 1970).

Unpublished Works

Holm et al., 'Historical Marine Footprint for Atlantic Europe, 1500–2019', paper in progress (2023), pp. 1–26.

Ludlow, Francis et al., 'Explosive Volcanism as a Driver of North Atlantic Marine Productivity, Fisheries and Markets, 1400–1850 CE', paper in progress (2023), pp. 1–27.

Scherer, Cordula et al., 'A Historical Plankton Index: Zooplankton Abundance in the North Sea since 800 CE', paper in review (2023), pp. 1–36.

Online Secondary Sources

Allen, Robert C., and Richard W. Unger (eds), 'Allen-Unger Global Commodity Prices Dataset Introduction' (https://iisg.nl/hpw/allen-unger_about.pdf) (accessed 28 January 2023).

Byrne, Patricia M., 'Went, Arthur Edward James' (*DIB*, 2009) (https://doi.org/10.3318/dib.008966.v1) (accessed 29 January 2023).

Canny, Nicholas, 'Boyle, Richard' (*DIB*, 2009) (https://doi.org/10.3318/dib.000853.v1) (accessed 29 July 2022).

Clavin, Terry, 'Fitton (Fytton), Sir Edward' (*DIB*, 2009) (https://doi.org/10.3318/dib.003115.v1) (accessed 29 January 2023).

—. 'Weston, Nicholas' (*DIB*, 2009) (https://doi.org/10.3318/dib.008977.v1) (accessed 29 July 2022).

Crooks, Peter, 'The Irish Chancery Rolls, Introduction' (TCD, 2020) (https://chancery.tcd.ie/content/irish-chancery-rolls) (accessed 2 February 2020).

Curtis, Matthew, and Peter H. Lindert (eds), 'Global Price and Income History Group' (Davis: University of California, 2019) (http://gpih.ucdavis.edu/index.html) (accessed 2 June 2020).

'EMODnet DTM for European Sea Regions' (European Commission, 2018) (https://www.emodnet-bathymetry.eu/data-products) (accessed January 2020).

Environmental Protection Agency, 'EPA Irish Water' (https://www.epa.ie/irelandsenvironment/water/) (accessed 14 August 2020).

European Environmental Agency, 'Copernicus Land Monitoring Service – EU-DEM' (2017) (https://www.eea.europa.eu/data-and-maps/data/copernicus-land-monitoring-service-eu-dem) (accessed 8 January 2020).

Eurostat, European Commission, 'Aquaculture Production in Tonnes and Value' (https://ec.europa.eu/eurostat/web/products-datasets/-/tag00075) (accessed 12 May 2019).

Fiontar & Scoil na Gaeilge and The Placenames Branch [Department of Culture, Heritage and the Gaeltacht], 'Placenames Database of Ireland' (https://www.logainm.ie/en/) (accessed 8 March 2020).

Froese, R., and D. Pauly (eds), 'FishBase' (www.fishbase.org) (accessed 3 February 2023).

'High Court of Admiralty Research Guide' (http://www.nationalarchives.gov.uk/help-with-your-research/research-guides/high-court-admiralty-records/) (accessed 8 October 2019).

Marine Institute of Ireland, 'Ireland's Marine Atlas' (http://atlas.marine.ie/#?c=53.9108:-15.9082:6) (accessed 7 August 2019).

—. 'Irish Groundfish Survey' (https://shiny.marine.ie/igfs/) (accessed 6 July 2020).

—. 'Salmon Life Cycle' (https://www.marine.ie/Home/site-area/areas-activity/fisheries-ecosystems/salmon-life-cycle) (accessed 7 August 2019).

—. 'Species Dashboard' (https://shiny.marine.ie/speciesdash/) (accessed 30 September 2020).

—. 'The Stock Book' (https://shiny.marine.ie/stockbook/) (accessed 30 April 2020).

McCormack, Anthony M., 'Fitzgerald, Gerald fitz James 15th earl of Desmond' (*DIB*, 2009) (https://doi.org/10.3318/dib.003149.v1) (accessed 29 July 2022).

The Measuring Worth Foundation, 'Measuring Worth' (https://www.measuringworth.com/calculators/ukcompare/relativevalue.php) (accessed 7 August 2021).

The National Archives of the United Kingdom, 'Introduction: State Papers Ireland 1509–1782: Government Papers on Irish Affairs' (https://www.nationalarchives.gov.uk/help-with-your-research/research-guides/state-papers-ireland-1509-1782/) (accessed 8 December 2019).

O'Dowd, Mary, 'O'Malley, Gráinne [Grace] (fl. 1577–1597)' (*Oxford Dictionary of National Biography*, 2008) (https://doi.org/10.1093/ref:odnb/20753) (accessed 13 June 2020).

Sea Fisheries Protection Authority, '2019 All Landings into Ireland' (https://www.sfpa.ie/Statistics/Annual-statistics/Annual-Statistics/2019-Statistics) (accessed 1 May 2020).

—. '2020 Quota Uptake – Landings' (https://www.sfpa.ie/Statistics/Quotas) (accessed 15 June 2020).

Valler, Cindy, 'Baltimore, Ireland 20 June 1631' (http://www.cindyvallar.com/Baltimore.html) (accessed 1 June 2020).

Index

References to figures, plates, maps and tables are given in **bold**.

Irish Historical Monographs previous volumes

I. Ruling Ireland 1685–1742: Politics, Politicians and Parties, *D. W. Hayton*, 2004

II. Lord Broghill and the Cromwellian Union with Ireland and Scotland, *Patrick Little*, 2004

III. Irish Migrants in New Zealand, 1840–1937: 'The Desired Haven', *Angela McCarthy*, 2005

IV. The Fenian Ideal and Irish Nationalism, 1882–1916, *M. J. Kelly*, 2006

V. Loyalism in Ireland, 1789–1829, *Allan Blackstock*, 2007

VI. Constructing the Past: Writing Irish History, 1600–1800, *edited by Mark Williams and Stephen Forrest*, 2010

VII. The Making of the Irish Protestant Ascendancy: The Life of William Conolly, 1662–1729, *Patrick Walsh*, 2010

VIII. The Militia in Eighteenth-Century Ireland: In Defence of the Protestant Interest, *Neal Garnham*, 2012

IX. The Anglo-Irish Experience, 1680–1730: Religion, Identity and Patriotism, *D. W. Hayton*, 2012

X. The Presbyterians of Ulster, 1680–1730, *Robert Whan*, 2013

XI. The Welsh and the Shaping of Early Modern Ireland, 1558–1641, *Rhys Morgan*, 2014

XII. The Irish in the Spanish Armies in the Seventeenth Century, *Eduardo de Mesa*, 2014

XIII. Consumption and Culture in Sixteenth-Century Ireland: Saffron, Stockings and Silk, *Susan Flavin*, 2014

XIV. Elite Women in Ascendancy Ireland, 1690–1745: Imitation and Innovation, *Rachel Wilson*, 2015

XV. The Stuart Restoration and the English in Ireland, *Danielle McCormack*, 2016

XVI. Charity Movements in Eighteenth-Century Ireland: Philanthropy and Improvement, *Karen Sonnelitter*, 2016

XVII. Hugh de Lacy, First Earl of Ulster: Rising and Falling in Angevin Ireland, *Daniel Brown*, 2016

XVIII. Catholic Survival in Protestant Ireland, 1660–1711: Colonel John Browne, Landownership and the Articles of Limerick, *Eoin Kinsella*, 2018

XIX. Popular Protest and Policing in Ascendancy Ireland, 1691–1761, *Timothy D. Watt*, 2018

XX. The Old English in Early Modern Ireland: The Palesmen and the Nine Years' War, 1594–1603, *Ruth A. Canning*, 2019

XXI. Political Conflict in East Ulster, 1920–22: Revolution and Reprisal, *Christopher Magill*, 2020

XXII. Aristocratic Women in Ireland, 1450–1660: The Ormond Family, Power and Politics, *Damien Duffy*, 2021

XXIII. The Jacobite Duchess: Frances Jennings, Duchess of Tyrconnell, c.1649–1731, *Frances Nolan*, 2021

XXIV. Ireland's English Pale, 1470–1550: The Making of a Tudor Region, *Steven G. Ellis*, 2021

XXV. Irish Women in Religious Orders, 1530–1700: Suppression, Migration and Reintegration, *Bronagh Ann McShane*, 2022

XXVI. Ireland and Empire in the Late Nineteenth Century, *Fergal O'Leary*, 2023

Printed and bound by CPI Group (UK) Ltd, Croydon, CR0 4YY

15/09/2026

14962418-0001